AF553718

BIOCHEMISTRY OF PROTEINS

BIOCHEMISTRY
OF
PROTEINS

By
Dr. S.K. Prasad
School of Studies of Zoology & Biotechnology
Vikram University
Ujjain

DISCOVERY PUBLISHING HOUSE PVT. LTD.
NEW DELHI-110 002

Published by:
Tilak Wasan
DISCOVERY PUBLISHING HOUSE PVT. LTD.
4383/4B, Ansari Road, Darya Ganj
New Delhi-110 002 (India)
Phone : +91-11-23279245, 43596064-65
Fax : +91-11-23253475
E-mail : discoverypublishinghouse@gmail.com
namitwasan9@gmail.com
sales@discoverypublishinggroup.com
web : www.discoverypublishinggroup.com

Reprinted: 2019
First Edition: 2010

ISBN: 978-81-8356-611-7

Biochemistry of Proteins

Printed at:
Infinity Imaging Systems
Delhi

Preface

Biochemistry today has made spectacular progress in unraveling the mysteries of animate nature. This progress has allowed us to gain deeper insight into the principles of vital activity and has to a very significant extent stimulated the development of applied disciplines, especially. medicine. The present title *"Biochemistry of Proteins"* is intended for those who wish to understand living organisms, especially man. Biochemistry is essential for this purpose, but it would be almost impossible for a student to survey on his own the massive body of existing knowledge, constantly augmented by a remarkable torrent of brilliat discoveries. The purpose of the book, then is to organize our knowledge into something that can be comprehended in a relatively short time and still convey a reasonable complete picture of the chemical structure and function of man. Readability without sacrifice of coverage has been a prime goal. An important device in gaining that goal is to keep attention constantly focused on function, with repeated use of rationalization to show that the chemical facts are not isolated, but part of a whole.

Biochemistry has two major goals as a fundamental science, it treats the vital functions from the standpoint of physical chemistry, and as an applied discipline, it points out practical applications for the wealth of scientific knowledge it has acquired. The dual purpose has been, as far as possible, take into account in the writing of this book. We have also tried to summarize our pedagogical experience in teaching biochemistry to students specializing in medicine and pharmacy. The material of this book has been organized according to the principle of functionally in order to trace the close relationship between the functions of the living organism and the structures of its constituents molecules as well as the chemical and physico-chemical process in which they are involved.

The aim of this book is to present a core of biochemical knowledge that is desirable for undergraduate and postgraduate students and also those involved in the field of medical, microbiology, biotechnology and pharmaceutical. Every attempt has been made to keep abrest of the advances in the subject and at the same time to include the fundamentals.

To make the work more comprehensive and informative, the author has consulted many authoritative books, research journals, abstracts, monographs etc. He is grateful to all those great scholars whose work are cited or substantially reproduced.

There can be no claim to originality except in the manner of treatment and much of the information has been obtained from the books and scientific journals available in the different libraries.

The author expresses his thanks to his friends and colleagues whose continue inspirations have initiated him to bring out this book.

The author expresses his gratitude to Mr. Wasan and Staff of M/s Discovery Publishing House Pvt. Ltd., for their whole hearted co-operation in the publication of this book.

In the mean time, the author will remain sincerely responsible for any shortcomings of the book and be grateful to the readers for their suggestions and constructive criticism for the continuous betterment of the book. He takes this opportunity to appeal to the readers to send their suggestions straightway to his publisher.

Author

CONTENTS

1 AMINO ACIDS

Amino acids are organic molecules of low relative molecular mass (approximately 100-200) which contain at least one carboxyl (COOH) and one amino (NH_2) group and are essential constituents of plant and animal tissues. The variations that occur between the different amino acids lie in the nature of their R groups (side chains) a feature that is of fundamental importance and confers individuality upon each amino acid. From a knowledge of the chemical nature of the R group, predictions can be made regarding the properties of a particular amino acid and, conversely, a knowledge of the properties of an amino acid under investigation will assist in the identification of the R group and hence of the amino acid.

There are several methods available for the quantitation of amino acids which only give information about the total amino acid content of the sample regardless of whether one or several amino acids are present and cannot differentiate between the individual components. When it is necessary to detect a particular amino acid in the presence of others, in theory a method is chosen which makes use of a chemical or physical characteristic that is specific to the amino acid in question. In practice this is usually difficult to achieve and many such methods show varying degrees of specificity. Thus, some of the most useful and widely employed methods involve the separation of the various amino acids in the sample by a chromatographic or electrophoretic technique followed by the qualitative or quantitative determination of each component using one of the general colorimetric or fluorimetric methods. However, if the aim of the analysis is to detect and identify the various amino acid components without prior separation, a more specific reagent may be used where the reaction depends on the presence of a particular type of amino acid. Such reagents are also often used as visualization agents with paper and thin-layer chromatographic or electrophoretic techniques.

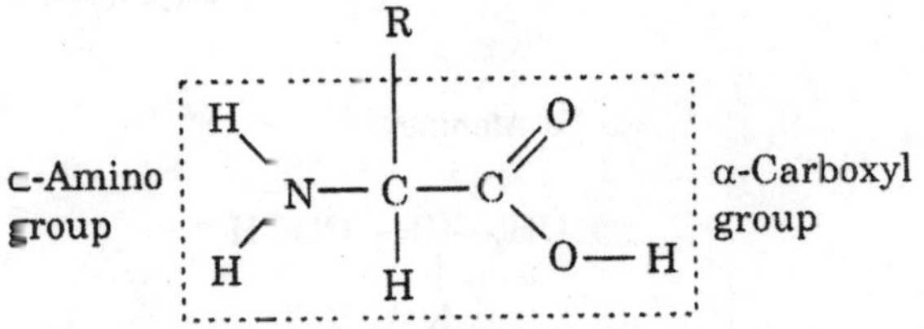

α-Amino group | R | H, H — N — C — C(=O)—O—H | H | α-Carboxyl group

Fig. 1.1. General structure of an α-amino acid. The part of the molecule shown inside the box is common to all α-amino acids while the R represents the side chain, which is different for each amino acid.

GENERAL STRUCTURE AND PROPERTIES

There are approximately 20 amino acids found in proteins, all of which are α-amino acids with the exception of the two α-imino acids proline and hydroxyproline, which for the purpose of this discussion will be considered with the amino acids because of their similarity. The α-amino acids are so called because the amino group is attached to the α-carbon of the chain which is, by convention, the carbon atom adjacent to the carboxyl group. Succeeding carbon atoms are designated β, γ, δ, and ε. Hence in α-amino acids both the amino group and the carboxyl group are attached to the same carbon atom. Many naturally occurring amino acids not found in protein, but which are of importance either in metabolism or as constitutents of plants and antibiotics, have structures that differ from the α-amino acids. In these compounds the amino group is attached to a carbon atom other than the α-carbon atom and they are called β, γ, δ, or ε amino acids accordingly.

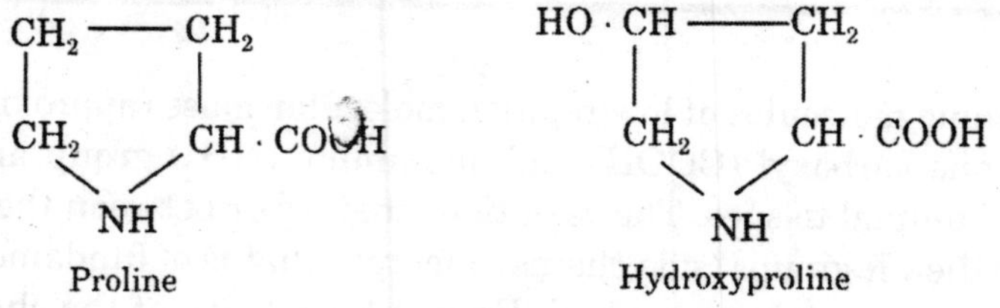

```
CH2 —— CH2                HO · CH ——— CH2
 |      |                      |        |
CH2    CH · COOH              CH2      CH · COOH
  \    /                        \      /
    NH                             NH
  Proline                    Hydroxyproline
```

Fig. 1.2. α-Imino acids.

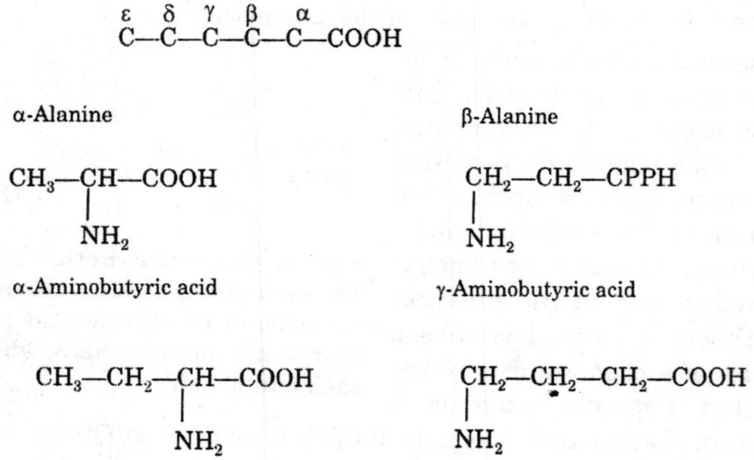

```
ε   δ   γ   β   α
C—C—C—C—C—COOH
```

α-Alanine β-Alanine

```
CH3—CH—COOH          CH2—CH2—CPPH
     |               |
    NH2              NH2
```

α-Aminobutyric acid γ-Aminobutyric acid

```
CH3—CH2—CH—COOH      CH2—CH2—CH2—COOH
         |           |
        NH2          NH2
```

Fig. 1.3. Structure of alternative forms of amino acids.

Classification

It is helpful when considering the principles and applications of methods for the determination of amino acids to be able to appreciate the characteristics of these compounds. Although it is not always essential to know the exact structural formula of individual amino acids it is useful to be able to remember particular properties or the presence of functional groups.

Table 1.1. Some naturally accurring amino acids not found in proteins

Amino acid	*Metabolic significance or tissue source*	*Formula*
α-Aminobutyric acid	Plant and animal tissues	$CH_3—CH_2—CH(NH_2)—COOH$
α, γ-Diaminobutyric acid	Antibiotics	$CH_2(NH_2)—CH_2—CH(NH_2)—COOH$
β-Alanine	Coenzyme A	$CH_2(NH_2)—CH_2—COOH$
γ-Aminobutyric acid	Brain tissue	$CH_2(NH_2)—CH_2—CH_2—COOH$
α,ε-Diaminopimelic acid	Bacterial cell wall	$COOH—CH(NH_2)—(CH_2)_3—CH(NH_2)—COOH$

Table 1.2. Classification of amino acids based on chemical structure

Chemical nature of R group	*Examples*
Aliphatic	Gly, Ala, Val, Leu
Aromatic	Phe, Tyr, Trp
Hydroxylic	Ser, Thr
Carboxylic	Asp, Glu
Sulphur containing	Cys, Met
Imino	Pro, Hyp
Amino	Lys, Arg
Amide	Asn, Gln

It is convenient to group together animo acids with similar R groups that show common chemical or physical characteristics. The classification may be based on the chemical nature of the R group and such a system makes it easier to remember the general properties of each amino acid. However, a classification system based on the polarity of the R group may also be useful and this, together with the structural formula and abbreviated forms of the α-amino acids.

Isomerism

The α-carbon of all amino acids, with the exception of glycine, has four different substituent groups and is therefore an asymmetric carbon atom. Such an atom can exist in two different

Table 1.3. Amino acids

Common name	*Abbreviation*	*Structure*	*pI*	*Comments*
Uncharged polar R groups				
Glycine	Gly	$H-C(NH_2)(COOH)-H$	6.0	Hydrogen atom on the side chain—not optically active
Serine	Ser	$H-C(NH_2)(COOH)-CH_2OH$	5.7	Hydroxyl group in side chain
Thereonine	Thr	$H-C(NH_2)(COOH)-C(H)(OH)-CH_3$	6.5	Hydroxyl group in side chain—has two asymmetric carbon atoms
Hydroxyproline	Hyp	$H-C(COOH)$ in ring with $HN-CHOH-CH_2-CH_2$	5.8	The hydroxyl group is added to praline after synthesis into protein and is only found in collagen and gelatine—has two asymmetric carbon atoms
Cysteine	Cys	$H-C(NH_2)(COOH)-CH_2-SH$	5.0	Cotains sulphur—forms disulphide bonds
Tyrosine	Tyr	$H-C(NH_2)(COOH)-CH_2-C_6H_4-OH$	5.7	Phenolic aromiatic side chain
Asparagine	Asn	$H-C(NH_2)(COOH)-CH_2-CONH_2$	5.4	Amide of aspartic acid

Common name	Abbreviation	Structure	pI	Comments
Glutamine	Gln	$H-C(NH_2)(COOH)-CH_2-CH_2-CONH_2$	5.7	Amide of glutamic acid
Non-polar R groups				
Alanine	Ala	$H-C(NH_2)(COOH)-CH_3$	6.0	Aliphatic side chain
Valine	Val	$H-C(NH_2)(COOH)-CH(CH_3)_2$	6.0	Branched aliphatic side chain
Leucine	Leu	$H-C(NH_2)(COOH)-CH_2-CH(CH_3)_2$	6.0	Branched aliphatic side chain
Isoleucine Ile		$H-C(NH_2)(COOH)-CH(CH_3)-CH_2-CH_3$	6.[illegible]	Branched aliphatic side chain—has two asymmetric carbon atoms
Proline	Pro	$HN-CH_2-CH_2-CH_2-C(H)(COOH)$ (ring)	[illegible].1	Imino acid—distorts the regular α-helix structure
Phenylalanine	Phe	$H-C(NH_2)(COOH)-CH_2-C_6H_5$	6.0	Aromatic side chain

Common name	*Abbreviation*	*Structure*	*pI*	*Comments*
Tryptophan	Typ	NH_2 / H—C—CH_2—(indole ring, NH) / COOH	5.9	Heterocyclic aromatic side chain
Methionine	Met	NH_2 / H—C—CH_2—CH_2—S—CH_3 / COOH	5.8	Aliphatic side chain contains sulphur
Polar with an extra carboxyl group				
Aspartic acid	Asp	NH_2 / H—C—CH_2—COOH / COOH	2.9	Dicarboxylic
Glutamic acid	Glu	NH_2 / H—C—CH_2—CH_2—COOH / COOH	3.2	Dicarboxylic
Polar with an extra ionizable N-containing group				
Lysine	Lys	NH_2 / H—C—CH_2—CH_2—CH_2—CH_2—NH_2 / COOH	9.7	Diamino
Hydroxylysine	Hyl	NH_2 / H—C—CH_2—CH_2—CHOH—CH_2—NH_2 / COOH	9.2	Hydroxy group is added after synthesis into protein—only found in collagen and gelatine
Arginine	Arg	NH_2 / H—C—CH_2—CH_2—CH_2—NH—C(=NH)—NH_2 / COOH	10.8	Guanidino group—important in the urea cycle
Histidine	His	NH_2 / H—C—CH_2—C═CH_2 (ring: HN, NH, C—H) / COOH	7.6	Heterocyclic imidazole group in side chain

spatial arrangements which are mirror images of each other. These structural forms of molecules are known as stereoisomers and the common notation of D and L forms is used, a nomenclature that refers to their absolute spatial configuration when compared with that of glyceraldehyde.

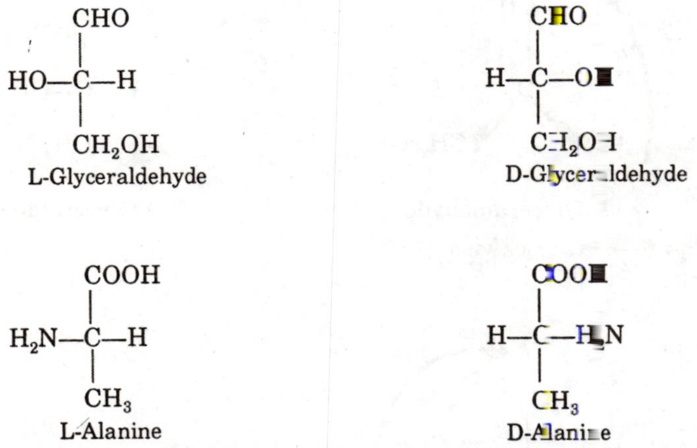

$$\begin{array}{c} CHO \\ | \\ HO—C—H \\ | \\ CH_2OH \end{array} \qquad\qquad \begin{array}{c} CHO \\ | \\ H—C—OH \\ | \\ CH_2OH \end{array}$$

L-Glyceraldehyde — D-Glyceraldehyde

$$\begin{array}{c} COOH \\ | \\ H_2N—C—H \\ | \\ CH_3 \end{array} \qquad\qquad \begin{array}{c} COOH \\ | \\ H—C—H_2N \\ | \\ CH_3 \end{array}$$

L-Alanine — D-Alanine

Fig. 1.4. The stereochemical relationship between amino acids and glyceraldehyde. The designation of D or L to an amino acid refers to its absolute configuration relative to the structure of L-glyceraldehyde respectively. The D and L forms of a particular compound are called enatiomers.

All amino acids except glycine exist in these two different isomeric forms but only the L isomers of the α-amino acids are found in proteins, although many D amino acids do occur naturally, for example in certain bacterial cell walls and polypeptide antibiotics. It is difficult to differentiate between the D and the L isomers by chemical methods and when it is necessary to resolve a racemic mixture, an isomer-specific enzyme provides a convenient way to degrade the unwanted isomer, leaving the other isomer intact. Similarly in a particular sample, one isomer may be determined in the presence of the other using an enzyme with a specificity for the isomer under investigation.

The other isomer present will not act as a substrate for the enzyme and no enzymic activity will be demonstrated. The enzyme L-amino acid oxidase, for example, is an enzyme that shows activity only with L amino acids and will not react with the D amino acids. Although the D and L notation is still commonly used in amino acid and carbohydrate terminology, modern nomenclature employs a system that permits the configuration of an asymmetric atom to be specified. This is called the *R—S* convention or the sequence rule and involves assigning a priority *(a>b>c>d)* to the four different substituent atoms attached to the asymmetric atom on the basis of the atomic number of each. The sequence of substituent groups relative to the axis between the asymmetric carbon atom and the substituent group with the lowest priority *(d)* is used to designate the atom as either *R (rectus,* right) or *S (sinister,* left).

Ionic Properties

Amino acids contain both acidic (COOH) and basic (NH_2) groups. As a result they can act as both weak acids and weak bases and are therefore called ampholytes. Their behaviour is termed amphiprotic because they can either accept or donate a proton, a reaction that can be represented by the following equation:

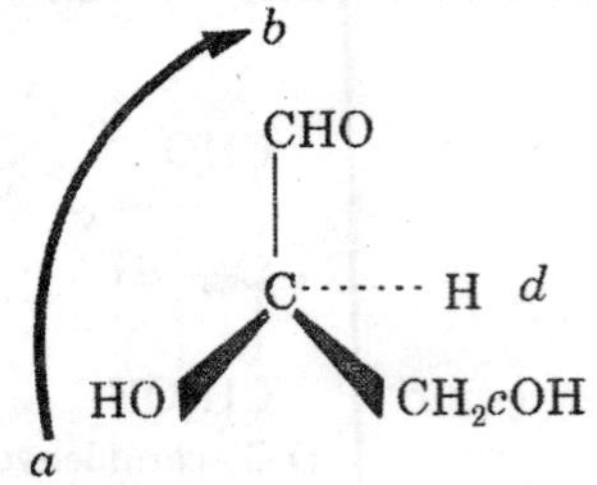

b

CHO

C······ H d

HO CH$_2$cOH

a

R-Glyceraldehyde

$a \rightarrow b \rightarrow c$ is clockwise

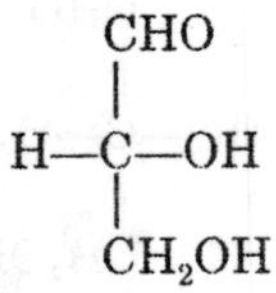

CHO

H—C—OH

CH_2OH

D-Glyceraldehyde

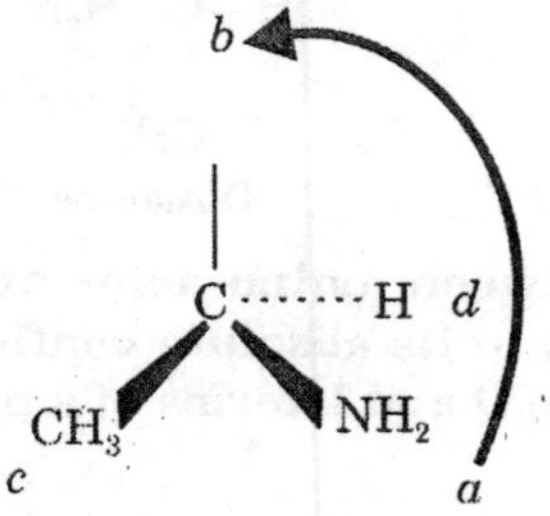

b

C······ H d

CH_3 NH_2

c a

S-Alanine

$a \rightarrow b \rightarrow c$ is anticlockwise

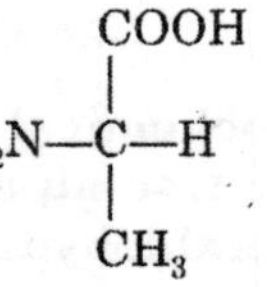

COOH

H_2N—C—H

CH_3

L-Alanine

Fig. 1.5. Systematic naming of an amino acid using the *R—S* convention. The compound is named by designating the configuration of the asymmetric carbon atom as either *R* or 5. For amino acids containing more than one asymmetric centre, *e.g.* threonine and isoleucine, the configuration about each asymmetric atom is specified.

$$R.NH_3^+.COOH \rightleftharpoons R.NH_2COOH \rightleftharpoons R.NH_2.COO^- + H^+$$

Even this representation is not completely true, because it implies that an amino acid exists in an uncharged form ($R.NH_2.COOH$), whereas the molecule in this state carries one negative and one positive charge and as a result shows no net charge. This is known as the dipolar form or **'Zwitterion'**of the amino acid.

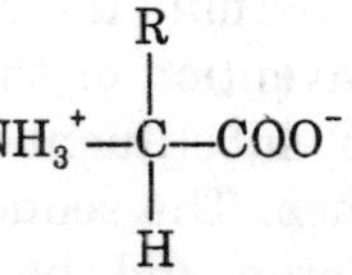

R

NH_3^+—C—COO^-

H

Fig. 1.6. Dipolar or zwitter ionic form of an amino acid. Amino acids exist in a charged form in aqueous solution, the carboxyl group being dissociated and the amino group associated. Some amino acids also have an extra lonizable group present in their side chain (R group). The ionization of each group is pH-dependent and for each amino acid there is a pH at which the charges are equal and opposite and the molecule bears no net charge. This is called the isoionic pH (pI).

In solution the dissociation of each ionizable group in the molecule may be represented as follows:

$$COOH \rightleftharpoons COO^- + H^+$$
$$NH_3^+ \rightleftharpoons NH_2 + H^+$$

The dissociated and undissociated forms of each group exist in equilibrium with each other and the position of the equilibrium (or the tendency of each of the groups to dissociate) may be expressed in terms of the equilibrium (dissociation constant) K, often termed K_a because it refers to the dissociation of groups that liberate protons, *i.e.* acids. The actual values for K_a are often very small and are conventionally expressed as the negative logarithm of the value, a term known as the pK_a value:

$$pK_a = -\log K_c$$

This function results in a numerical value which is less cumbersome to use and is comparable with the method of expressing the hydrogen ion concentration of a solution, the pH value:

$$pH = -\log [H^+]$$

The concentration of hydrogen ions liberated by the dissociation of an acid is related to the dissociation constant for that acid and this relationship can be expressed by the Henderson-Hasselbalch equation:

$$pH = pK_a + \log \frac{[salt]}{[acid]}$$

where the square brackets indicate the molar concentration of the named substance.

An examination of this equation reveals the fact that when the concentrations of salt and undissociated acid are equal, then the *p*H of the solution is numerically equal to the pK_a for that acid. The lower the value of pK_a for an acid, the greater is the ability of the acid to dissociate, yielding hydrogen ions, a characteristic known as the strength of the acid. Amino acids with two ionizable groups, an α-carboxyl and an α-amino group, will be characterized by a pK_a value for each group and the actual value will give an indication of the strength of the acidic or basic group concerned. The ionization of an amino acid is most easily demonstrated in a titration curve, which can be prepared by titrating a solution of the amino acid in the fully protonated form with a solution of sodium hydroxide and plotting the amount of alkali added against the resulting *p*H of the solution.

The titration curve for a simple amino acid will show two regions where the addition of alkali results in only a small change in the *p*H value of the mixture. The buffering action of an amino acid is most significant over these *p*H ranges. The first end-point in such a titration is due to the carboxyl group an the pK_a value for this is called pK_{a1} while the second pK value is for the amino group and is called pK_{a2}.

In practice each acid and its salt will act as a buffer over a *p*H range of approximately one unit on either side of its pK_a value. For the amino acid alanine where pK_{a1} is 2.4 and pK_{a2} is 9.6, the most effective buffering action occurs over the *p*H ranges 2.4 ± 1.0 and 9.6 ± 1.0. In addition to the α-amino and α-carboxyl groups those amino acids with an extra ionizable group will also have a pK_a value. Glutamic acid is an example of an amino acid with an extra acidic group (COOH) on the γ-carbon, and lysine is an example of an amino acid with an extra amino group on the ε-carbon atom. As a result they each have three ionizable groups and three pK_a values can be demonstrated the pK_{a3} value being for the extra group.

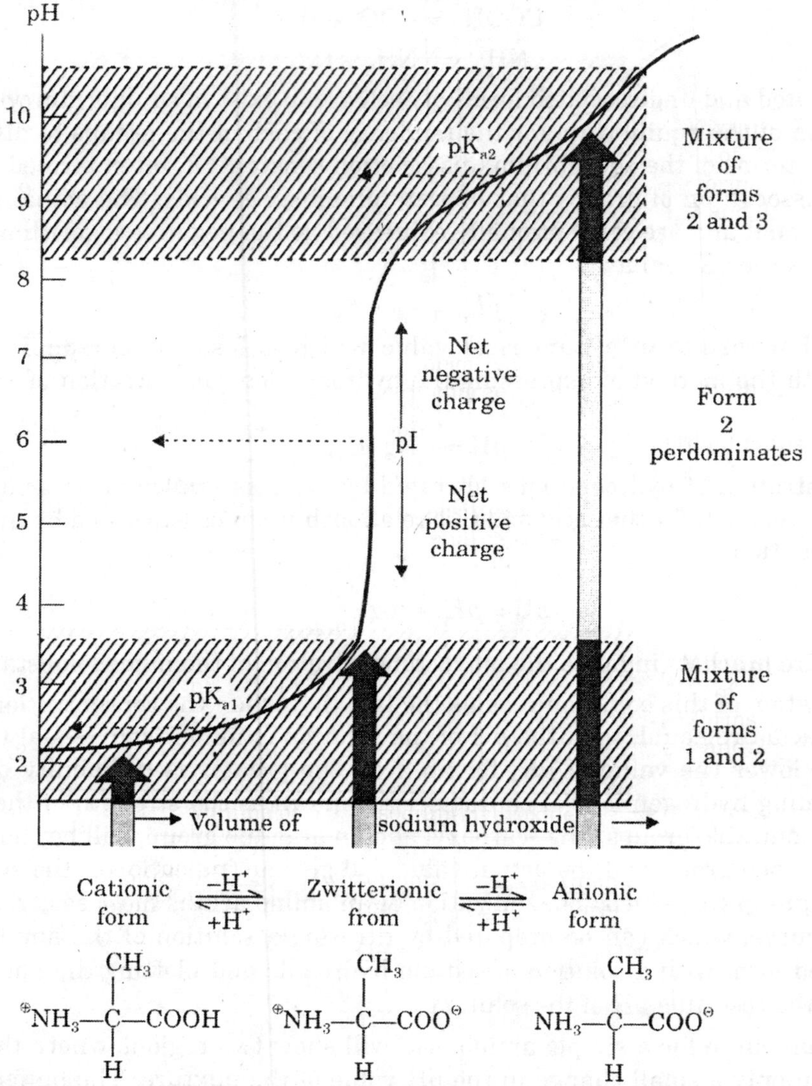

Fig. 1.7. Titration curve of alanine. A solution of alanine (0.1 mol l^{-1}) in the fully protonated form at pH 2.0 is titrated with 0.1 mol l^{-1} sodium hydroxide. The volumes of sodium hydroxide added are recorded and plotted against the resulting pH values to give a titration curve which is typical of an amino acid with only two ionizable groups (one carboxyl and one amino). The two shaded areas show the pH range over which the addition of alkali results in only a very small change in pH and where the amino acid exhibits its most significant buffering action. At a pH equivalent to pK_{a1}, there are equal amounts of forms 1 and 2, while at a pH equivalent to pK_{a2}, forms 2 and 3 are in equal concentrations. The pI value for alanine is 6.0 and is the mean of pK_{a2} (2.4) and pK_{a2} (9.6). At a pH below its pI value, an amino acid will carry a net positive charge but it will carry a net negative charge at pH values greater than its pI.

Table 1.4.

Amino acid	*Formula*	*pK_a values*
Glutamic acid (extra COOH group)	COOH*	4.1
	|	
	CH_2	
	|	
	CH_2	
	|	
	H—C—NH_2*	9.5
	|	
	COOH*	2.1
Lysine (extra NH_2 group)	NH_2*	10.8
	|	
	CH_2	
	|	
	CH_2	
	|	
	CH_2	
	|	
	CH_2	
	|	
	H—C—NH_2*	9.2
	|	
	COOH*	2.2

The asterisks (*) show where proton gain or loss can occur.

Table 1.5. pK_a values for some amino acids

Amino acid	*Extra ionizable group*	*α-Carboxyl* pK_{a1}	*α-Amino* pK_{a2}	*Extra group* pK_{a3}
Arginine	Guanidinium	[illegible].8	8.9	12.5
Cysteine	Sulphydryl	[illegible].7	10.8	8.3
Tyrosine	Pheolic	2.0	9.1	10.1
Histidine	Imidazole	1.8	9.2	6.0

Other functional groups present in an amino acid may also be ionizable and will have characteristic pK_a values. Such amino acids result in complex titration curves. The overall charge carried by an amino acid depends upon the pH of the solution and the pK_a values of the ionizable groups present. If the pH is greater than the pK_a value for a group, a proton will be lost and the molecule will carry a negative charge, but if the pH is less than the pK_a value, a positive charge will predominate. The fact that at different pH values different amino acids will be present in different ionic forms and will carry different net charges is utilized in many analytical methods, *e.g.* electrophoresis and ion-exchange chromatography. The **iso-ionic point** of a molecule is the pH at which the number of negative charges due to proton loss equal the number of positive charges due to proton gain and the zwitterionic form predominates.

The iso-electric point (p*I*) is the pH of the solution at which the molecules show no migration in an electric field and can be determined experimentally by electrophoresis: for amino acids it is equal to the iso-ionic point. The iso-ionic point of an amino acid with one carboxylic and one amino group is the mean of the two pK_a values. However, when three ionizable groups are present, the effect of an extra acid group will be to reduce the ionic character of the other acid group and hence the pI value will not be the mean of the three separate pK_a values but will more closely approximate to the mean of the closest pK_a values.

Peptides

When two amino acids are linked together by the condensation of the α-amino group from one amino acid and the α-carboxyl group from another to form a peptide bond, the resulting compound is called a dipeptide. The ionic character of the constituent amino acids will be modified due to the loss of either an amino or a carboxyl group and the properties of the dipeptide will depend not only on the terminal amino and carboxyl groups but also on any ionizable R groups. For peptides containing increasing numbers of amino acids the significance of the two terminal groups (COOH and NH_2) becomes less important and the ionic nature of the R groups becomes more important. Molecules containing many amino acids linked in such a manner are known as polypeptides and are generally only classed as proteins when they are composed of more than 50 amino acids and their relative molecular mass exceeds 5000.

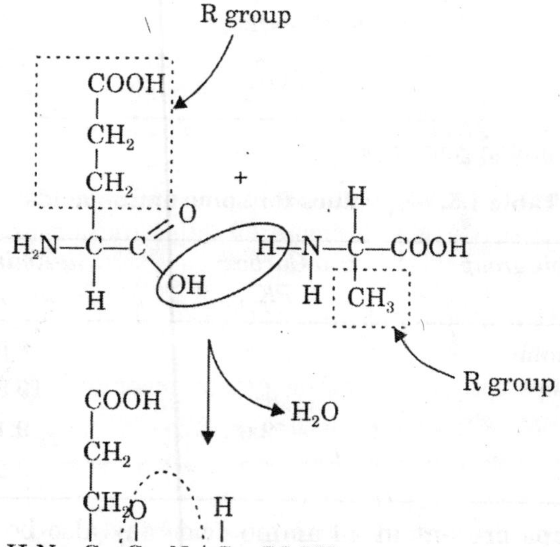

R group

COOH
|
CH_2
|
CH_2
|
$H_2N—C—C(=O)OH$ + $H—N(H)—C(H)(CH_3)—COOH$
|
H

R group

H_2O

COOH
|
CH_2
|
CH_2
|
$H_2N—C(H)—C(=O)—N(H)—C(H)(CH_3)—COOH$

Peptide bond

Fig. 1.8. Structure of a dipeptide. The peptide bond joins glutamic acid and alanine by condensation of the α-carboxyl group of glutamic acid and the α-amino group of alanine. The resulting dipeptide is called glutamylalanine, which can be abbreviated to NH_2-Glu-Ala-COOH or Glu-Ala.

The peptide bonds in proteins are between the α-amino and the α-carboxyl groups but peptides do occur naturally where the peptide linkage involves a carboxyl or amino group which is attached to a carbon atom other than the α-carbon. The amino acid glutamic acid contains two carboxyl groups attached to the α- and γ-carbon atoms and either may be involved in peptide linkages. A dipeptide formed between the γ-carboxyl group of glutamic acid and the amino group of alanine is called γ-glutamylalanine.

Linkages may also be formed between the ε-amino group of lysine and other amino acids. Peptides are also found whose constituents are amino acids other than the ε-amino acids; carbosine, for instance, is a dipeptide found in muscle and consists of β-alanine and histidine. The only residual evidence of the amino acids that make up polypeptides is their R group and the individual components are now termed amino acid residues.

It is conventional when representing peptides to show the terminal amino acid residue with the free amino group to the left of any diagram and to designate it Residue 1 (the N-terminal residue) and the one with the free carboxyl group (the C-terminal residue) on the right of any diagram. Many naturally occurring hormones and antibiotics are polypeptides and investigation into both the amino acid constituents and their sequence in the polypeptide chain are important areas of research. These investigations may reveal information regarding the biologically active part of the molecule, a fact that may then be used in the commercial production of a synthetic peptide containing only that small part of the original polypeptide but showing a physiological activity comparable with the whole molecule.

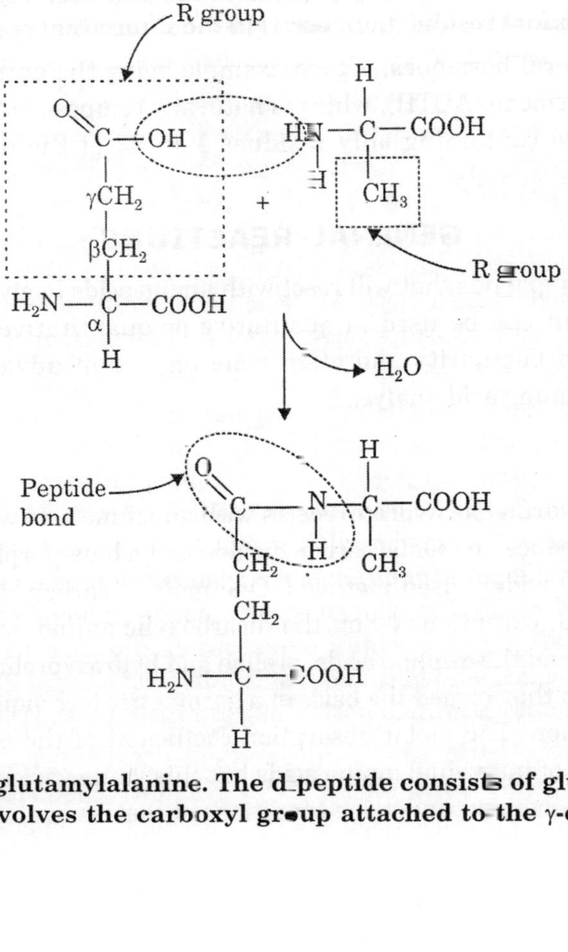

Fig. 1.9. Structure of γ-glutamylalanine. The dipeptide consists of glutamic acid linked by a peptide bond which involves the carboxyl group attached to the γ-carbon atom and the α-amino group of alanine.

N-terminal residue

1 Ser — Tyr — Ser — Met — 5 Glu — His — Phe — Arg — Trp — 10 Gly —

Lys — Pro — Val — Gly — 15 Lys — Lys — Arg — Arg — Pro — 20 Val —

Lys — Val — Tyr — Pro — 25 Asn — Gly — Ala — Glu — Asp — 30 Glu —

Ser — Ala — Glu — Ala — 35 Phe — Pro — Leu — Glu — 39 Phe

C-terminal residue

Fig. 1.10. Amino acid sequence of human adrenocorticotrophin. The amino acid residues in this polypeptide hormone are linked by peptide bonds and each residue is given a number starting with the N-terminal residue (number 1) to the C-terminal residue.

This is true for several hormones, a good example being the anterior pituitary hormone, adrenocorticotrophic hormone (ACTH), which is naturally composed of 39 amino acid residues, but a synthetic peptide containing only residues 1 to 23 of the original hormone shows comparable physiological activity.

GENERAL REACTIONS

There are several compounds that will react with amino acids to give coloured or fluorescent products and as a result can be used in qualitative or quantitative methods. Fluorimetric methods are gaining in popularity and offer some important advantages over absorption spectrophotometry for amino acid analysis.

Ninhydrin

Ninhydrin (triketohydrindene hydrate) reacts with an amino acid when heated under acidic conditions (pH 3-4) to produce ammonia, carbon dioxide and a blue-purple complex. This reaction forms the basis of many widely used methods. One mole of carbon dioxide is liberated from each mole of amino acid, exceptions being the dicarboxylic amino acids, which produce two moles of carbon dioxide, and the α-imino acids, proline and hydroxyproline, which do not produce carbon dioxide. Although this formed the basis of a gasometric technique, colorimetric methods are now the most common. The molar absorption coefficient of the coloured product can be used in the quantitation of individual amino acids but this value varies from one amino acid to another and must be determined under the conditions of the assay.

An accepted value, however, must be used in the quantitation of the total amino acids in a mixture when absorbance readings are normally taken at 570 nm. Amines other than α-amino acids will also give a colour reaction with ninhydrin but without the production of carbon dioxide. Thus *β*-, γ-, δ- and ε-amino acids and peptides react more slowly than α-amino acids, to give the blue complex, while imino acids result in the formation of a yellow-coloured product which can be measured at 440 nm.

Removal of substances such as protein, ammonia and urea from biological samples may be necessary in quantitative work because they also react in a similar manner. The ninhydrin colour reaction has proved very useful in qualitative work and is widely used in the visualization of amino acid bands after electrophoretic or chromatographic separation of mixtures. The reagent used in such circumstances is usually prepared in ethanol and, if 2,4,6-collidine is added, the variations in colour produced by different amino acids will aid their identification.

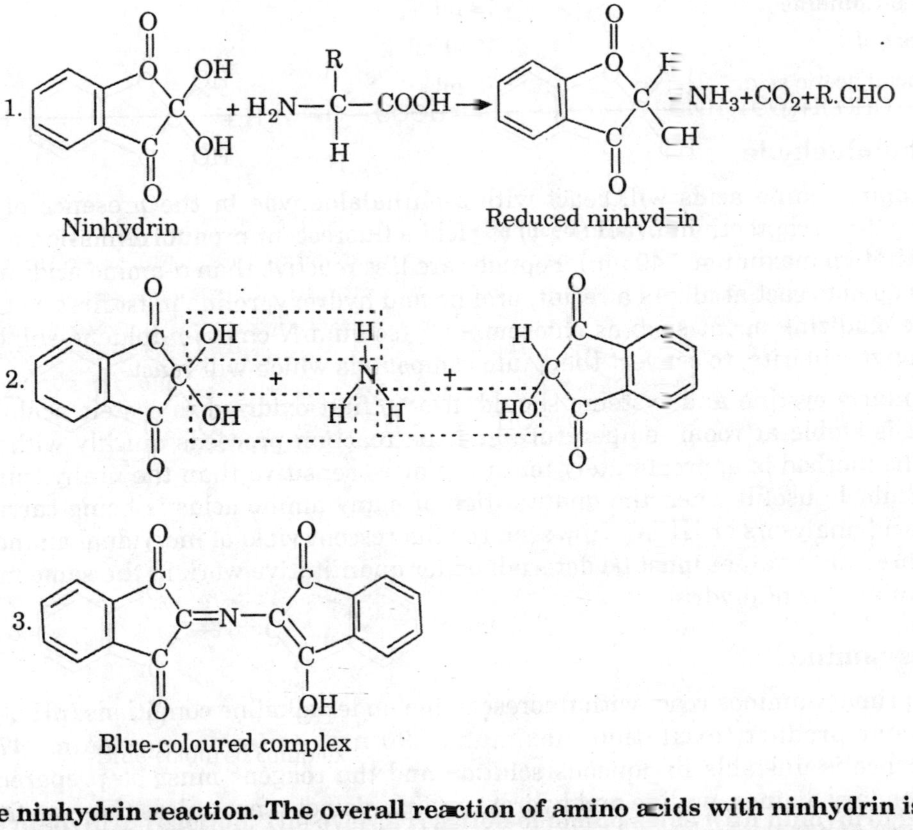

1. Ninhydrin + $H_2N-C(R)(H)-COOH \rightarrow$ Reduced ninhydrin + $NH_3 + CO_2 + R.CHO$

2. Ninhydrin + NH_3 + Reduced ninhydrin

3. Blue-coloured complex

Fig. 1.11. The ninhydrin reaction. The overall reaction of amino acids with ninhydrin is:

1. **oxidative decarboxylation of the amino acid and the production of reduced ninhydrin, ammonia and carbon dioxide;**
2. **reduced ninhydrin reacts with more ninhydrin and the liberated ammonia;**
3. **a blue-coloured complex is formed.**

Table 1.6. Modified ninhydrin reagent

Amino acid		*Colour produced*
Histidine		Brown
Phenylalanine		Brown/grey
Glycine		Blue
Glutamic acid		Bright blue
Lysine		Grey/blue
Tyrosine		Grey
Proline		Orange
Hydroxyproline		Orange
Aspartic acid		Orange/yellow
Reagent composition		
Ninhydrin	2.5 g	
2,4,6-Collidine	73 ml	
Ethanol	1750 ml	
Glacial acetic acid	73 ml	

o-Phthalaldehyde

Primary amino acids will react with *o*-phthalaldehyde in the presence of the strongly reducing 2-mercaptoethanol (pH 9-1 1) to yield a fluorescent product (emission maximum, 455 nm; excitation maximum, 340 nm). Peptides are less reactive than α-amino acids and secondary amines do not react at all. As a result, proline and hydroxyproline must first be treated with a suitable oxidizing agent such as chloramine T (sodium *N*-chloro-*p*-toluene-sulphonamide) or sodium hypochlorite, to convert them into compounds which will react.

Similarly cystine and cysteine should also be first oxidized to cysteic acid. The aqueous reagent is stable at room temperature and the reaction proceeds quickly without requiring heat. The method is approximately ten times more sensitive than the ninhydrin method and is particularly useful when the quantitation of many amino acids is being carried out using amino acid analysers or HPLC. However, the fluorescent yield of individual amino acids varies and fluorescence values must be determined for quantitative work in the same manner as the colour values for ninhydrin.

Fluorescamine

All primary amines react with fluorescamine under alkaline conditions (pH 9-11) to form a fluorescent product (excitation maximum, 390 nm; emission maximum, 475 nm). The fluorescence is unstable in aqueous solution and the reagent must be prepared in acetone. The secondary amines, proline and hydroxyproline, do not react unless they are first converted to primary amines, which can be done using *N*-chlorosuccinimide. Although the reagent is of interest because of its fast reaction rate with amino acids at room temperature, it does not offer any greater sensitivity than the ninhydrin reaction.

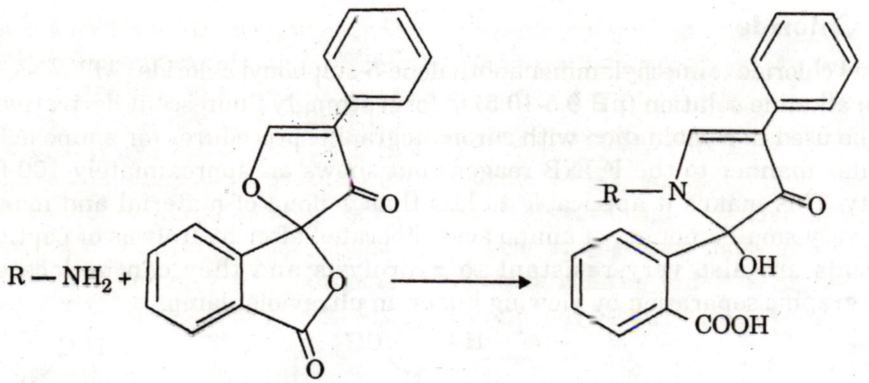

Primary amine Fluorescamine (non-fluorescent) Fluorescent product

Fig. 1.12. The reaction of fluorescamine with a primary amino group. The reaction of an amino acid containing a primary amino group with a solution of fluorescamine in acetone at pH 9.0 results in the conversion of the non-fluorescent fluorescamine to a fluorescent product.

N-TERMINAL ANALYSIS

1-Fluoro-2,4-Dinitrobenzene

1-Fluoro-2,4-dinitrobenzene (FDNB) reacts in alkaline solution (pH 9.5) with the free amino group of an amino acid or a peptide to form a yellow dinitrophenyl (DNP) derivative. This reaction cannot be used for the accurate quantitation of mixtures of amino acids because the molar absorption coefficients of the DNP derivatives of different amino acids vary, but it is very useful in qualitative methods of analysis. The yellow DNP derivatives of free amino acids in a sample can be clearly seen after separation by paper or thin layer chromatography and, for identification purposes, comparison of the R_F values is made with those of known amino acids treated in an identical manner. FDNB is a hazardous substance and should be used only in exceptional circumstances and under strict safety conditions.

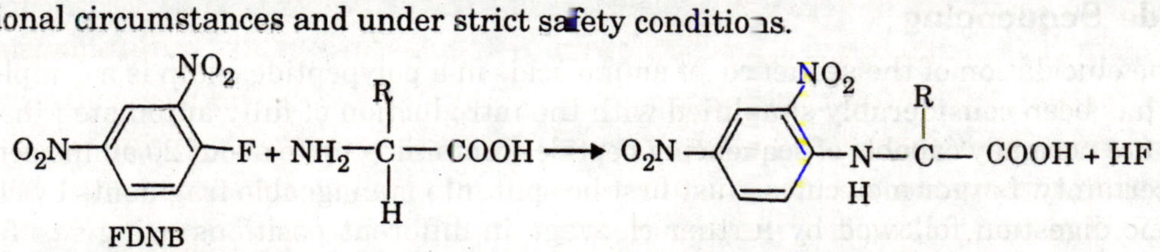

$O_2N-C_6H_3(NO_2)-F + NH_2-CH(R)-COOH \longrightarrow O_2N-C_6H_3(NO_2)-NH-CH(R)-COOH + HF$

FDNB

Fig. 1.13. The reaction of FDNB with compounds containing a free amino group. The reaction at pH 9.5 between FDNB and amino acids or peptides results in the formation of yellow-coloured dinitrophenyl derivatives.

The amino group of the N-terminal amino acid residue of a peptide will react with the FDNB reagent to form the characteristic yellow DNP derivative, which may be released from the peptide by either acid or enzymic hydrolysis of the peptide bond and subsequently identified. This is of historic interest because Dr F. Sanger first used this reaction in his work on the determination of the primary structure of the polypeptide hormone insulin and the reagent is often referred to as Sanger's reagent.

Dansyl Chloride

Dansyl chloride (dimethylaminonaphthalene-5-sulphonyl chloride) will react with free amino groups in alkaline solution (pH 9.5-10.5) to form strongly fluorescent derivatives. This method can also be used in combination with chromatographic procedures for amino acid identification in a similar manner to the FDNB reagent but shows an approximately 100-fold increase in sensitivity. This makes it applicable to less than 1 nmol of material and more amenable for use with very small amounts of amino acids liberated after hydrolysis of peptides. The dansyl amino acids are also very resistant to hydrolysis and they can be located easily after chromatographic separation by viewing under an ultraviolet lamp.

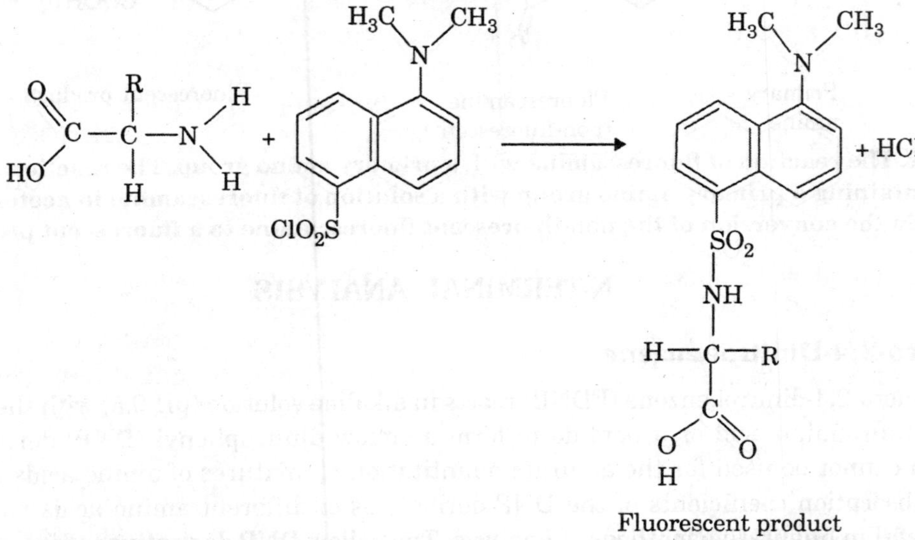

H_3C CH_3 N R O H C—C—N + HO H H ClO_2S → H_3C CH_3 N + HCl SO_2 NH H—C—R C O O H

Fluorescent product

Fig. 1.14. The reaction of dansyl chloride with compounds containing a free amino group. At an alkaline pH, the reaction results in the formation of fluorescent derivatives of free amino acids and the N-terminal amino acid residue of peptides.

Peptide Sequencing

The elucidation of the sequence .of amino acids in a polypeptide chain is a complex process but it has been considerably simplified with the introduction of fully automated instruments. These are normally capable of sequencing peptides containing up to about 20 amino acid residues with certainty. Larger molecules must first be split into manageable fragments by chemical or enzymic digestion followed by further cleavage in different positions using residue-specific enzymes in order to produce the overlapping fragments required for full analysis of a protein or long polypeptide.

Peptide sequencers automatically carry out all the reactions of the Edman degradation procedure under controlled conditions, and a typical scheme is described below.

The released N-terminal derivatives are then analysgd by reverse-phase HPLC. The sample to be sequenced is first applied to a solid support within an enclosed chamber. A variety of support media are available but glass fibre discs offer the advantage of high reaction efficiencies owing to their large surface area. Pre-treatment of the disc with polybrene confers a slight charge to the surface, which attracts the proteins but does not produce the problems which were experienced when peptides were covalently bound to a glass support. The programmed cycle begins with the removal of the oxygen in the chamber by purging with argon.

Trimethylamine is added to give the alkaline conditions required for the coupling of the phenylisothiocyanate to the N-terminus of the peptide. Trifluoroacetic acid (100%) is then used to cleave the N-terminal derivative, leaving the N-terminus of the remaining peptide ready for the next cycle. The anilinothiazolozone (ATZ) derivative released is eluted from the disc with butyl chloride into a separate chamber where it is converted to the more stable phenylthiohydantoin (PTH) derivative by the addition of trifluoroacetic acid (25%). After drying with argon and reconstitution in approximately 100 μl of solvent, the sample is ready for analysis by reverse-phase HPLC.

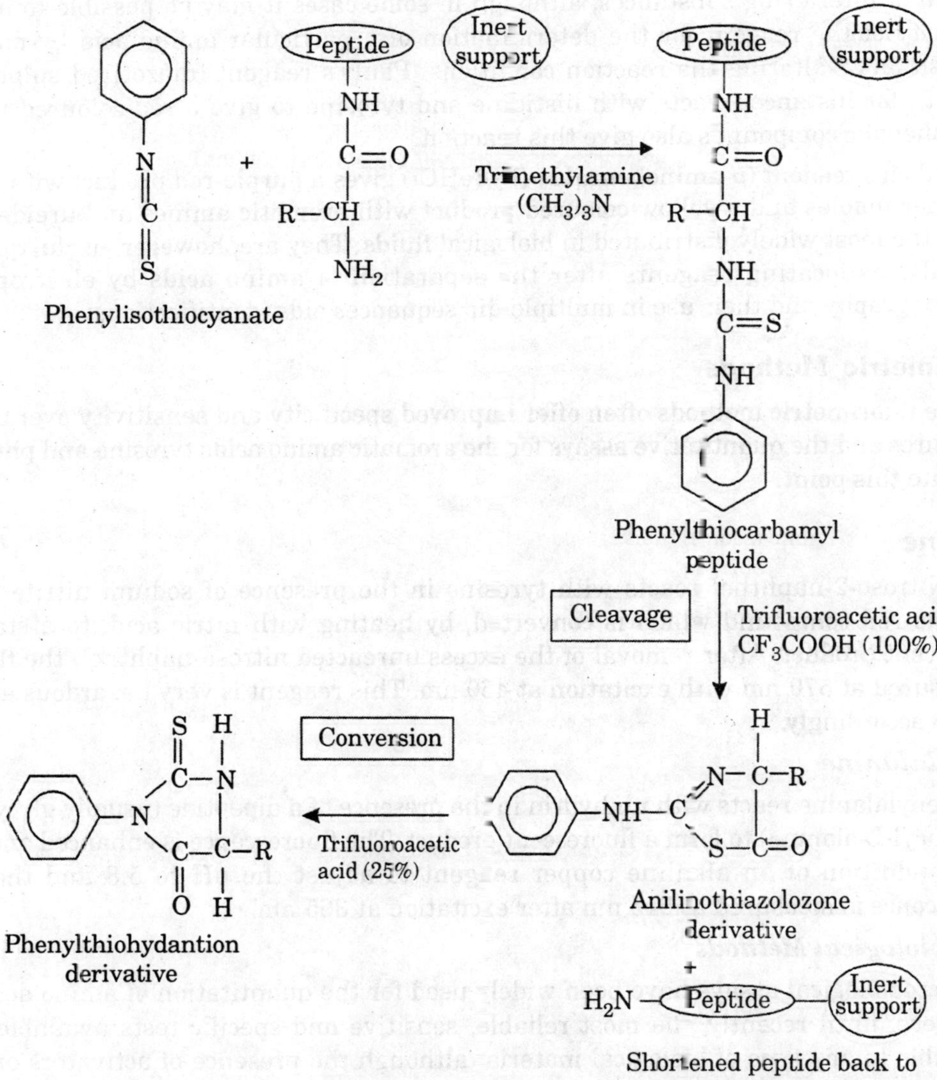

Phenylisothiocyanate + Peptide, Inert support

Trimethylamine $(CH_3)_3N$

Phenylthiocarbamyl peptide

Cleavage — Trifluoroacetic acid CF_3COOH (100%)

Anilinothiazolozone derivative + H_2N — Peptide — Inert support

Shortened peptide back to start for next cycle

Conversion — Trifluoroacetic acid (25%)

Phenylthiohydantion derivative

Fig. 1.15. Solid-phase peptide sequencing procedure based on Edman degradation technique.

Reactions of Specific Amino Acids

It is often difficult to quantitate one particular amino acid in the presence of others because of chemical similarities. Interference from substances other than amino acids is also a problem in many reputedly specific methods. Ultraviolet spectroscopy is of little value in the detection of aromatic amino acids because they have similar absorbance maxima and considerably different molar absorption coefficients.

Colorimetric Methods

There are several colour reagents which are of little quantitative value without the prior removal of interfering substances, although in some cases it may be possible to increase the specificity of the reagent for the determination of a particular amino acid by modifying its composition or altering the reaction conditions. Pauly's reagent (diazotized sulphanilic acid reagent), for instance, reacts with histidine and tyrosine to give a red-coloured product but other phenolic compounds also give this reaction.

Erhlich's reagent (*p*-aminobenzalde-[hydeHCl) gives a purple-red product with tryptophan and other indoles and a yellow-coloured product with aromatic amines and ureides, of which urea is the most widely distributed in biological fluids. They are, however, useful qualitatively, especially as locating reagents after the separation of amino acids by electrophoresis or chromatography and their use in multiple-dip sequences aids identification.

Fluorimetric Methods

The fluorimetric methods often offer improved specificity and sensitivity over colorimetric procedures and the quantitative assays for the aromatic amino acids tyrosine and phenylalanine illustrate this point.

Tyrosine

1-Nitroso-2-naphthol reacts with tyrosine in the presence of sodium nitrite to form an unstable red compound which is converted, by heating with nitric acid, to a stable yellow fluorescent product. After removal of the excess unreacted nitroso-naphthol, the fluorescence is measured at 570 nm with excitation at 460 nm. This reagent is very hazardous and must be treated accordingly.

Phenylalanine

Phenylalanine reacts with ninhydrin in the presence of a dipeptide (usually glycyl-L-leucine or L-leucyl-L-alanine) to form a fluorescent product. The fluorescence is enhanced and stabilized by the addition of an alkaline copper reagent to adjust the pH to 5.8 and the resulting fluorescence is measured at 515 nm after excitation at 365 nm.

Microbiological Methods

Microbiological assays have been widely used for the quantitation of amino acids because they were, until recently, the most reliable, sensitive and specific tests available. They are applicable to any type of biological material although the presence of activators or inhibitors sometimes causes problems. Such methods lend themselves to the analysis of large batches of samples and are inexpensive, but they are time consuming and not suitable for all amino acids.

They have now been largely superseded by ion-exchange chromatography for the quantitation of specific amino acids and will only be discussed briefly. Certain microorganisms require amino acids for growth and without them they cannot replicate.

Procedure : Fluorimetric determination of phenylalanine

Reagents

Copper regent

Sodium carbonate	1.6 g l^{-1}
Sodium potassium tartrate	65 g l^{-1}
Copper sulphate	60 mg l^{-1}

Buffer pH 5.8

Sodium succinate (0.3 mol l^{-1})

Dipeptide reagent

L-Leucyl-L-alanine (5 m mol l^{-1})

or Glycyl-L-leucine (5 m mol l^{-1})

Ninhydrin reagent

Triketohydrindene hydrate (30 m mol l^{-1})

NB. This reagent is hazardous and must be handled in accordance with approved procedures.

Method

1. Mix

   20 μl sample

   20 μl succinate buffer

   80 μl ninhydrin reagent

   40 μl dipeptide reagent.
2. Heat at 60°C for 2 h and then cool to 20°C
3. Add 2 ml copper reagent.
4. Measure the fluorescence at 515 nm after excitation at 365 nm.

Calculation

The fluorescence resulting from an identical treatment of several standard solutions of phenylalanine (0.1–1.0 mmol 1^{-1}) is used to calculate the test concentration.

Many strains of such microorganisms have been produced which show dependence on a particular amino acid. Hence attempts to culture such a microorganism in the presence of only a small amount of that amino acid will result in a limited degree of growth. This can be assessed using turbidimetry or by measuring the increase in lactic acid production by either microtitration or pH change. A modification of the microbiological assay which utilizes diffusion in gels has been successfully introduced into clinical biochemistry laboratories for the mass screening of blood samples for the raised phenylalanine levels found in phenylketonuria (PKU), which is an inherited disorder of amino acid metabolism.

It is often called the 'Guthrie test' after its originators, Guthrie and Susi, and is the most extensively used microbiological assay for the measurement of an amino acid. It is a bacterial inhibition assay and is based on the ability of phenylalanine to counteract the effects of a competitive metabolic antagonist *β*-2-thienylalanine on the growth of a special strain of *Bacillus subtilis* which requires phenylalanine as a growth factor. The assay is performed on a layer of agar in which is incorporated a mixture of the suspension of *Bacillus subtilis* spores, the minimum amount of growth nutrients and a fixed amount of the metabolic antagonist β-2-thienylalanine.

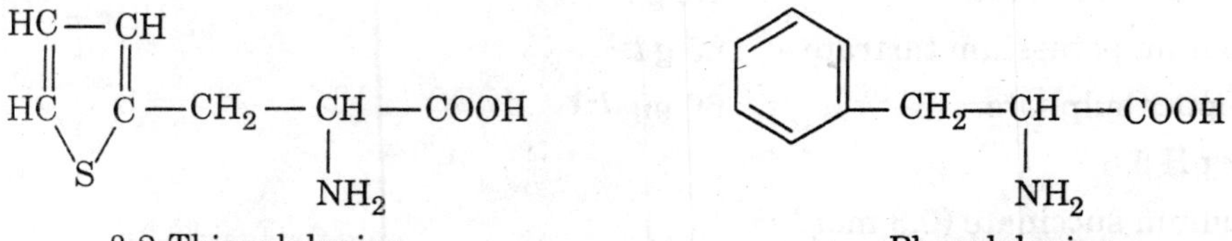

HC—CH
‖ ‖
HC C—CH_2—CH—COOH
\S/ |
NH_2

β-2-Thienylalanine

(benzene ring)—CH_2—CH—COOH
|
NH_2

Phenylalanine

Fig. 1.16. The Guthrie test. The similarity in structure between phenylalanine and its metabolic antagonist, β-2-thienylalanine, provides the basis for a microbiological assay for phenylalanine.

Blood-soaked filter paper discs of identical diameter (approximately mm) are placed on the surface of the agar together with a range of phenylalanine standards also in the form of blood discs and the agar plates are incubated overnight at 37°C. Bacterial growth will occur only when the concentration of phenylalanine in the blood discs is sufficient to overcome the effects of the metabolic antagonist, resulting in zones of growth around each disc. The following day the diameter of bacterial growth around each disc is measured and is related to phenylalanine concentration.

Enzymic Methods

There are several enzymes that, in theory, may be used for quantitation but because they react with more than one amino acid cannot be used to measure an individual amino acid in a mixture. The specificity of an enzyme for a particular isomer may be used for measurement of D amino acids in the presence of the L isomers, or vice versa, and the amino acid oxidases are useful in this respect. They catalyse the oxidative deamination of amino acids:

$$\text{Amino acid} + O_2 \xrightarrow{\text{amino acid oxidase}} \text{oxo acid} + H_2O_2 + NH_3$$

D-Amino acid oxidase (EC 1.4.3.3) extracted from sheep kidney possesses low selectivity and at pH 8-9 will oxidise many D amino acids, whereas L-amino acid oxidase (EC 1.4.3.2) from snake venom *(Crotalus adamanteus)* at pH 8-9 catalyses the oxidation of many L amino acids. However, as these enzymes show different reactivity towards different amino acids, the results for a sample that contains several D and L amino acids may be difficult to interpret.

The use of these enzymes is therefore only recommended for the measurement of one isomer of an isolated amino acid. They may also be used to remove an unwanted isomer from a sample containing both to allow subsequent measurement of the other. One approach to the measurement of amino acids using an amino acid oxidase is to measure the amount of ammonia formed during the reaction either using an ion-selective electrode or by linking it to the oxidation of NADH by the enzyme glutamate dehydrogenase (EC 1.4.1.3).

In an alternative method the amount of hydrogen peroxide formed is measured either using an ion-selective electrode or by the oxidation of a suitable chromogen using a peroxidase. These procedures are common to other assays employing oxidases (for example, glucose oxidase). L-Amino acid oxidase has been used to measure L-phenylalanine and involves the addition of a sodium arsenate-borate buffer, which promotes the conversion of the oxidation product, phenylpyruvic acid, to its enol form, which then forms a borate complex having an absorption maximum at 308 nm.

Tyrosine and tryptophan react similarly but their enol-borate complexes have different absorption maxima at 330 and 350 nm respectively. Thus by taking absorbance readings at these wavelengths the specificity of the assay is improved. The assay for L-alanine may also be made almost completely specific by converting the L-pyruvate formed in the oxidation reaction to L-lactate by the addition of lactate dehydrogenase (EC 1.1.1.27) and monitoring the oxidation of NADH at 340 nm. A group of enzymes which may be employed in the measurement of L amino acids are the L-amino acid decarboxylases (EC 4.1.1) of bacterial origin, many of which are substrate specific. They catalyse reactions of the type:

$$NH_2\text{—}\underset{}{\overset{\displaystyle R\\ |}{CH}}\text{—}COOH \longrightarrow R\text{—}CH_2\text{—}NH_2 + CO_2$$

Some enzymes with improved single amino acid specificity are commercially available.

An example is phenylalanine dehydrogenase (EC 1.4.1.1), derived from bacterial sources, which acts on phenylalanine with the simultaneous conversion of NAD to NADH. Quantitation of the phenylalanine is based on determining the amount of NADH produced using standard procedures. In the direct methods, the absorbance at 340 nm is measured, whereas in the colorimetric methods, the reaction is coupled to an electron acceptor such as *l*-methoxy-5-methyl-phenazium methylsulphate (1-MPMS) or alternatively to a tetrazolium salt. Variations in the substrate specificity of enzymes derived from different sources does occur and cross-reactivity should always be checked when developing an enzymic assay. This includes an investigation of the interference from a variety of substances that may be present in the sample in addition to studies on amino acid specificity.

SEPARATION OF AMINO ACID MIXTURES

The identification and quantitation of the individual amino acids in a mixture is often required in metabolic studies and investigations of protein structure. The use of thin-layer chromatography or electrophoresis may be adequate to indicate the relative amounts and number of different amino acids in a sample but the use of gas-liquid chromatography or an amino acid analyser is essential for quantitative analysis.

Paper and Thin-layer Chromatography

Paper chromatography has been used successfully for many years and is still a useful tool despite the fact that thin-layer techniques, especially with readily available commercially prepared plastic or foil-backed plates, offer advantages of speed, resolution and easier handling. Larger volumes of sample can be applied to paper, permitting the subsequent elution of a particular amino acid for further purification and analysis, and this may be of particular importance in the identification of an unknown sample constituent. Prior to chromatography, it may be necessary to remove interfering substances such as protein carbohydrates and salts and this may be done using an ion-exchange resin.

A small column containing a cation-exchange resin, *e.g.* Zeo-Karb 225, is prepared in the acid form by treating it with hydrochloric acid (2 mol l^{-1}). After the resin column has been washed with water, the acidified sample is applied and the interfering substances washed through with water and discarded. The amino acids are retained on the resin and may be subsequently eluted by adding a small volume of ammonia solution (2 mol l^{-1}) to the column and washing through with distilled water. The alkaline eluent is collected and reduced in volume using a rotary evaporator. Other techniques such as solvent extraction, dialysis or protein precipitation may also be used to separate the amino acids from the other components of the sample.

Table 1.7. Solvents for both paper and thin-layer chromatography

Solvent	*Proportions*	*Comments*
n-Butanol Glacial acetic acid Water	12 3 5	Very widely used for one-way runs or as a first solvent in two-way chromatography
n-Butanol Acetone Glacial acetic acid Water	7 7 2 4	Very widely used for one-way runs or as a first solvent in two-way chromatography
Phenol Water	160 g 40 ml	Gives a large spread of R_F values for a range of amino acids. Must always be used second in any two-way runs. The removal of the phenol is time consuming. Precautions must be exercised in its use because of the toxic and corrosive nature of phenol
Phenol Water Ammonia	160 g 40 ml 1 ml	Useful for the separation of basic amino acids and must always be used second in any two-way runs. The addition of ammonia causes discoloration of the solvent and so should be prepared frequently. Other comments as above for phenol
n-Butanol Acetone Diethylamine Water	10 10 2 5	This can often replace phenol as the solvent in some two-way runs on paper. For thin-layer work it should be used as the first solvent. The diethylamine must be removed from the paper before locating agents are used
Isopropanol Formic acid Water	20 1 5	This gives very good separation if used second for thin-layer chromatography

The identification of an amino acid is achieved by comparison of R_F values with those of reference solutions and the use of at least three different solvent systems is recommended before its identity can be established with any degree of certainty. The nature of the amino acids is an important factor in the choice of a solvent and different solvents will permit better resolution of acidic, basic or neutral components. In general, increasing the proportion of water in the solvent will increase all R_F values and the introduction of small amounts of ammonia will increase the R_F of the basic amino acids.

Some solvents contain noxious chemicals, *e.g.* phenol, and this may restrict their routine use. The chemical composition may also limit the range of locating reagents which can be satisfactorily applied. For example, sulphanilic acid reagent cannot be used with phenolic solvents. The resolving power of paper or thin-layer chromatography can be increased by the use of two-dimensional techniques, which involve the use of two different solvent systems.

A larger volume (×3) of sample than is normally used for one-dimensional chromatography separation is applied to one corner of the support medium and separation in the first dimension is carried out. The chromatogram is then air dried thoroughly, turned through an angle of 90° and run in the second solvent. After further drying, it is dipped in the chosen locating reagent.

In two-dimensional chromatography the composition of the two solvents will determine the order in which they are used. Two-dimensional separations permit the resolution of large numbers of amino acids present in a sample and those having a similar mobility in one dimension will usually be separated from each other in the second. This is especially useful in the detection of components that are present only in low concentrations and might be obscured in one dimension by other amino acids that are present in higher concentrations.

Locatmg Reagents

Reagents used for the visualisation of amino acids on the dried chromatogram may be applied either by spraying or dipping. Those commonly used produce intensely coloured bands with approximately 20 n mol of each amino acid for paper chromatography and 5 n mol for thin-layer separations, although smaller amounts can be detected.

Ninhydrin is the most commonly used reagent. If the composition is modified from the original 2.0 g l^{-1} in acetone by the addition of acetic acid and 2,4,6-collidine, the colours produced vary from different amino acids and this greatly aids interpretation and identification. All the α-amino acids will react in the cold within a few hours and if heat is applied for 10 min at about 100 °C, all compounds containing a primary or secondary amino group attached to an aliphatic carbon atom will react with the formation of various colours. Thus, if a compound yields a colour on heating but not in the cold, it is almost certainly not an α-amino acid.

The ninhydrin colours will slowly fade, especially in the presence of strong acid fumes, but are more stable if the chromatogram is stored in the dark at 4 °C. Preservation of the spots can also be achieved by treatment with a solution of a copper or nickel salt but the colours are altered to deep pink. Alternatively cadmium acetate (1.0 g l^{-1}) may be incorporated into the ninhydrin reagent. Other reagents that are more specific for particular amino acids have been described and their use significantly assists in the identification process.

The different locating reagents may be applied either to separate chromatograms or as part of a multi-dip sequence, when they should be used in the recommended order to prevent interference of one reagent by another. Many reagents are hazardous and must be handled in accordance with approved safety procedures. The formation of **DNP or dansyl amino acid derivatives** followed by chromatography or electrophoresis is a useful technique in certain circumstances.

Table 1.8. Reagents for the colorimetric detection of amino acids

Common Name	*Composition and reaction conditions**	*Amino acids detected***	*Colour produced*	*Comments*
Isatin	2.0 gl^{-1} in acetone. Heat at 105°C for 2–3 min	**Pro** **Hyp** Asp Glu Ala Phe Tyr βAla Gln	Dark blue blue/green Light blue Light brown Grey/blue Blue blue/grey Light blue Light red	Non-specific, mainly for proline but certain sulphur-containing amino acids and some aromatic acids react. Use first in multiple-dip procedures.
Ehrilich	*p*-Dimethylaminobenzaldehyde (100 gl^{-1}) in conc. HCL. Mix 1 volume with 4 volumes of acetone. No heat required. Reacts within 20 min	**Trp** Citrulline	Pink/red Yellow	Some indoles, aromatic amines and ureides react. Use after ninhydrin in multiple-dip sequences
Pauly	Sulphanilic acid (9 gl^{-1}) in conc. HCl. Mix 1 volume with 10 volumes of water, 1 volume sodium nitrite (50 gl^{-1}) and 1 volume Sodium carbonate (100 gl^{-1})	**His** **Tyr**	Red Light orange	Some imidazoles and phenolic compounds and ammonium salts react. Colours vary – red, brown, yellow. Use after ninhydrin, isatin or Ehrlich in multiple-dip sequences
Diacetyl	10 gl^{-1} α-naphthol in 80 gl^{-1} NaOH plus an Equal volume of diacetyl (1 ml per litre water.) Mix before use. Heat at 100°C for 2-3 min	**Arg**	Purple/red	Mono-and di-substituted guanidines, *e.g.* creatine, creatinine, also react
Cyanide nitro prusside	100 gl^{-1} of each of the following: sodium hydroxide, sodium nitroprusside, potassium ferricyanide. Mix 1 volume of each solution with 3 volumes of water and sstand for 30 min before use. No heat required	**Cys** **Cystine** Homocys Homocys-tine	Purple Purple Purple Purple	Some sulphur-containing amino acids react. Colours fade in aobut 30 min

*Details given are for chromatographic or electrophoretic location and will differ for use with liquid samples.
**Amino acids customarily detected are in bold type.

The preparation of DNP derivatives may be indicated when the sample for analysis contains a variety of other substances, removal of which would be complicated, leading possibly to considerable analytical errors. However, the derivative formation and extraction is time consuming and itself can introduce inaccuracies into the analysis and should be used only when it offers an advantage over the separation of untreated amino acids. The use of dansyl derivatives is not recommended for routine analysis of free amino acids but is very suitable in the identification of an unknown amino acid that has been selectively extracted from the original sample and is present in small quantities.

Both kinds of derivative can be easily separated by chromatography or electrophesis and no locating reagent is required for either because the DNP derivatives are themselves yellow in colour and the dansyl derivatives are fluorescent.

Electrophoresis

The fact that different amino acids carry different net charges at any particular pH permits mixtures to be separated using low or high voltage electrophoresis. The most frequently used supporting media are paper or thin-layer sheets (cellulose or silica gel) and the locating reagents already described for chromatography may be used for visualization of the spots. Separations at high voltages can be achieved more quickly than at low voltages and one of the principal advantages of the former is that salts and other substances that may be present in the sample affect the quality of electrophoretogram to a lesser extent.

This permits the separation of amino acids in relatively crude extracts and untreated fluids, whereas prior to low voltage electrophoresis it is necessary to remove interfering substances such as proteins, carbohydrates and salts using the same methods as described for chromatograpy. Although electrophoretic separations can be achieved using buffers over a wide range of pH values, in practice the pH values chosen are either pH 2.0 or pH 5.3. At pH 2.0 all amino acids will carry a positive charge and the basic amino acids, having the highest positive charge, will migrate furthest towards the cathode whereas at pH 5.3 migration will occur towards both electrodes depending on the charge carried.

Separations at pH 5.3 are particularly useful to determine the acidic or basic nature of an unknown amino acid or dipeptide.

A two-dimensional technique involving initial separation by high voltage electrophoresis at pH 2.0 followed by chromatography is a useful means of separating similar amino acids and short peptides and does not require desalting or excessive purification of the sample.

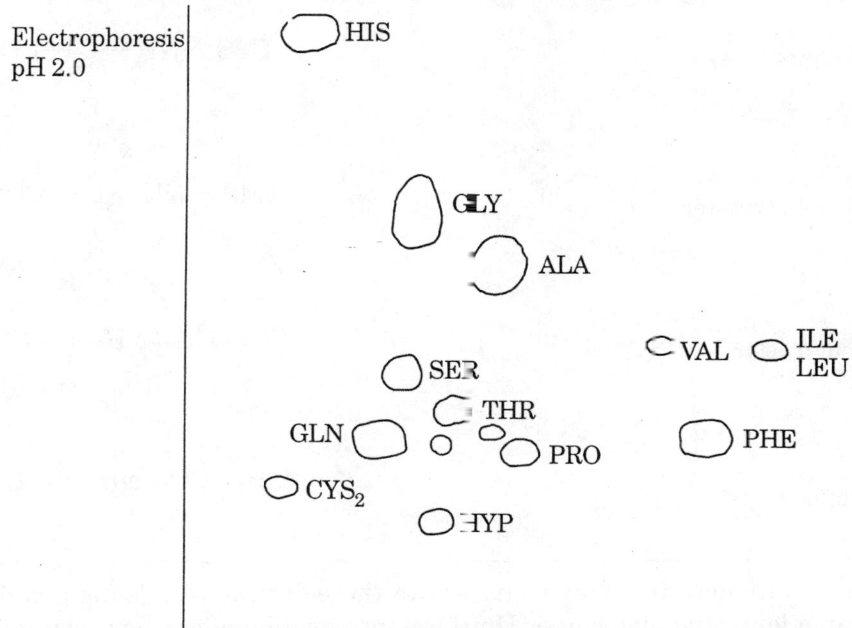

Fig. 1.17. Two-dimensional high voltage electrophoresis and chromatography of amino acids. Paper high voltage electrophoresis (4000 V) in an acetic acid-formic acid buffer at pH 2.0 in the first dimension followed by descending chromatography in the second dimension in an n-butanol-acetic acid-water solvent (12 : 3 : 5). The spots were visualized with a ninhydrin-collidine reagent.

Gas-liquid Chromatography

Derivatives of amino acids are required because amino acids are not themselves sufficiently volatile for gas-liquid chromatography and difficulties may be encountered in the choice and method of derivatization. In the past no single column was normally capable of resolving the derivatives of such a diverse group of compounds but the introduction of fused silica capillary columns has resulted in considerably improved resolution. The advantage of trimethylsilyl (TMS) derivatives lies in the simplicity of the derivatization procedure, which is carried out by the addition of *N,O*- bis (trimethylsilyl) trifluoroacetamide (BSTFA) in acetonitrile and heating for approximately 2 h at 150°C under anhydrous conditions in a sealed tube.

However, there may be problems owing to the formation of multiple derivatives of each amino acid. Another technique involves the formation of *n*-butyl esters of the amino acids and their subsequent trimethylsilylation by a similar procedure. The *n*-butyl esters are formed by heating the amino acids for 15 min in *n*-butanol and HCl and these are then converted to the *N*-TMS-*n*-butyl ester derivatives. *N*-acyl amino acid alkyl esters are commonly used. Acetylation of the butyl, methyl or propyl esters of amino acids, to give trifluorocetyl (TFA) or heptafluorobutyryl (HFB) derivatives can be performed by reacting them with either TFA or HFB anhydrides in methylene dichloride at 150°C for 5 min.

Table 1.9. Derivatives of Amino Acids suitable for Gas-liquid Chromatography

Derivative	*Formula*
N—TMS—TMS ester	TMS—NH—C(R)(H)—COOTMS
N—TMS—*n*-butyl ester	TMS—NH—C(R)(H)—$COOC_4H_9$
TFA—*n*-butyl ester	CF_3—CO—NH—C(R)(H)—$COOC_4H_9$
HFB—*n*-propyl ester	C_3H_7—CO—NH—C(R)(H)—$COOC_3H_7$

Electron capture detectors may be used with these derivatives, giving greater sensitivity than with flame ionization detectorsr. Unitl the introduction of capillary columns, it was not possible to separate all the amino acids found in proteins on one column. The choice of stationary phase will depend upon the types of derivatives that have been prepared and in some situations it still may be preferable to use two different columns simultaneously. Another difficulty in the gas chromatographic separation of amino acids is the choice of detector and it may be necessary to split the gas stream and use two different detectors.

The flame ionization detector, which is commonly used, is non-specific and will detect any non-amino acid components of the sample unless purification has been performed prior to derivatization. In addition the relative molar response of the flame ionization detector varies for each amino acid, necessitating the production of separate standard curves. As a consequence, although gas chromatography offers theoretical advantages, its practical application is mainly reserved for special circumstances when a nitrogen detector may be useful to increase the specificity.

High performance liquid chromatography

The use of reverse-phase columns with pre-column derivatization of the amino acids offers an acceptable alternative to the dedicated instrumentation of an amino acid analyser or separation by HPLC followed by post-column derivatization. Buffered mobile phases are used and the proportions of polar solvents (*e.g.* methanol, tetrahydrofuran) depend upon the type of derivative employed. Gradient elution is required for the resolution of complex mixtures and analysis times are less than 1 h. Ultraviolet, fluorescence or electrochemical detectors are used depending upon the nature of the amino acid derivatives. Derivatization of primary amino acids with *o*-phthalaldehyde (OPA) is simple and the poor reproducibility due to the instability of the reaction product can be improved by automation and the use of alternative thiols, *e.g.* ethanthiol in place of the 2-mercaptoethanol originally used.

An alternative fluorimetric method using 9-fluoroenylmethylchloroformate (FMOC-CL) requires the removal of excess unreacted reagent prior to column chromatography. This procedure is more difficult to automate fully and results are less reproducible. However, sensitivity is comparable with the OPA method with detection at the low picomole or femtomole level, and it has the added advantage that both primary and secondary amino acids can be determined. Dansyl chloride (5-dimethylaminonaphthlene-1-sulphonyl chloride) and the related dabsyl chloride (4-dimethyl-aminoazobenzene-4′-sulphonyl chloride) have also been used.

Sensitivity of the latter method is poor and both suffer problems from interference with reagent excess. Reaction with phenylisothiocyanate (PITC) in alkaline conditions produces stable phenylthiocarbamyl (PTC) adducts which can be detected either in the ultraviolet below 250 nm or electrochemically. However, this method involves a complex derivatization procedure and offers poorer sensitivity than the alternatives available for individual amino acids. It is useful, however, in conjunction with the automated analysis of peptides when single derivatized residues can be cleaved and analysed after conversion in acidic conditions to phenylthiohydantoins.

Ion-exchange Chromatography

Ion-exchange resins may be used to isolate amino acids from other substances prior to analysis, but ion-exchange chromatography is also used to separate mixtures of amino acids. By using a cation-exchange resin and varying the pH of the eluting buffer, it is possible to separate a wide range of amino acids. The amino acid analyser, which also quantifies each amino acid, is based on this technique. Anion-exchange resins tend to be used only in special circumstances for the separation of strongly acidic amino acids.

AMINO ACID ANALYSER

Although the chromatographic separation of amino acids on starch columns was introduced in 1941 it was more than 10 years later that Spackman, Stein and Moore developed the first amino acid analyser based on separation by ion-exchange chromatography and quantitation of each component in the column effluent using the ninhydrin reaction. The instruments currently available are based on their original design, permitting buffers of varying pH or ionic strength to be pumped through a thermostatically controlled resin column, although many modifications resulting in improved performance are constantly being introduced.

The most significant changes have been the advent of high quality resins, sophisticated automation and increased sensitivity of the detection systems. These have contributed to a reduction in analysis time from days to hours and extended the analytical range to below the nanomole (10^{-9} mol) level.

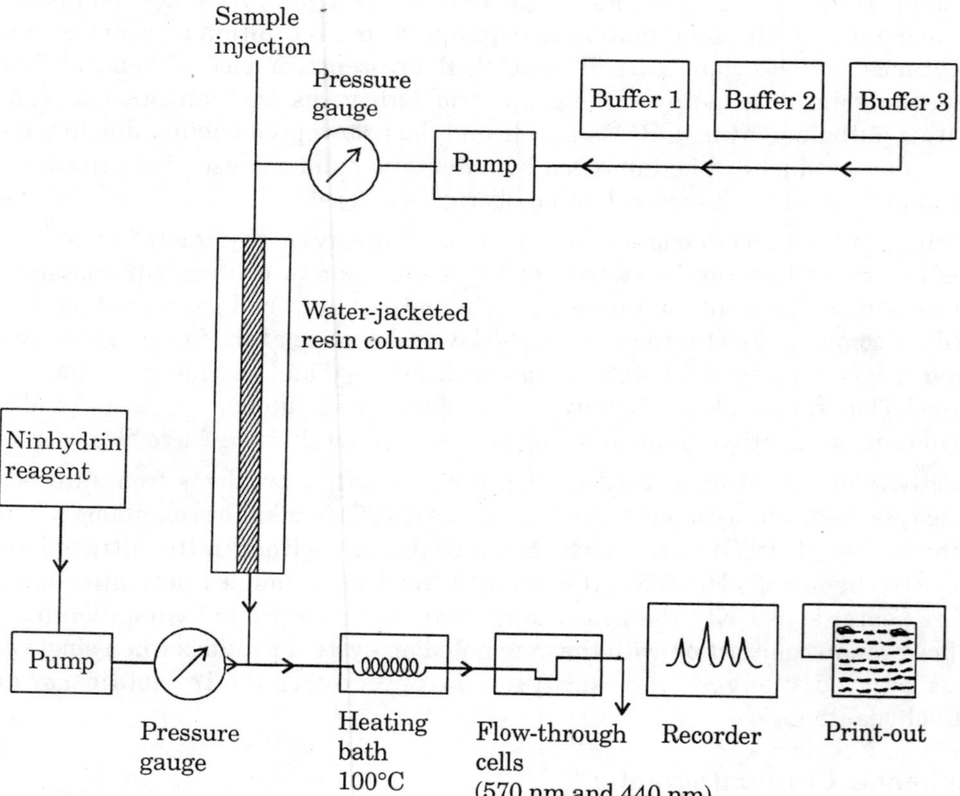

Fig. 1.18. Schematic diagram of an amino acid analyser using the ninhydrin reagent for quantitation.

Principles of Separation

The separation takes place in a column of sulphonated cross-linked polystyrene resin, which is a strong cation exchanger. The matrix of the resin is strongly anionic in nature (SO_3^-) and at the low pH used initially, the amino acids will be positively charged and will be attracted to the negatively charged sulphonate groups. As the pH of the buffer passing through the column is raised, the amino acids will be differentially eluted as their net positive charge diminishes and they are less strongly attracted to the sulphonate sites on the resin matrix. They will emerge from the column in a sequence relating to their individual p*I* values.

At pH 3.25, the pH at which the analysis usually commences, the amino acids with an extra acidic group in their side chain will have very little affinity for the resin and will be the first to emerge from the column, whereas at the same pH value those amino acids whose side chain contains an extra ionizable group capable of carrying a positive charge, for example lysine and histidine, will be strongly held on the resin and will be eluted from the column only as the pH is raised substantially and their net positive charge reduced. However, it is not only the pH of the eluting buffer that determines the relative elution position of the amino acids, but also the cation concentration of the buffer. Sodium citrate buffer solutions are commonly used and the positive sodium ions compete with the positively charged amino acids for the sulphonic acid sites on the resin:

$$\text{resin—}\ SO_3^- \ldots {}^+AA + Na^+ \rightleftharpoons \text{resin—}\ SO_3^- \ldots {}^+Na + AA^+$$

Although the amino acids have a considerable affinity for the resin, the sodium ions are constantly present in a much higher concentration and, as a result, the equilibrium of the above equation is shifted to the right and the amino acids are displaced from the resin. Thus the molarity of the eluting buffer affects elution and when the ionic concentration of the buffer is increased, the amino acids are eluted more rapidly from the column. Non-ionic interactions between the amino acids and the resin also influence the elution sequence, allowing amino acids with identical pI values to be eluted separately.

The polarity of the amino acid is important in this respect and those with a non-polar, hydrophobic R group interact with the strongly hydrophobic resin matrix. The degree of cross-linking also affects separation because it influences the rate of diffusion of the charged species to the exchange sites. Thus an accurate prediction of the positions of the amino acids on a chromatogram is difficult and may vary considerably from that determined inpractice. The quantitative resolution of amino acid mixtures is achieved by varying the composition of the buffer flowing through the column by either increasing the pH and maintaining a constant molarity (cation concentration) or by keeping the pH constant but varying the molarity, or by using a combination of both. Other factors such as column temperature, buffer flow rate and type of buffer cation play important, but less significant, parts in the quality of the separation.

Practical Aspects

The column

In the prototype analysers, two columns were often needed to achieve complete separation of all the amino acids. A 50-100 cm column was used to separate the acidic and neutral amino acids and a 5-10 cm column for the basic amino acids, each with a diameter of 1 or 2 cm, but today's instruments use single columns with narrower diameters. As peak width is proportional to the square root of the column length, these glass or stainless steel columns give narrow peaks and improved separation of closely related amino acids.

Buffers

The composition and pH of the buffer should be accurate to 0.001 mol l^{-1} and 0.01 pH units. Most methods rely on the sequential application of a series of buffer solutions of increasing pH and molarity, with the initial pH around 3.2. Sodium citrate or, preferably, lithium citrate buffers are used, which incorporate a detergent (BRIJ 35), an antioxidant (thiodiglycol) and a preservative (caprylic acid) and may be used to perform either stepwise or gradient elution. When stepwise elution is being carried out, each buffer is pumped through the column for varying lengths of time and these are chosen with reference to the types of amino acid to be separated and the column dimensions. Buffers often used are pH 3.25, pH 4.25, pH 5.25 and pH 10.0 with either constant or varying molarity if a separation of acidic, neutral and basic amino acids is to be achieved.

However, when groups of amino acids with similar ionic characteristics are under investigation, for example those with an acidic R group, it is often necessary to use only one or two of the buffers. When the technique of gradient elution is employed the buffers are mixed in a predetermined manner to give gradual changes in pH and ionic strength. A two-buffer gradient elution system is used in some models where an acidic buffer, pH 2.2, and an alkaline buffer, pH 11.5, are mixed in varying proportions to achieve increasing pH values.

Gradient elution gives improved separation with decreased analysis time and eliminates dramatic baseline fluctuations due to sudden changes in the buffer. A less common practice is to use an 'Iso-pH' system which employs a stepwise or gradient elution using buffers with increasing cation concentration, which may be from 0.2 to 1.6 mol^{-1} and an almost constant pH value between 3.25 and 3.65.

Temperature

The temperature of the resin column must be carefully maintained to avoid changes in both the pH of the buffers and ionization of the amino acids. Although increasing temperature usually results in faster elution, the effect may be variable for different amino acids and the relative elution positions can be altered, making interpretation of results difficult. The temperature often chosen is 60 °C although lower temperatures are sometimes required to resolve two similar amino acids. Temperature programming, which entails an alteration in temperature at a specified time in the separation procedure, is widely used.

Column Flow Rate

Successful and reproducible separations require a steady buffer flow rate and this is achieved with either a constant pressure or a constant displacement pump. These pumps are designed to deliver a constant rate of fluid independent of the resistance to the flow and recent developments in pump design permit the production of a precise and pulseless flow; this has contributed towards the increased analytical precision and sensitivity that can now be achieved with amino acid analysers. The choice of flow rate is dependent upon the type of resin, the dimensions of the column and overall design of the instrument and this varies between models.

Sample preparation

The volume of sample which can be applied to the resin column will vary for the different instruments commercially available. With progressive refinements in instrumentation the tendency has been towards a decreased sample volume, of 50 μl or less. Correct sample preparation is very important if reproducible results are to be obtained along with trouble-free operation. This will vary according to the nature of the sample and its constituents but the fluid applied to the column must be clean and free from protein and other large molecules. Failure to comply with this requirement will result in a slow clogging the resin column and may necessitate the removal, cleaning and re-packing of the resin, or purchase of a new column.

In general, liquid samples need only filtration or centrifugation to remove artifacts but any protein present must be removed by precipitation with picric acid or salicylsulphonic acid or by dialysis or ultraflltration. Solid samples, such as foodstuffs and animal or plant tissue, may require homogenization, extraction and deproteinization. Any peptides or proteins under investigation for their amino acid composition must first be hydrolysed to release the free amino acids. The prepared sample may be applied in pH 2.2 buffer directly to the top of the column and the analysis sequence started, and after elution of all the amino acids and regeneration of the resin column, the next sample can be applied.

However, many newer models incorporate an automatic loading device which enables several samples to be stored ready for analysis either in sample cups or in small Teflon coils. After the completion of an analysis the next stored sample is automatically applied to the resin and the buffer cycle restarted.

Detection

A second pump is required to deliver a constant flow of reagent to meet the column effluent. When the reaction has taken place the stream is monitored continuously using a flow-through cell in either a colorimeter or a fluorimeter. Ninhydrin reagent is the most widely used and after the solution has passed through a coil in the heating bath at 100°C, the absorbance is monitored at 570 and 440 nm to detect amino and imino acids respectively.

Ninhydrin reagent should be prepared accurately using high quality chemicals. The ninhydrin is dissolved in peroxide-free ethyleneglycol, monomethyl ether (methylcello-solve) and buffered with acetate at pH 5.5. Nitrogen gas is bubbled through during the preparation of the reagent to exclude air and a small quantity of a reducing agent, stannous chloride or titanious chloride, is added to ensure the production of a limited but accurate amount of reduced ninhydrin. Some analysers introduce argon or nitrogen to the column effluent to produce a gas-segmented stream and in these instruments either sodium cyanide or hydrazine is used as the reducing agent.

The prepared reagent should be a pale straw colour and must be stored in a dark bottle under nitrogen pressure because it is sensitive to light and oxygen.. The ninhydrin reaction requires heat and therefore the stream of column eluate plus ninhydrin reagent must pass through a coil of narrow-bore tubing (approximate diameter 1 mm) held in a 100°C heating bath? It is important to ensure that the flow is not restricted because excessive heating may cause the ninhydrin to precipitate in these micro-bore tubes, resulting in complete stoppage of the analyzer. The *o*-phthalaldehyde reaction compares favourably with the ninhydrin reaction in several respects.

The reagent is stable and is in an aqueous form, which eliminates the use of potentially toxic chemicals and storage under nitrogen. Because the reaction proceeds quickly at room temperature there is no need for the 100 °C heating bath with all its inherent problems, and the increased sensitivity permits detection at the picomole level.

Quantitation

The colour or fluorescence produced per mole of amino acid varies slightly for different amino acids and this must be determined for each one to be quantitated. This is done by loading a mixture of amino acids containing the same concentration of each amino acid including the chosen internal standard and from the areas of the peaks on the recorder trace calculating each response factor in the usual way. These values are noted and used in subsequent calculations of sample concentrations. An internal standard should always be used for every analysis carried out. This is an amino acid that is known to be absent from the sample under investigation.

For instance in blood plasma analysis either of the non-physiological amino acids, nor-leucine or α-amino-β-guanidinobutyric acid, may be used. This should be added in a known amount to the sample prior to any sample pre-treatment (for example, removal of protein). If the amount of internal standard which was added to the sample is known, the concentration of the unknown amino acid can be determined using peak area relationships. These calculations must take the various response factors into account. In this way losses due to sample preparation will be allowed for as will variation in the intensity of the colour or fluorescence produced with different preparations of reagent and changes in analysis conditions.

Catabolism of Amino Acids

The assembly of new proteins requires a source of amino acids. These building blocks are generated by the digestion of proteins in the intestine and the degradation of proteins within the cell. Many cellular proteins are constantly degraded and resynthesized. To facilitate this recycling, a complex system for the controlled turnover of proteins has evolved. Damaged or unneeded proteins are marked for destruction by the covalent attachment of chains of a small protein, *ubiquitin.* Polyubiquitinated proteins are subsequently degraded by a large, ATP-dependent complex called the *proteasome.* The primary uses of amino acids are as building blocks for protein and peptide synthesis and as a source of nitrogen for the synthesis of other amino acids and other nitrogenous compounds such as nucleotide bases.

Amino acids in excess of those needed for biosynthesis cannot be stored, in contrast with fatty acids and glucose, nor are they excreted. Rather, surplus amino acids are used as metabolic fuel. *The χ-amino group is removed, and the resulting carbon skeleton is converted into a major metabolic intermediate.* Most of the amino groups of surplus amino acids are converted into urea through the *urea cycle,* whereas their carbon skeletons are transformed into acetyl CoA, aceroacetyl CoA, pyruvate, or one of the intermediates of the citric acid cycle. *Hence, fatty acids, ketone bodies, and glucose can be formed from amino acids.* Several coenzymes play key roles in amino acid degradation, foremost among them is *pyridoxal phosphate.* This coenzyme forms Schiff-base intermediates that allow χ-amino groups to be shuttled between amino acids and ketoacids. We will consider several genetic errors of amino acid degradation that lead to brain damage and mental retardation unless remedial action is initiated soon after birth.

Phenylketonuria, which is caused by a block in the conversion of phenylalanine into tyrosine, is readily diagnosed and can be treated by removing phenylalanine from the diet. The study of amino acid metabolism is especially rewarding because it is rich in connections between basic biochemistry and clinical medicine.

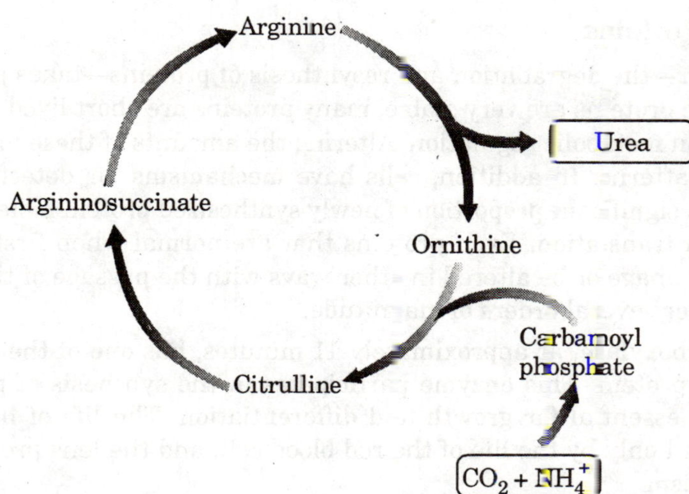

Degradation of cyclin B. This important protein in cell cycle regulation is visible as the green areas in the images above (the protein was fused with green fluorescent protein). Cyclin B is prominent during metaphase, but is degraded in anaphase to prevent premature initiation of another cell cycle. A large protease complex called the proteasome digests the protein into amino acids. These are either reused, or further processed by the urea cycle, which removes the nitrogen as urea.

Proteins Are Degraded to Amino Acids

Dietary protein is a vital source of amino acids. Proteins ingested in the diet are digested into amino acids or small peptides that can be absorbed by the intestine and transported in the blood. Another source of amino acids is the degradation of defective or unneeded cellular proteins.

DIGESTION AND ABSORPTION OF PROTEINS

Protein digestion begins in the stomach, where the acidic environment favors protein denaturation. Denatured proteins are more accessible as substrates for proteolysis than are native proteins. The primary proteolytic enzyme of the stomach *is pepsin,* a nonspecific protease that, remarkably, is maximally active at pH 2. Thus, pepsin can be active in the highly acidic environment of the stomach, even though other proteins undergo denaturation there. Protein degradation continues in the lumen of the intestine owing to the activity of proteolytic enzymes secreted by the pancreas. These proteins, are secreted as inactive zymogens and then converted into active enzymes.

The battery of enzymes displays a wide array of specificity, and so the substrates are degraded into free amino acids as well as di- and tripeptides. Digestion is further enhanced by proteases, such as aminopeptidase N, that are located nn the plasma membrane of the intestinal cells. Aminopeptidases digest proteins from the amino-terminal end. Single armno acids, as well as di- and tripeptides, are transported into the intestinal cells from the lumen and subsequently released into the blood for absorption by other tissues.

Degradation of Proteins

Protein turnover—the degradation and resynthesis of proteins—takes place constantly in cells. Although some proteins are very stable, many proteins are short lived, particularly those that are important in metabolic regulation. Altering the amounts of these proteins can rapidly change metabolic patterns. In addition, cells have mechanisms for detecting and removing damaged proteins. A significant proportion of newly synthesized protein molecules are defective because of errors in translation. Even proteins that are normal when first synthesized may undergo oxidative damage or be altered in other ways with the passage of time. The half-lives of proteins range over several orders of magnitude.

Ornithine decarboxylase, at approximately 11 minutes, has one of the shortest half-lives of any mammalian protein. This enzyme participates in the synthesis of polyamines, which are cellular cations essential for growth and differentiation. The life of hemoglobin, on the other hand, is limited only by the life of the red blood cell, and the lens protein, crystallin, by the life of the organism.

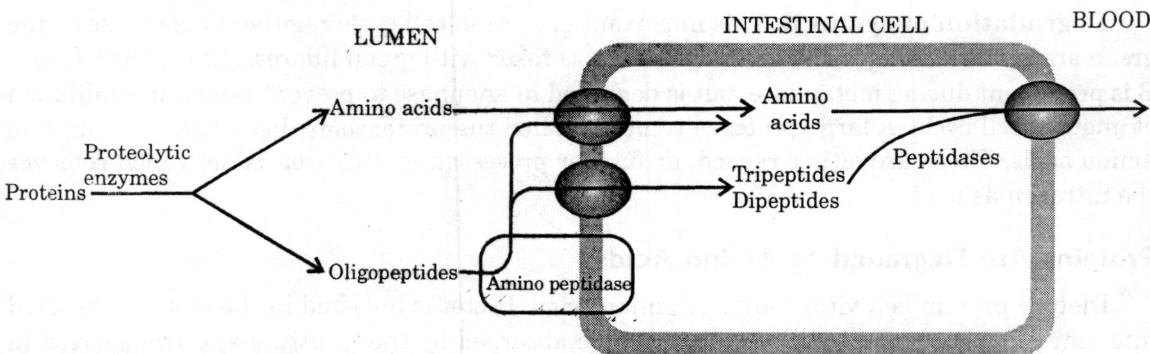

Fig. 1.19. Digestion and Absorption of Proteins. Protein digestion is primarily a result of the activity of enzymes secreted by the pancreas. Aminopeptidases associated with the intestinal epithelium further digest proteins. The amino acids and di- and tripeptides are absorbed into the intestinal cells by specific transporters. Free amino acids are then released into the blood for use by other tissues.

Table 1.12. Half-lives of cytosolic yeast proteins

Highly stabilizing residues			
($t_{1/2}$ > 20 hours)			
Ala	Cys	Gly	Met
Pro	Ser	Thr	Val
Intrinsically destabilizing residues			
($t_{1/2}$ = 2 to 30 minutes)			
Arg	His	Ile	Leu
Lys	Phe	Trp	Tyr
Destabilizing residues after chemical modification			
($t_{1/2}$ = 3 to 30 minutes)			
Asn	Asp	Gln	glu

REGULATED PROTEIN TURNOVER

How can a cell distinguish proteins that are meant for degradation? *Ubiq-uitin,* a small (8.5-kd) protein present in all eukaryotic cells, is the tag that marks proteins for destruction. Ubiquitin is the cellular equivalent of the "black spot" of Robert Louis Stevenson's *Treasure Island:* the signal for death.

Effects of Ubiquitin

Ubiquitin is highly conserved in eukaryotes: yeast and human ubiquitin differ at only 3 of 76 residues. The carboxyl-terminal glycine residue of ubiquitin (Ub) becomes covalently attached to the γ-amino groups of several lysine residues on a protein destined to be degraded. The energy for the formation of these *isopeptide bonds (iso* because γ-rather than χ-amino groups are targeted) comes from ATP hydrolysis.

Ub
O
HN
Isopeptide bond
Lys
O
H
H
N
O
N
H
Peptide bond
O
Peptide bond

Three enzymes participate in the attachment of ubiquitin to each protein: ubiquitin-activating enzyme, or E1, ubiquitin-conjugating enzyme, or E2, and ubiquitin-protein ligase, or E3. First, the terminal carboxylate group of ubiquitin becomes linked to a sulfhydfyl group of E1 by a thioester bond. This ATP-driven reaction is reminiscent of fatty acid activation.

An adenylate is linked to the C-terminal carboxylate of ubiquitin with the release of pyrophosphate, and the ubiquitin is transferred to a sulfhydryl group of a key cysteine residue in E1. Second, activated ubiquitin is shuttled to a sulfhydryl group of E2. Finally, E3 catalyzes the transfer of ubiquitin from E2 to an γ-amino group on the target protein. The attachment of a single molecule of ubiquitin is only a weak signal for degradation. However, the ubiquitination reaction is processive: chains of ubiquitin can be generated by the linkage of the γ-amino group of lysine residue 48 of one ubiquitin molecule to the terminal carboxylate of another. Chains of four or more ubiquitin molecules are particularly effective in signaling degradation. The use of chains of ubiquitin molecules may have at least two advantages. First, the ubiquitin molecules interact with one another to form a binding surface distinct from that created by a single ubiquitin molecule. Second, individual ubiquitin molecules can be cleaved off without loss of the degradation signal. Although most eukaryotes have only one or a small number of distinct E1 enzymes, all eukaryotes have many distinct E2 and E3 enzymes. Moreover, there appears to be only a single family of evolutionary related E2 proteins but many distinct families of E3 proteins.

Although the E3 component provides most of the substrate specificity for ubiquitination, the multiple combinations of the E2-E3 complex allow for more finely tuned substrate discrimination. What determines whether a protein becomes ubiquitinated? One signal turns out to be unexpectedly simple. *The half-life of a cytosolic protein is determined to a large extent by its ammo-terminal residue.* This dependency is referred to as the *N-terminal rule.* A yeast protein with methionine at its N terminus typically has a half-life of more than 20 hours, whereas one with arginine at this position has a half-life of about 2 minutes. A highly destabilizing N-terminal residue such as arginine or leucine favors rapid ubiquitination, whereas a stabilizing residue such as methionine or proline does not. E3 enzymes are the readers of N-terminal residues.

Other signals thought to identify proteins for degradation include *cyclin destruction boxes,* which are amino acid sequences that mark cell-cycle proteins for destruction, and proteins rich in proline, glutamic acid, serine, and threonine (PEST sequences). Some pathological conditions vividly illustrate the importance of the regulation of protein turnover. For example, human papilloma virus (HPV) encodes a protein that activates a specific E3 enzyme. The enzyme ubiquitinates the tumor suppressor p53 and other proteins that control DNA repair, which are then destroyed. The activation of this E3 enzyme is observed in more than 90% of cervical carcinomas. Thus, the inappropriate marking of key regulatory proteins for destruction can trigger further events, leading to tumor formation.

Digestion by Proteasomes

If ubiquitin is the mark of death, what is the executioner? A large protease complex called the *proteasome* or the *26S proteasome* digests the ubiquitinated proteins. This ATP-driven multisubunit protease spares ubiq-uitin, which is then recycled. The 26S proteasome is a complex of two components: a 20S proteasome, which contains the catalytic activity, and a 19S regulatory subunit. The 20S complex is constructed from two copies each of 14 subunits and has a mass of 700 kd. All 14 subunits are homologous and adopt the same overall structure. The subunits are arranged in four rings of 7 subunits that stack to form a structure resembling a barrel. The components of the two rings at the ends of the barrel are called the χ subunits and those of the two central rings the δ subunits.

The active sites of the protease are located at the N-termini of certain δ subunits on the interior of the barrel—specifically, those δ chains having an N-terminal threonine or serine residue. The hydroxyl groups of these amino acids are converted into nucleophiles with the assistance of their own amino groups. These nucleophilic groups then attack the carbonyl groups of peptide bonds and form acyl-enzyme intermediates. The structure of the complex sequesters the proteolytic active sites from potential substrates until they are directed into the barrel. Substrates are degraded in a processive manner without the release of degradation intermediates, until the substrate is reduced to peptides ranging in length from seven to nine residues. The 20S proteasome is a sealed barrel. Access to its interior is controlled by a 19S regulatory complex, itself a 700-kd complex made up of 20 subunits. This complex binds to both ends of the 20S proteasome core to form the complete 26S proteasome. The 19S subunit binds specifically to polyubiquitin chains.

Key components of the 19S complex are six distinct ATPases of the AAA class (ATPase associated with various cellular activities) characterized by a conserved 230 amino acid ATP-binding domain of the P-loop NTPase family. This class of ATPase found in all kingdoms, is associated with a variety of cell functions including cell-cycle regulation and organelle biogenesis. Although the exact role of the ATPase remains uncertain, ATP hydrolysis may assist the 19S complex to unfold the substrate and induce conformational changes in the 20S proteasome so that the substrate can be passed into the center of the complex. Finally, the 19S subunit also contains an isopeptidase that cleaves off intact ubiquitin molecules. Thus, the ubiquitinization pathway and the proteasome cooperate to degrade unwanted proteins. The ubiquitin is recycled and the peptide products are further degraded by other cellular proteases to yield individual amino acids.

Regulative Degradation

Lists a number of physiological processes that are controlled at least in part by protein degradation. This control is exerted by dynamically altering the stability and abundance of regulatory proteins. Consider, for example, control of the inflammatory response. A transcription factor called NF-μ**B.** (NF for nuclear factor) initiates the expression of a number of the genes that take part in this response. This important factor is itself activated by the degradation of an inhibitory protein, I-μ**B.** NF-μ**B** is maintained in the cytoplasm in its inactive state by association with I-μB (I for inhibitor). In response to inflammatory signals, I-μδ is phosphorylated at two serine residues, creating an E3 binding site. The binding of E3 leads to the ubiquitination and degradation of I-μ**B** and thereby disrupts the inhibitor's association with NF-μ**B.** The liberated transcription factor migrates to the nucleus to stimulate transcription of the target genes. The NF-μ**B-I**-μ**B** system illustrates the interplay of several key regulatory motifs: receptor-mediated signal transduction, phosphorylation, compartmentalization, controlled and specific degradation, and selective gene expression.

Ubiquitin Pathway

Both the ubiquitin pathway and the proteasome appear to be present in all eukaryotes. Homologs of the proteasome are found in prokaryotes, although the physiological roles of these homologs have not been well established. The proteasomes of some archaea are quite similar in overall structure to their eukaryotic counterparts and similarly have 28 subunits. In the archaeal proteasome, however, all χ subunits and all δ subunits are identical; in eukaryotes, each χ or δ subunit is one of seven different isoforms. This specialization provides distinct substrate specificity. Ubiquitin, however, has not been found in prokaryotes. Indeed, the high level of sequence similarity between the human and yeast proteins suggests that ubiquitin, in its present form, diverged relatively recently in evolutionary terms. Ubiquitin's molecular ancestors were recently identified in prokaryotes.

Remarkably, these proteins take part not in protein modification but in coenzyme biosynthesis. The biosynthesis of thiamine begins with a sulfide ion derived from cysteine. This sulfide is added to the C-terminal carboxylate of the protein ThiS, which had been activated as an acyl adenylate. The activation of ThiS and the addition of sulfide are catalyzed by the enzyme ThiF. Human E1 includes two tandem regions of 160 amino acids that are 28% identical in amino acid sequence with a region of ThiF from *E. coli.* The evolutionary relationships between these two pathways were cemented by the determination of the three-dimensional structure of ThiS, which revealed a structure very similar to that of ubiquitin, despite being only 14% identical in amino acid sequence. Thus, a eukaryotic system for protein modification evolved from a preexisting prokaryotic pathway for coenzyme biosynthesis.

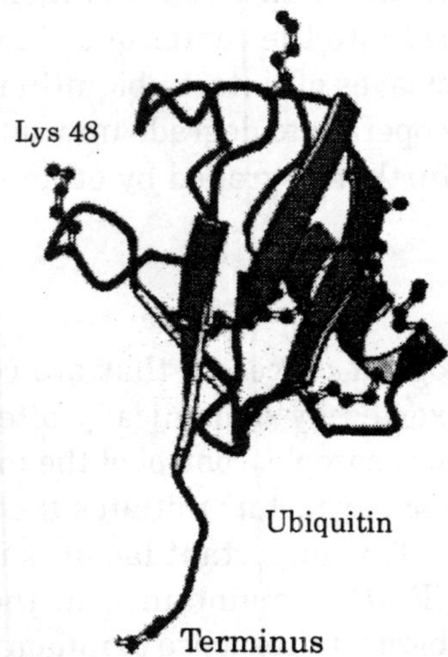

Fig. 1.20. Ubiquitin. The structure of ubiquitin reveals an extended carboxyl terminus that is activated and linked to other proteins. Lysine residues also area shown, including lysine 48, the major site for linking additional ubiquitin molecules.

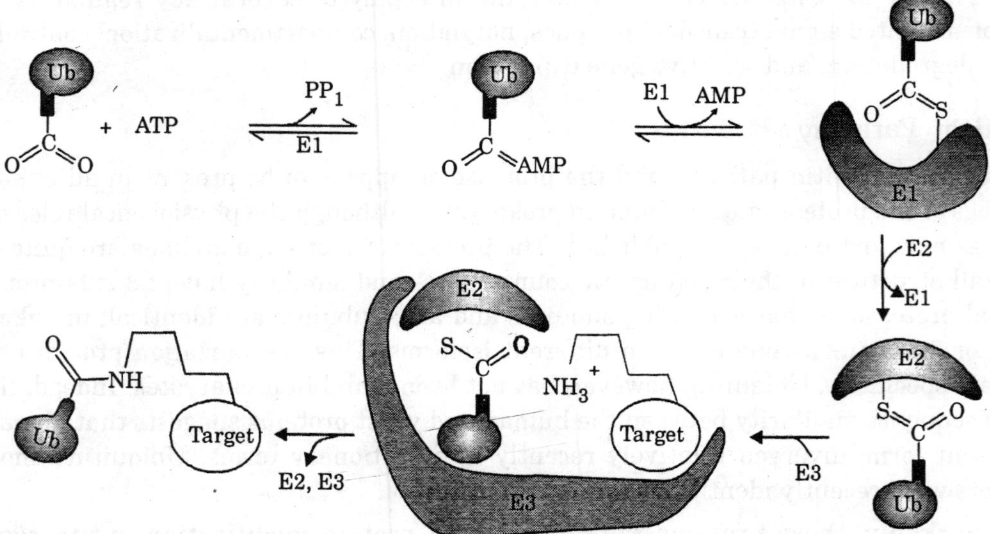

Fig. 1.21. Ubizuitin conjugation. The ubiquitin-activating enzyme E1 adenylated ubiquitin (Ub) and transfers the ubiquitin to one of its own cysteine residues. Ubiquitin is then transferred to a cysteine residue in the ubiquitin-conjugating enzyme E2. Finally, the ubiquitin-protein ligase E3 transfers the ubiquitin to a lysine residue on the target protein.

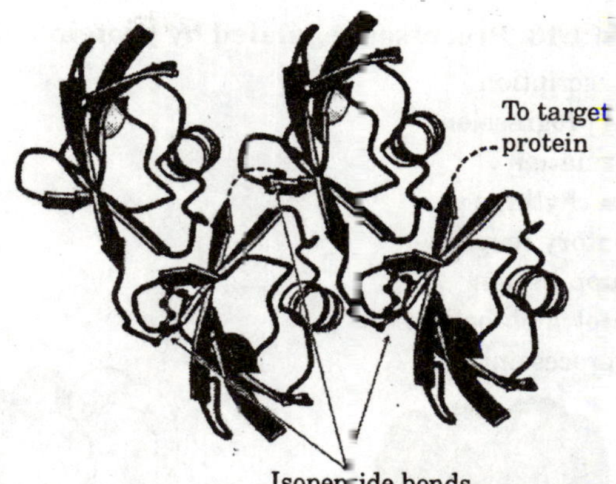

Fig. 1.22. Tetra Ubiquitin. Four ubiquitin molecules are linked by isopeptide bonds. This unit is the primary signal for degradation when linked to a target protein.

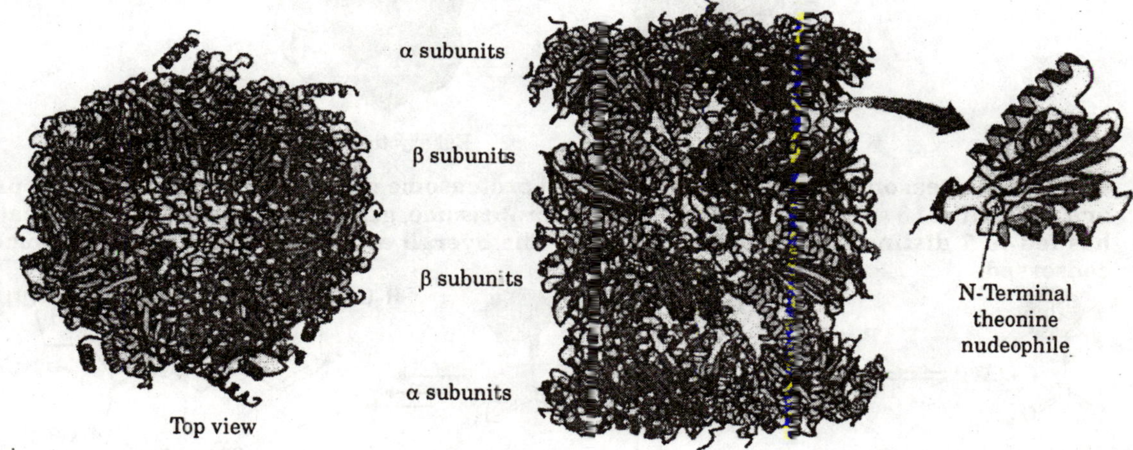

Fig. 1.23. 20S Proteasome. The 20S proteasome comprises 28 homologous subunits (χ, red; δ, blue), arranged in four rings of 7 subunits each. Some of the δ subunits (highlighted in yellow) include protease active sites at the amino termini. The top view shows the approximate seven-fold symmetry of the structure.

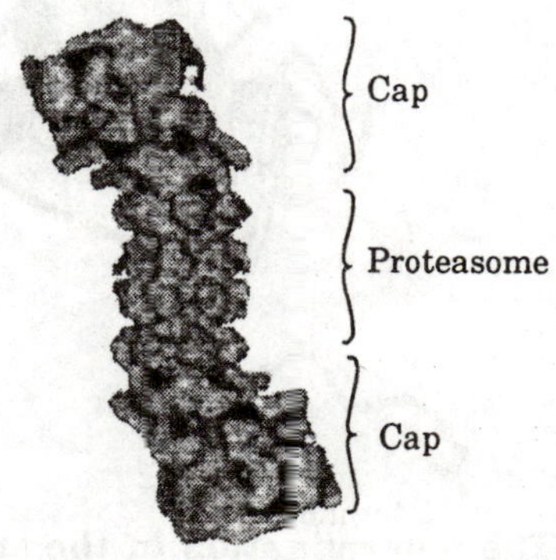

Fig. 1.24. 26S Protcasome. A 19S cap is attached to each end of the 20S catalytic unit.

Table 1.13. Processes regulated by protein degradation

Processes
Gene transcription
Cell-cycle progression
Organ formation
Circadian rhythms
Inflammatory response
Tumor suppression
Cholesterol metabolism
Antigen processing

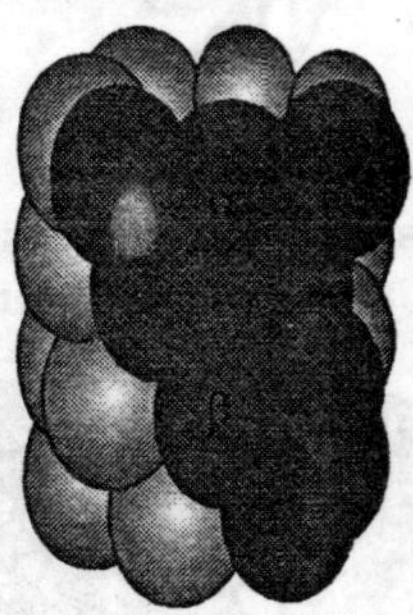

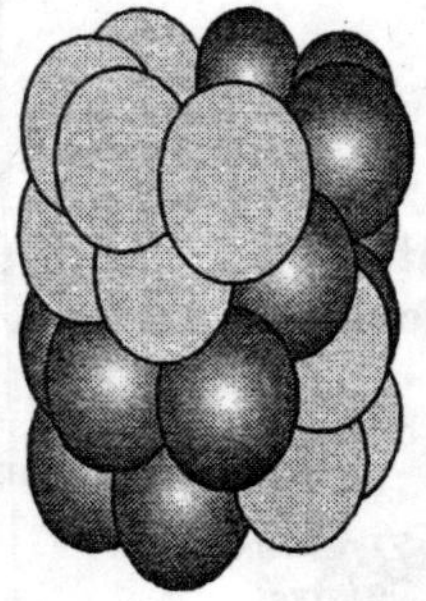

Fig. 1.25. Proteasome Evolution. The archaeal proteasome consists of 14 identical χ subunits and 14 identical δ subunits. In the eukaryotic proteasome, gene duplication and specialization has led to 7 distinct subunits of each type. The overall architecture of the proteasome is conserved.

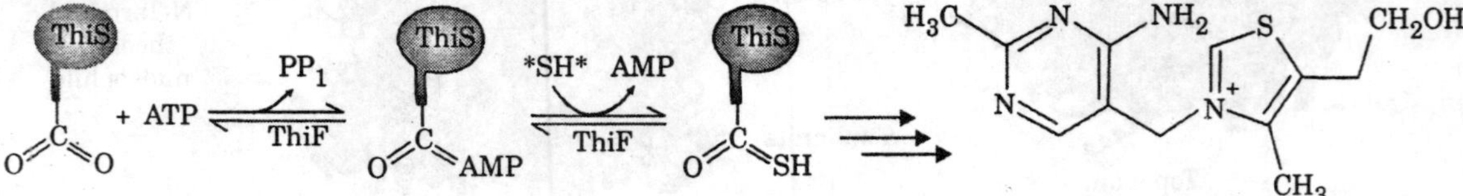

Fig. 1.26. Biosynthesis of Thiamine. The biosynthesis of thiamine begins with the addition of sulfide to the carboxyl terminus of the protein ThiS. This protein is activated by adenylation and conjugated in a manner analogous to the first steps in the ubiquitin pathway.

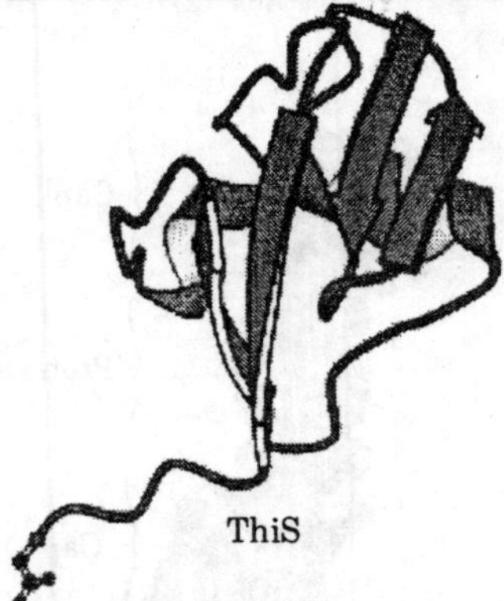

Fig. 1.27. Structure of This. The determination of the structure of ThiS revealed it to be structurally similar to ubiquitin despite only 14% sequence identity. This observation suggests that a prokaryotic protein such as ThiS evolved into ubiquitin.

REMOVAL OF NITROGEN

What is the fate of amino acids released on protein digestion or turnover? Any not needed as building blocks are degraded to specific compounds. The major site of amino acid degradation in mammals is the liver. The amino group must be removed, inasmuch as there are no nitrogenous compounds in energy-transduction pathways. The χ-ketoacids that result from the deamination of amino acids are metabolized so that the carbon skeletons can enter the metabolic mainstream as precursors to glucose or citric acid cycle intermediates. The fate of the χ-amino group will be considered first, followed by that of the carbon skeleton.

Oxidative Deamination of Glutamate

The χ-amino group of many amino acids is transferred to χ-ketoglutarate to form *glutamate,* which is then oxidatively deaminated to yield ammonium ion (NH_4^+).

^+H_3N–CH(R)–COO^- ⟶ ^+H_3N–CH($CH_2CH_2COO^-$)–COO^- ⟶ NH_4^+

Amino acid **Glutamate**

Aminotransferases catalyze the transfer of an χ-amino group from an χ-amino acid to an χ-ketoacid. These enzymes, also called *transaminases,* generally funnel χ-amino groups from a variety of amino acids to χ-keto-glutarate for conversion into NH_4^+.

^-OOC–CH(^+H_3N)–R_1 + ^-OOC–C(=O)–R_2 ⇌ (Aminotransferase) ^-OOC–C(=O)–R_1 + ^-OOC–CH(^+H_3N)–R_2

Aspartate aminotransferase, one of the most important of these enzymes, catalyzes the transfer of the amino group of aspartate to χ-ketoglutarate.

Aspartate + α-ketoglutarate ⇌ oxaloicetate + glutamate

Alanine aminotransferase catalyzes the transfer of the amino group of alanine to χ-ketoglutarate.

Alanine + α-ketoglutarate ⇌ pyruvate + glutamate

These transamination reactions are reversible and can thus be used to synthesize amino acids from χ-ketoacids.

The nitrogen atom that is transferred to χ-ketoglutarate in the transamination reaction is converted into free ammonium ion by oxidative deamination. This reaction is catalyzed by *glutamate dehydrogenase.* This enzyme is unusual in being able to utilize either NAD^+ or $NADP^+$, at least in some species. The reaction proceeds by dehydrogenation of the C-N bond, followed by hydrolysis of the resulting Schiff base.

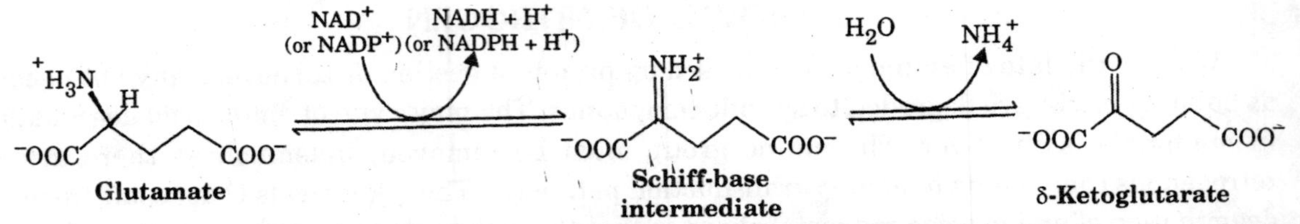

NAD$^+$ (or NADP$^+$) NADH + H$^+$ (or NADPH + H$^+$) H_2O NH_4^+

^+H_3N H ^-OOC COO^- **Glutamate** ⇌ NH_2^+ ^-OOC COO^- **Schiff-base intermediate** ⇌ O ^-OOC COO^- **δ-Ketoglutarate**

The equilibrium for this reaction favors glutamate; the reaction is driven by the consumption of ammonia. Glutamate dehydrogenase is located in mitochondria, as are some of the other enzymes required for the production of urea. This compartmentalization sequesters free ammonia, which is toxic.

In vertebrates, the activity of glutamate dehydrogenase is allosterically regulated. The enzyme consists of six identical subunits. Guanosine triphosphate and adenosine triphosphate are allosteric inhibitors, whereas guanosine diphosphate and adenosine diphosphate are allosteric activators. Hence, *a lowering of the energy charge accelerates the oxidation of amino acids.*

The sum of the reactions catalyzed by aminotransferases and glutamate dehydrogenase is

$$\alpha\text{-Amino acid} + NAD^+ \ (NADP^+) \rightleftharpoons \alpha\text{-ketoacid} + NH_4^+ + NADH + H^+ \ (\text{or } NADPH)$$

In most terrestrial vertebrates, NH_4^+ is converted into urea, which is excreted.

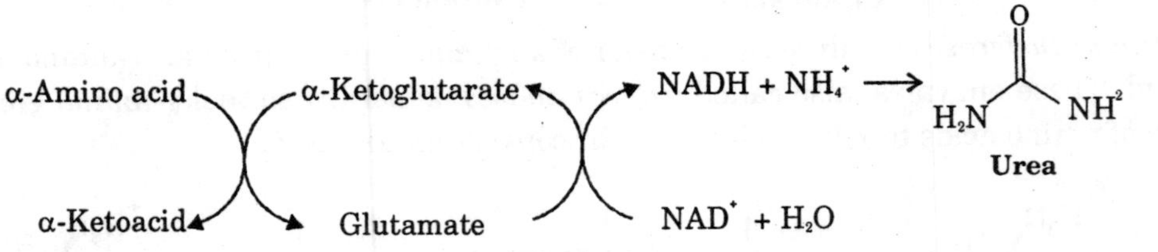

α-Amino acid → α-Ketoacid; α-Ketoglutarate → Glutamate; Glutamate + NAD$^+$ + H_2O → α-Ketoglutarate + NADH + NH_4^+ ⟶ O, H_2N, NH^2 **Urea**

Aminotransferases

All aminotransferases contain the prosthetic group *pyridoxal phosphate* (PLP), which is derived from *pyridoxine* (*vitamin* B_6). Pyridoxal phosphate includes a pyridine ring that is slightly basic as well as a phenolic hydroxyl group that is slightly acidic. Thus, pyridoxal phosphate derivatives can form stable tautomeric forms in which the pyridine nitrogen atom is protonated and, hence, positively charged while the hydroxyl group is deprotonated, forming a phenolate.

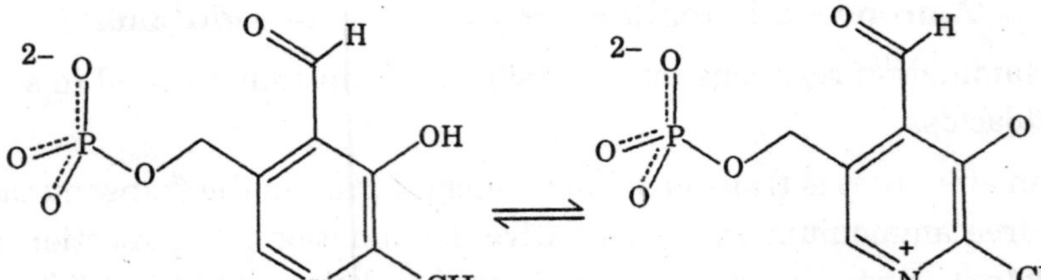

2– O O P O O O H OH N CH_3 **PLP** ⇌ 2– O O P O O O H O + N H CH_3 **PLP**

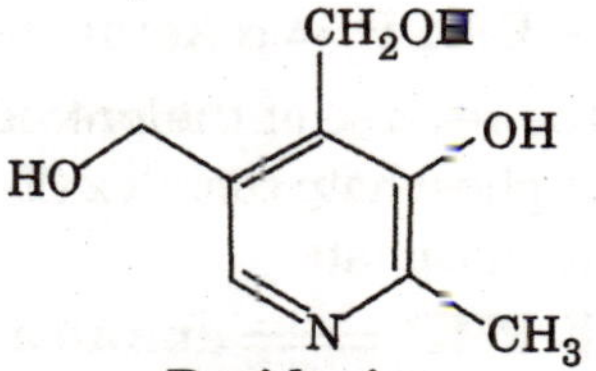

Pyridoxine (Vitamin B$_6$)

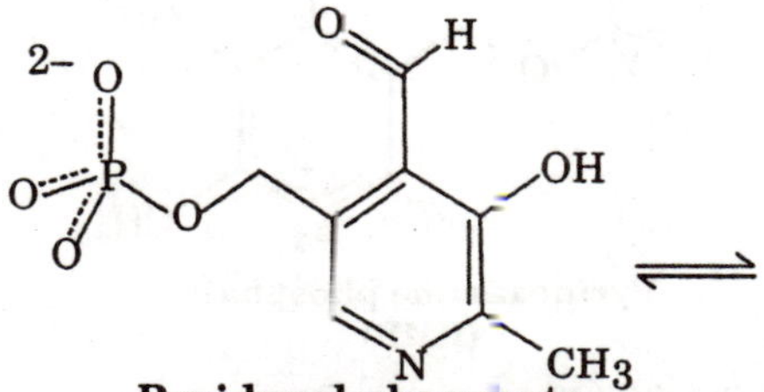

Pyridoxal phosphate PLP

The most important functional group on PLP is the aldehyde. This group allows PLP to form covalent Schiff-base intermediates with amino acid substrates. Indeed, even in the absence of substrate, the aldehyde group of PLP usually forms a Schiff-base linkage with the γ-amino group of a specific lysine residue of the enzyme. A new Schiff-base linkage is formed on addition of an amino acid substrate. These Schiff-base linkages are often protonated, with the positive charge stabilized by interaction with the negatively charged phenolate group of PLP.

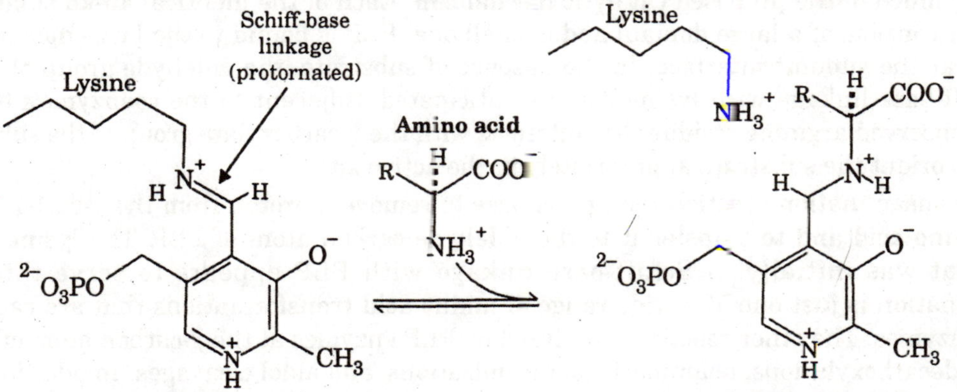

Internal aldimine **External aldimine**

The χ-amino group of the amino acid substrate displaces the γ-amino group of the active-site lysine residue. In other words, an internal aldimine becomes an external aldimine. The amino acid-PLP Schiff base that is formed remains tightly bound to the enzyme by multiple noncovalent interactions.

The Schiff base between the amino acid substrate and PLP, the external *aldimine,* loses a proton from the χ-carbon atom of the amino acid to form a *quinonoid* intermediate.

The negative charge that is left on the amino acid is stabilized by delocalization into the pyridinium ring. Reprotonation of this intermediate at the aldehyde carbon atom yields a *ketimine.* The ketimine is then hydrolyzed to an χ-ketoacid and *pyridoxamine phosphate (PMP).* These steps constitute half of the transamination reaction.

Amino acid$_1$ + E-PLP $\rightleftharpoons$ α-ketoacidl + E-PMP

The second half takes place by the reverse of the preceding pathway. A second χ-ketoacid reacts with the enzyme-pyridoxamine phosphate complex (E-PMP) to yield a second amino acid and regenerate the enzyme-pyridoxal phosphate

α-Ketoacid$_2$ + E-PMP $\rightleftharpoons$ amino acid$_2$ + E-PLP

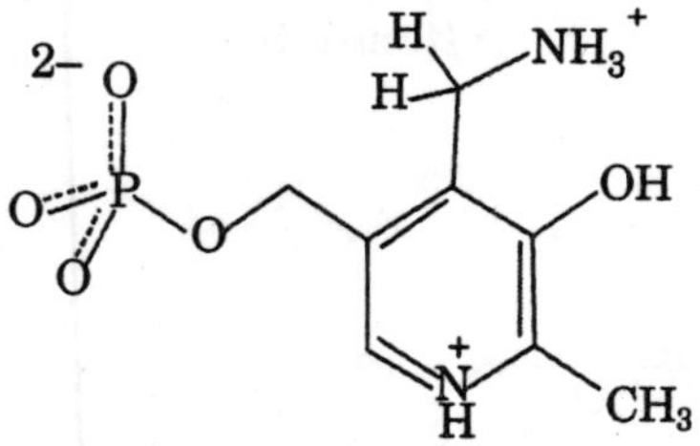

Pyridoxamine phosphate (PMP)

The sum of these partial reactions is

Amino acid$_1$ + α-ketoacid$_2$ $\rightleftharpoons$ amino acid$_2$ + α-ketoacid$_1$

Aspartate Aminotransferase

The mitochondrial enzyme aspartate aminotransferase provides an especially well studieq example of PLP as a coenzyme for transamination reactions. The results of X-ray crystallographic studies provided detailed views of how PLP and substrates are bound and confirmed much of the proposed catalytic mechanism. Each of the identical 45-kd subunits of this dimer consists of a large domain and a small one. PLP is bound to the large domain, in a pocket near the subunit interface. In the absence of substrate, the aldehyde group of PLP is in a Schiff-base linkage with lysine 258, as anticipated. Adjacent to the coenzyme's binding site is a conserved arginine residue tliat interacts with the χ-carboxylate group of the substrate, helping to orient the substrate appropriately in the active site.

The transamination reaction requires a base to remove a proton from the χ-carbon group of the amino acid and to transfer it to the aldehyde carbon atom of PLP. The lysine amino group that was initially in Schiff-base linkage with PLP appears to serve this role. Transamination is just one of a wide range of amino acid transformations that are catalyzed by PLP enzymes. The other reactions catalyzed by PLP enzymes at the χ-carbon atom of amino acids are decarboxylations, deaminations, racemizations, and aldol cleavages. In addition, PLP enzymes catalyze elimination and replacement reactions at the δ-carbon atom (*e.g.*, tryptophan synthetase; and the χ-carbon atom (*e.g.*, cytathionine δ-synthase, of amino acid substrates. Three common features of PLP catalysis underlie these diverse reactions.

1. A Schiff base is formed by the amino acid substrate (the amine component) and PLP (the carbonyl component).
2. The protonated form of PLP acts as an *electron sink to* stabilize catalytic intermediates that are negatively charged. Electrons from these intermediates can be transferred into the pyridine ring to neutralize the positive charge on the pyridinium nitrogen. In other words, PLP is an *electrophilic catalyst.*
3. The product Schiff base is cleaved at the completion of the reaction.

Many of the enzymes that catalyze these reactions, such as serine hydroxymethyl transferase, which converts serine into glycine, have the same fold as that of aspartate aminotransferase and are clearly related by divergent evolution. Others, such as tryptophan synthetase, have quite different overall structures. Nonetheless, the active sites of these enzymes are remarkably similar to that of aspartate aminotransferase, revealing the effects of convergent evolution.

How does a particular enzyme selectively favor the cleavage of one of three bonds at the χ-carbon atom of an amino acid substrate? An important principle is that *the bond being broken must be perpendicular to the ρ orbitals of the electron sink*. An aminotransferase, for example, achieves this goal by binding the amino acid substrate so that the C -H bond is perpendicular to the PLP ring. In serine hydroxymethyltransferase, the N-Cχ bond is rotated so that the Cχ -$C_δ$ bond is most nearly perpendicular to the plane of the PLP ring, favoring its cleavage. This means of choosing one of several possible catalytic outcomes is called *stereoelectronic control.*

Deamination of Serine and Threonine

Although the nitrogen atoms of most amino acids are transferred to χ-ketoglutarate before removal, the χ-amino groups of serine and threonine can be directly converted into NH_4^+. These direct deaminations are catalyzed by *serine dehydratase* and *threonine dehydratase,* in which PLP is the prosthetic group.

Serine ⟶ **Pyruvate + NH_4^+**

Threonine ⟶ **α-ketobutyrate + NH_4^+**

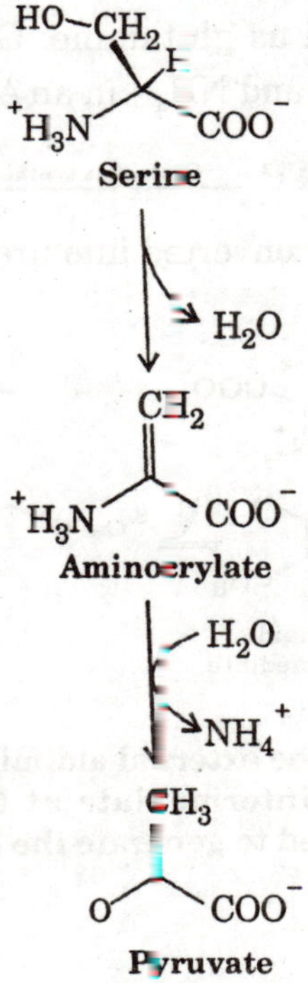

$HO-CH_2$, H, ^+H_3N, COO^-

Serine

↓ H_2O

CH_2, ^+H_3N, COO^-

Aminoacrylate

↓ H_2O, NH_4^+

CH_3, O, COO^-

Pyruvate

These enzymes are called *dehydratases* because *dehydration precedes deamination.* Serine loses a hydrogen ion from its χ-carbon atom and a hydroxide ion group from its δ-carbon atom to yield aminoacrylate. This unstable compound reacts with H_2O to give pyruvate and NH_4^+. Thus, the presence of a hydroxyl group attached to the δ-carbon atom in each of these amino acids permits the direct deamination.

Transport of Nitrogen to the Liver

Most amino acid degradation takes place in tissues other than the liver. For instance, muscle uses amino acids as a source of fuel during prolonged exercise and fasting. How is the nitrogen processed in these other tissues? As in the liver, the first step is the removal of the nitrogen from the amino acid. However, muscle lacks the enzymes of the urea cycle, so the nitrogen must be released in a form that can be absorbed by the liver and converted into urea.

Nitrogen is transported from muscle to the liver in two principal transport forms. Glutamate is formed by transamination reactions, but the nitrogen is then transferred to pyruvate to form alanine, which is released into the blood.

The liver takes up the alanine and converts it back into pyruvate by transamination. The pyruvate can be used for gluconeogenesis and the amino group eventually appears as urea. This transport is referred to as the *alanine cycle.* It is reminiscent of the Cori cycle discussed earlier and again illustrates the ability of the muscle to shift some of its metabolic burden to the liver.

Nitrogen can also be transported as glutamine. Glutamine synthetase, catalyzes the synthesis of glutamine from glutamate and NH_4^+ in an ATP-dependent reaction:

$$\mathbf{NH_4^+ + glutamate + ATP \xrightarrow{\text{Glutamine synthetase}} glutamine + ADP + P_i}$$

The nitrogens of glutamine can be converted into urea in the liver.

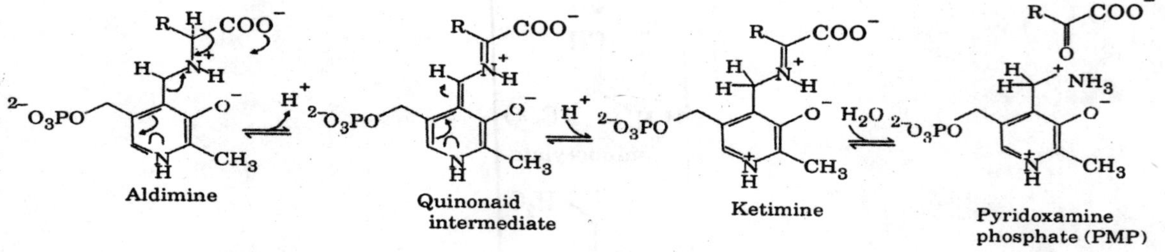

Aldimine

Quinonaid intermediate

Ketimine

Pyridoxamine phosphate (PMP)

Fig. 1.28. Transamination Mechanism. The external aldimine loses a proton to form a quinonoid intermediate. Reprotonation of this intermediate at the aldehyde carbon atom yields a ketimine. This intermediate is hydrolyzed to generate the δ-ketoacid product and pyridoxamine phosphate.

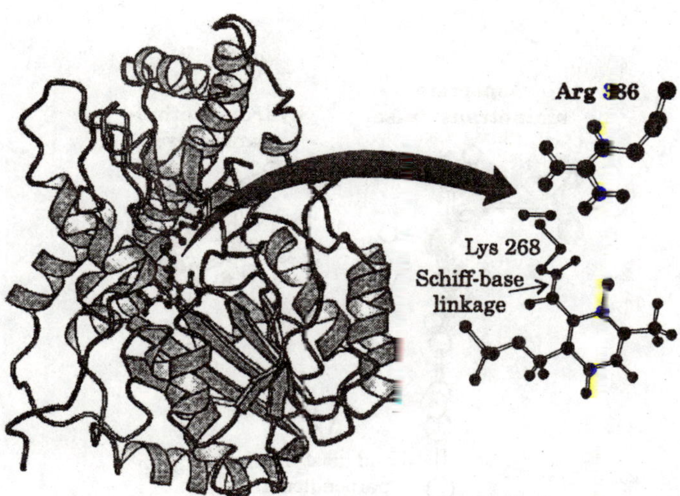

Fig. 1.29. Aspartate Aminotransferase. The active site of this prototypical PLP-dependent enzyme includes pyridoxal phosphate attached to the enzyme by a Schiff-base linkage with lysine 258. An arginine residue in the active site helps orient substrates by binding to their χ-carboxylate groups.

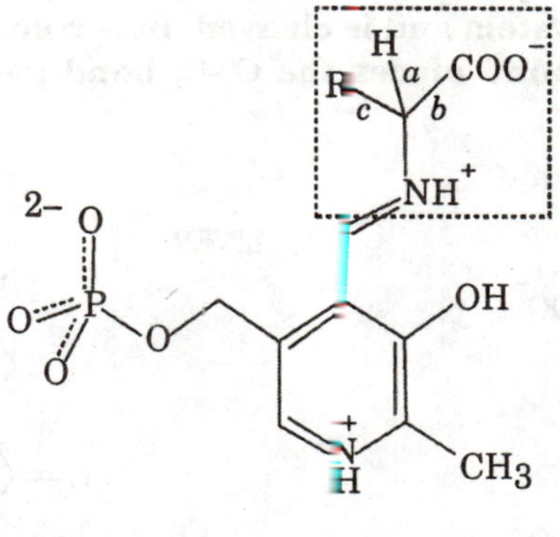

Fig. 1.30. Bond Cleavage by PLP Enzymes. Pyridoxal phosphate enzymes labilize one of three bonds at the χ-carbon atom of an amino acid substrate. For example, bond *a* is labilized by aminotransferases, bond *b* by decarboxylases, and bond *c* by aldolases (such as threonine aldolases). PLP enzymes also catalyze reactions at the δ-and χ-carbon atoms of amino acids.

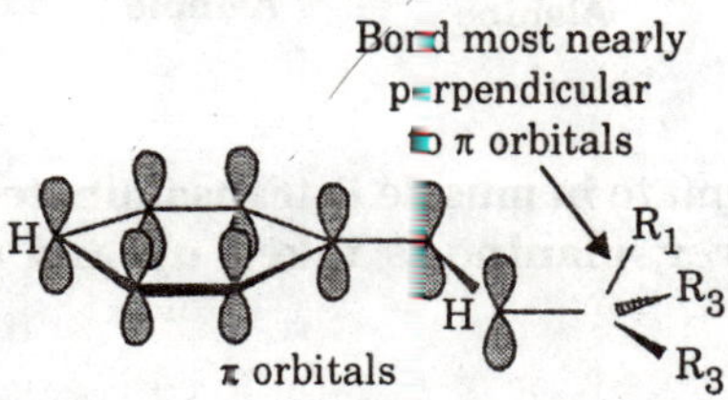

Fig. 1.31. Stereoelectronic Effect. The orientation about the NH-C_χ bond determines the most favored reaction catalyzed by a'pyridoxal phosphate enzyme. The bond that is most nearly perpendicular to the ρ orbitals of the pyridoxal phosphate electron sink is most easily cleaved.

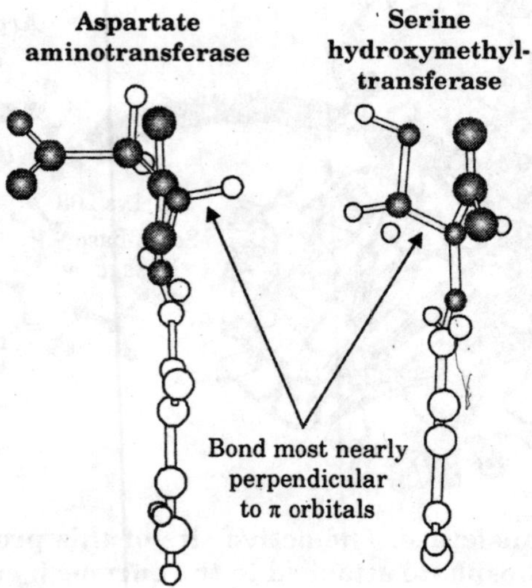

Fig. 1.32. Reaction Choice. In aspartate aminotransferase, the C_χ-H bond is most nearly perpendicular to the ρ-orbital system and is cleaved. In serine hydroxymethyltransferase, a small rotation about the N-C_χ bond places the C_χ-C_δ bond perpendicular to the ρ system, favoring its cleavage.

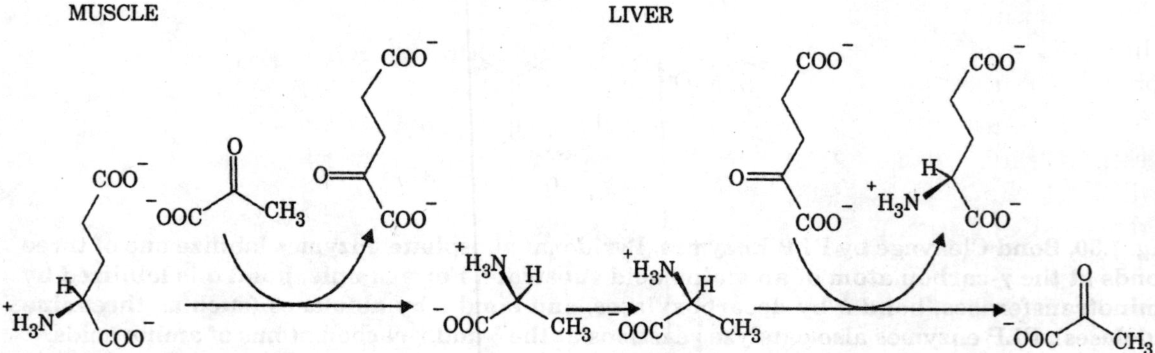

MUSCLE

LIVER

Glutamate

Alanine

Alanine

Pyruvate

Fig. 1.33. The Alanine Cycle. Glutamate in muscle is transaminated to alanine, which is released into the bloodstream. In the liver, alanine is taken up and converted into pyruvate for subsequent metabolism.

FORMATION OF UREA

Some of the NH_4^+ formed in the breakdown of amino acids is consumed in the biosynthesis of nitrogen compounds. In most terrestrial vertebrates, the excess NH_4^+ is converted into *urea* and then excreted. Such organisms are referred to as *ureotelic*.

In terrestrial vertebrates, urea is synthesized by the *urea cycle*. The urea cycle, proposed by Hans Krebs and Kurt Henseleit in 1932, was the first cyclic metabolic pathway to be discovered. One of the nitrogen atoms of the urea is transferred from an amino acid, aspartate. The other nitrogen atom is derived directly from free NH_4^+, and the carbon atom comes from HCO_3- (derived by hydration of CO_2)

$NH_4^+ \longrightarrow H_2N-C(=O)-NH_2$

Urea

Formation of Carbamoyl Phosphate

The urea cycle begins with the coupling of free NH_4^+ with HCO_3- to form carbamoyl phosphate. The synthesis of carbamoyl phosphate, though a simple molecule, is complex, comprising three steps, all catalyzed by *carbamoyl phosphate synthetase*.

Bicarbonate —(ATP → ADP)→ **Carboxyphosphate** —(NH_3 → P_1)→ **Carbamic acid** —(ATP → ADP)→ **Carbamoyl phosphate**

The reaction begins with the phosphorylation of HCO_3^- to form carboxyphosphate, which then reacts with ammonium ion to form carbamic acid. Finally, a second molecule of ATP phosphorylates carbamic acid to carbamoyl phosphate.

The consumption of two molecules of ATP makes this synthesis of carbamoyl phosphate essentially irreversible. The mammalian enzyme requires *N-acetyl-glutamate* for activity, as will be discussed shortly. The carbamoyl group of carbamoyl phosphate, which has a high transfer potential because of its anhydride bond, is transferred to *ornithine* to form *citrulline*, in a reaction catalyzed by *ornithine transcarbamoylase*.

Ornithine + **Carbamoyl phosphate** —(Ornithine transcarbamoylase; → P_1)→ **Citrulline**

Ornithine and citrulline are amino acids, but they are not used as building blocks of proteins. The formation of NH_4^+ by glutamate dehydrogenase, its incorporation into carbamoyl phosphate, and the subsequent synthesis of citrulline take place in the mitochondrial matrix. In contrast, the next three reactions of the urea cycle, which lead to the formation of urea, take place in the cytosol.

Citrulline is transported to the cytoplasm where it condenses with aspartate, the donor of the second amino group of urea. This synthesis of *argininosuccinate,* catalyzed by *argininosuccinate synthetase,* is driven by the cleavage of ATP into AMP and pyrophosphate and by the subsequent hydrolysis of pyrophosphate.

Citrulline + **Aspartate** —(ATP → AMP + PP_1; Argininosuccinate synthetase)→ **Argininosuccinate**

Argininosuccinase cleaves argininosuccinate into *arginine* and *fumarate.* Thus, the carbon skeleton of aspartate is preserved in the form of fumarate.

Argininosuccinate —(Argininosuccinase)→ **Arginine** + **Fumarate**

Finally, arginine is hydrolyzed to generate urea and ornithine in a reaction catalyzed by *arginase.* Ornithine is then transported back into the mitochondrion to begin another cycle. The urea is excreted. Indeed, human beings excrete about 10 kg (22 pounds) of urea per year.

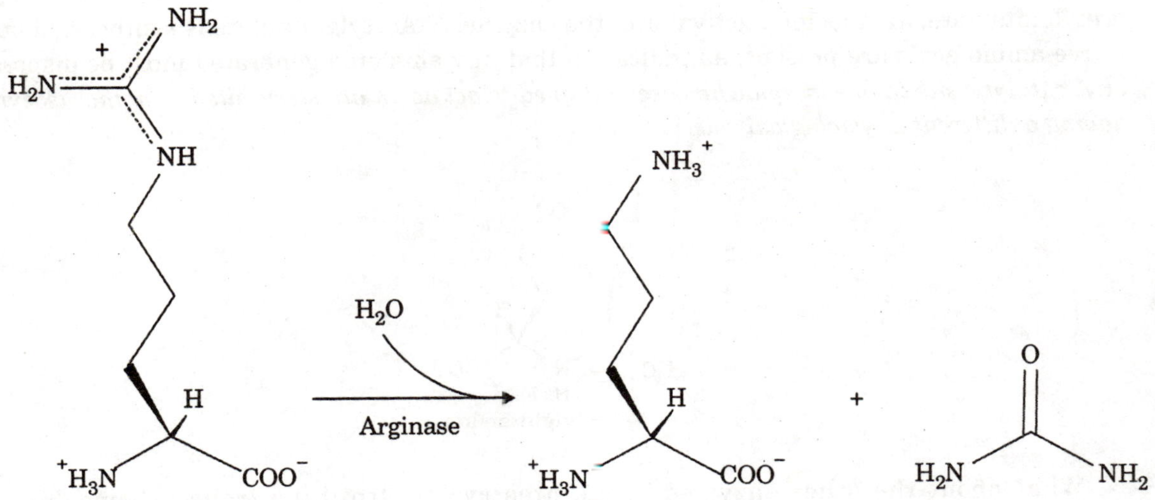

NH$_2$
+
H$_2$N
NH
H
$^+$H$_3$N COO$^-$
Arginine

H$_2$O
Arginase

NH$_3^+$
H
$^+$H$_3$N COO$^-$
Ornithine

+

O
H$_2$N NH$_2$
Urea

The Urea Cycle and the citric Acids Cycle

The stoichiometry of urea synthesis is

$$\mathbf{CO_2 + NH_4^+ + 3\,ATP + aspartate + 2\,H_2O \rightarrow}$$
$$\mathbf{urea + 2\,ADP + 2\,P_i + AMP + PP_i + fumarate}$$

Pyrophosphate is rapidly hydrolyzed, and so the equivalent of four molecules of ATP are consumed in these reactions to synthesize one molecule of urea. *The synthesis of fumarate by the urea cycle is important because it links the urea cycle and the citric acid cycle.* Fumarate is hydrated to malate, which is in turn oxidized to oxaloacetate. Oxaloacetate has several possible fates: (1) transamination to aspartate, (2) conversion into glucose by the gluconeogenic pathway, (3) condensation with acetyl CoA to form citrate, or (4) conversion into pyruvate.

The Evolution of the Urea Cycle

We have previously encountered carbamoyl phosphate as a sub strate for aspartate transcarbamoylase, the enzyme that catalyzes the first step in pyrimidine biosynthesis. Carbamoyl phosphate synthetase generates carbamoyl phosphate for both this pathway and the urea cycle. In mammals, two distinct carbamoyl phosphate synthetase isozymes are present.

As discussed earlier, the mitochondrial enzyme uses NH_4^+ as the nitrogen source, as is appropriate for its role in the urea cycle. In pyrimidine biosynthesis, carbamoyl phosphate synthetase differs in two important ways. First, this enzyme utilizes glutamine as a nitrogen source. The side chain amide is hydrolyzed within one domain of the enzyme and the ammonia generated moves through a tunnel in the enzyme to react with carboxyphosphate. Second, this enzyme is part of a large polypeptide called CAD that comprises three distinct enzymes: carbamoyl phosphate synthetase, aspartate transcarbamoylase, and dihydroorotase, all of which catalyze steps in pyrimidine biosynthesis.

Interestingly, the domain in which glutamine hydrolysis takes place is largely preserved in the urea-cycle enzyme, although that domain is catalytically inactive. This site binds *N*-

acetylglutamate, an allosteric activator of the enzyme. *N*-Acetylglutamate is synthesized only if free amino acids are present, an indication that any ammonia generated must be disposed of. *A catalytic site in one isozyme has been adapted to act as an allosteric site in another isozyme having a different physiological role.*

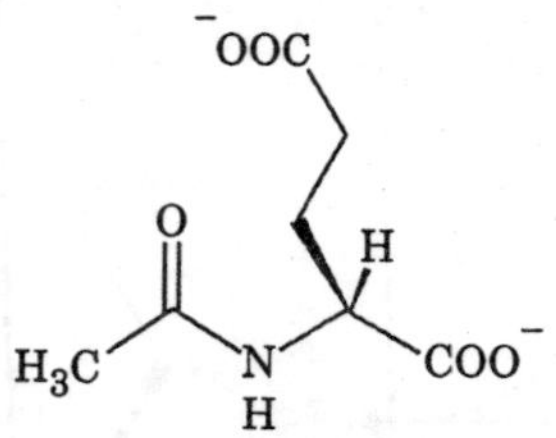

N-Acetylglutamate

What about the other enzymes in the urea cycle? Ornithine transcarbamoylase is homologous to aspartate transcarbamoylase and the structures of their catalytic subunits are quite similar. Thus, two consecutive steps in the pyrimidine biosynthetic pathway were adapted for urea synthesis. The next step in the urea cycle is the addition of aspartate to citrulline to form argininosuccinate, and the subsequent step is the removal of fumarate. These two steps together accomplish the net addition of an amino group to citrulline to form arginine. Remarkably, these steps are analogous to two consecutive steps in the purine biosynthetic pathway.

Aspartate **Fumarate**

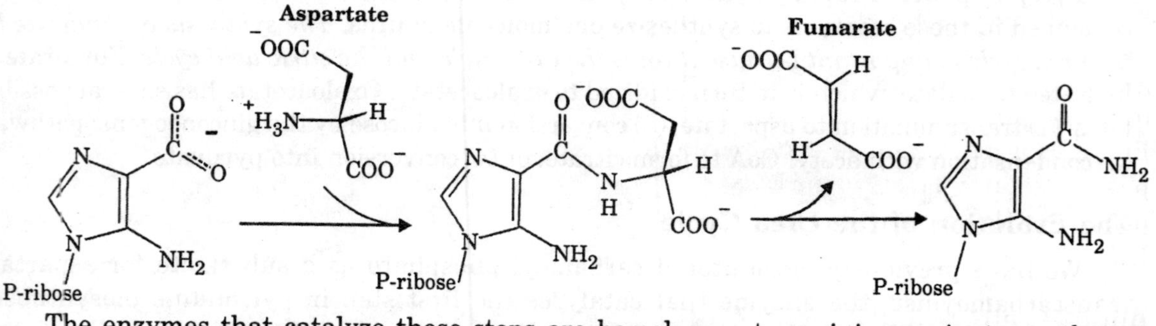

The enzymes that catalyze these steps are homologous to argininosuccinate synthetase and argininosuccinase, respectively. Thus, four of the five enzymes in the urea cycle were adapted from enzymes taking part in nucleotide biosynthesis. The remaining enzyme, arginase, appears to be an ancient enzyme found in all domains of life.

Hyperammonemia

The synthesis of urea in the liver is the major route of removal of NH_4^+. A blockage of carbamoyl phosphate synthesis or of any of the four steps of the urea cycle has devastating consequences because there is no alternative pathway for the synthesis of urea. *All defects in the urea cycle lead to an elevated level of NH_4^+ in the blood (hyperammonemia).* Some of these genetic defects become evident a day or two after birth, when the afflicted infant becomes lethargic and vomits periodically. Coma and irreversible brain damage may soon follow. Why are high levels of NH_4^+ toxic? The answer to this question is not yet known. One possibility is that elevated levels of glutamine, formed from NH_4^+ and glutamate, produce osmotic effects that lead directly to brain swelling.

$$\text{Glutamate} + NH_3 \xrightarrow[\text{Glutamine synthetase}]{ATP \rightarrow ADP + P_i} \text{Glutamine}$$

Ingenious strategies for coping with deficiencies in urea synthesis have been devised on the basis of a thorough understanding of the underlying biochemistry. Consider, for example, *argininosuccinase deficiency.* This defect can be partly bypassed *by providing a surplus of arginine in the diet and restricting the total protein intake.* In the liver, arginine is split into urea and ornithine, which then reacts with carbamoyl phosphate to form citrulline.

This urea-cycle intermediate condenses with aspartate to yield argininosuccinate, which is then excreted. Note that two nitrogen atoms—one from carbamoyl phosphate and another from aspartate—are eliminated from the body per molecule of arginine provided in the diet. In essence, *argininosuccinate substitutes for urea in carrying nitrogen out of the body.* The treatment *of carbamoyl phosphate synthetase deficiency* or *ornithine transcarbamoylase deficiency* illustrates a different strategy for circumventing a metabolic block.

Citrulline and argininosuccinate cannot be used to dispose of nitrogen atoms, because their formation is impaired. Under these conditions, excess nitrogen accumulates in glycine and glutamine. The challenge then is to rid the body of these two amino acids, which is accomplished by supplementing a protein-restricted diet with *large amounts of benzoate and phenylacetate.* Benzoate is activated to benzoyl CoA, which reacts with glycine to form hippurate. Likewise, phenylacetate is activated to phenylacetyl CoA, which reacts with glutamine to form phenylacetylglutamine. These conjugates substitute for urea in the disposal of nitrogen. Thus, *latent biochemical pathways can be activated to partly bypass a genetic defect.*

Ammonitelism and Uricotelism

As discussed earlier, most terrestrial vertebrates are ureotelic; they excrete excess nitrogen as urea. However, urea is not the only excretable form of nitrogen. *Ammoniotelic organisms, such as aquatic vertebrates and invertebrates, release nitrogen as NH_4^+* and rely on the aqueous environment to dilute this toxic substance. Interestingly, lungfish, which are normally ammoniotelic, become ureotelic in time of drought, when they live out of the water. Both ureotelic and ammoniotelic organisms depend on sufficient water, to varying degrees, for nitrogen excretion. *In contrast, uricotelic organisms, which secrete nitrogen as the purine uric acid, require little water.* Disposal of excess nitrogen as uric acid is especially valuable in animals, such as birds, that produce eggs having impermeable membranes that accumulate waste products. The pathway for nitrogen excretion developed in the course of evolution clearly depends on the habitat of the organism.

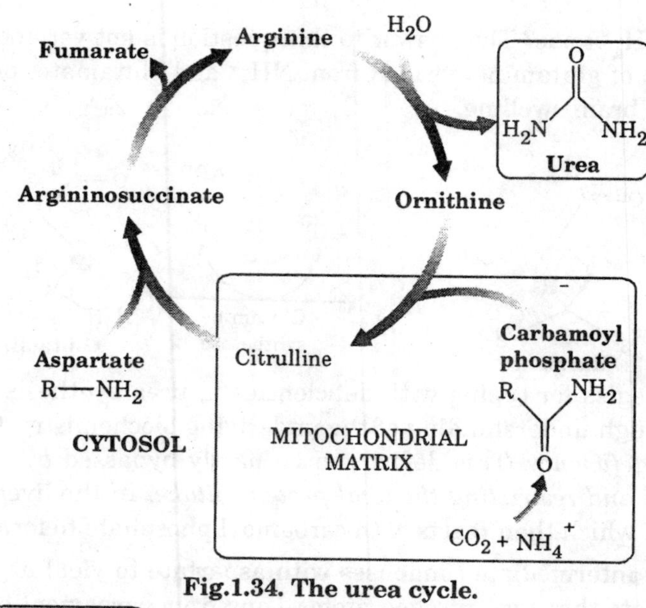

Fig.1.34. The urea cycle.

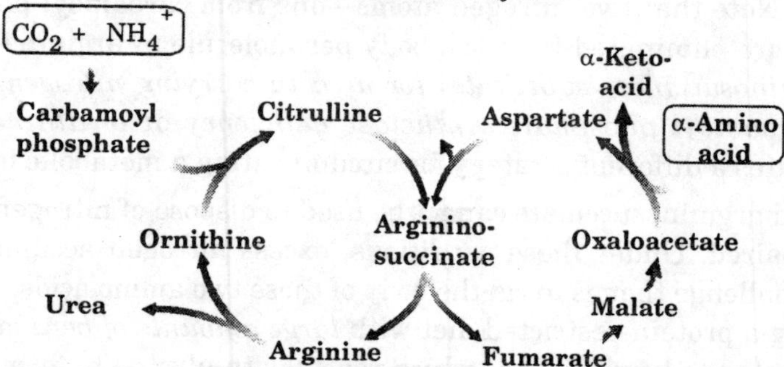

Fig. 1.35. Metabolic Integration of Nitrogen Metabolism. The urea cycle, the citric acid cycle, and the transaminaiton of oxaloacetate are linked by fumarate and aspartate.

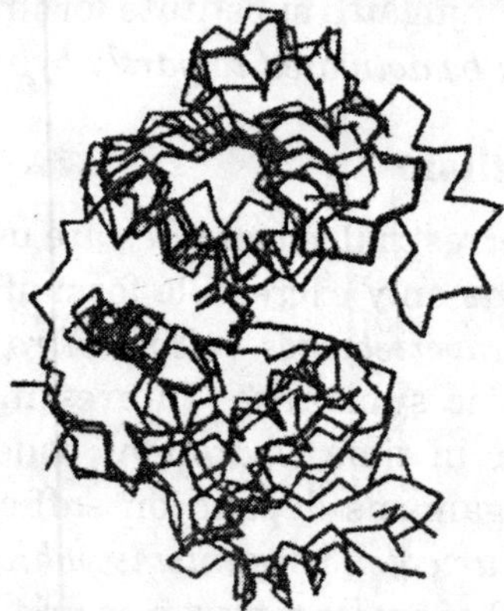

Fig. 1.36. Homologous Enzymes. The structure of the catalytic subunit of ornithine transcarbamoylase (blue) is quite similar to that of the catalytic subunit of aspartate transcarbamoylase (red), indicating that these two enzymes are homologs.

MAJOR METABOLIC INTERMEDIATES

We now turn to the fates of the carbon skeletons of amino acids after the removal of the χ-amino group. *The strategy of amino acid degradation is to transform the carbon skeletons into major metabolic intermediates that can be converted into glucose or oxidized by the citric acid cycle.* The conversion pathways range from extremely simple to quite complex. The carbon skeletons of the diverse set of 20 fundamental amino acids are funneled into only seven molecules: *pyruvate, acetyl CoA, acetoacetyl CoA, χ-ketoglutarate, succinyl CoA, fumarate,* and *oxaloacetate*. We see here a striking example of the remarkable economy of metabolic conversions, as well as an illustration of the importance of certain metabolites. Amino acids that are degraded to acetyl CoA or acetoacetyl CoA are termed *ketogenic* amino acids because they can give rise to ketone bodies or fatty acids.

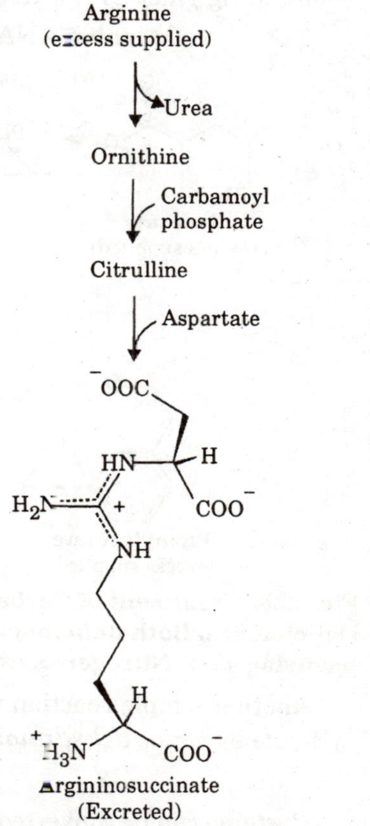

Arginine
(excess supplied)
→ Urea
Ornithine
Carbamoyl phosphate
Citrulline
Aspartate
^-OOC, HN, H, H_2N, +, COO^-, NH, H, ^+H_3N, COO^-
Argininosuccinate
(Excreted)

Fig. 1.37. Treatment of Argininosuccinase Deficiency. Argininosuccinase deficiency can be managed by supplementing the diet with arginine. Nitrogen is excreted in the form of argininosuccinate.

Amino acids that are degraded to pyruvate, χ-ketoglutarate, succinyl CoA, fumarate, or oxaloacetate are termed *glucogenic* amino acids. The net synthesis of glucose from these amino acids is feasible because these citric acid cycle intermediates and pyruvate can be converted into phosphoenolpyruvate and then into glucose. Recall that mammals lack a pathway for the net synthesis of glucose from acetyl CoA or acetoacetyl CoA. Of the basic set of 20 amino acids, only leucine and lysine are solely ketogenic. Isoleucine, phenylalanine, tryptophan, and tyrosine are both ketogenic and glucogenic.

Some of their carbon atoms emerge in acetyl CoA or acetoacetyl CoA, whereas others appear in potential precursors of glucose. The other 14 amino acids are classed as solely glucogenic. This classification is not universally accepted, because different quantitative criteria are applied. Whether an amino acid is regarded as being glucogenic, ketogenic, or both depends partly on the eye of the beholder. We will identify the degradation pathways by the entry point into metabolism.

Entry Point of Pyruvate

Pyruvate is the entry point of the three-carbon amino acids—alanine, serine, and cysteine—into the metabolic mainstream. The transamination of alanine directly yields pyruvate.

$$\text{Alanine} + \alpha\text{-ketoglutarate} \rightleftharpoons \text{pyruvate} + \text{glutamate}$$

As mentioned previously, glutamate is then oxidatively deaminated, yielding NH_4^+ and regenerating χ-ketoglutarate. The sum of these reactions is

$$\text{Alanine} + NAD(P)^+ + H_2O \rightarrow \text{pyruvate} + NH_4^+ \; NAD(P)H + H^+$$

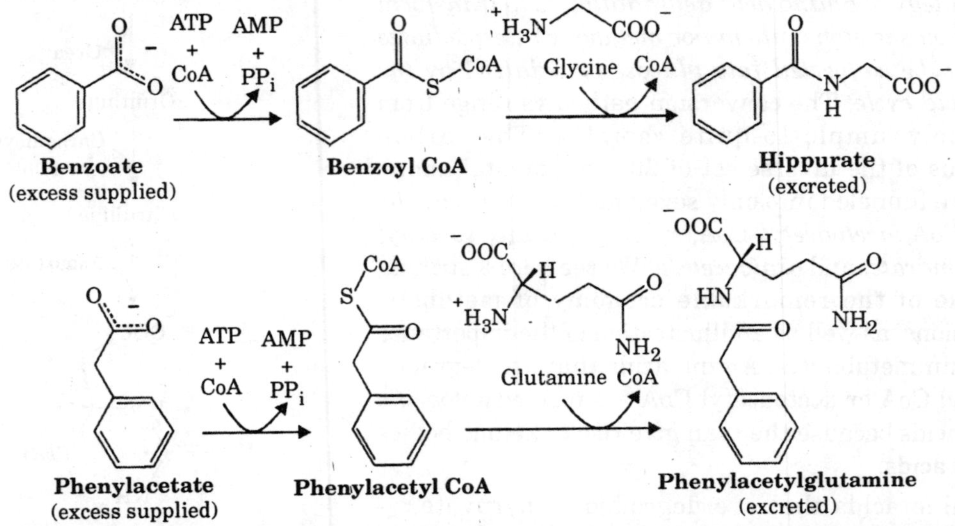

Benzoate (excess supplied) — ATP + CoA → AMP + PP_i → Benzoyl CoA — Glycine (^+H_3N–CH_2–COO^-) → CoA → Hippurate (excreted)

Phenylacetate (excess supplied) — ATP + CoA → AMP + PP_i → Phenylacetyl CoA — Glutamine → CoA → Phenylacetylglutamine (excreted)

Fig. 1.38. Treatment of Carbamoyl Phosphate Synthetase and Orninthine Transcarbamoylase Deficiencies. Both deficiencies can be treated by supplementing the diet with benzoate and phenylacetate. Nitrogen is excreted in the form of hippurate and phenylacetylglutamine.

Another simple reaction in the degradation of amino acids is the *deamination of serine to pyruvate* by *serine dehydratase.*

$$\text{Serine} \rightarrow \text{pyruvate} + NH_4^+$$

Cysteine can be converted into pyruvate by several pathways, with its sulfur atom emerging in H_2S, SCN^-, or SO_3^{2-}.

The carbon atoms of three other amino acids can be converted into pyruvate. *Glycine* can be converted into serine by enzymatic addition of a hydroxymethyl group or it can be cleaved to give CO_2, NH_4^+, and an activated one-carbon unit. *Threonine* can give rise to pyruvate through the intermediate aminoacetone. Three carbon atoms of *tryptophan* can emerge in alanine, which can be converted into pyruvate.

Oxaloacetate

Aspartate and asparagine are converted into oxaloacetate, a citric acid cycle intermediate. *Aspartate,* a four-carbon amino acid, is directly *transaminated to oxaloacetate.*

$$\text{Aspartate} + \alpha\text{-ketoglutarate} \rightleftharpoons \text{oxaloacetate} + \text{glutamate}$$

Asparagine is hydrolyzed by *asparaginase* to NH_4^+ and aspartate, which is then transaminated.

Recall that aspartate can also be converted *intofumarate* by the urea cycle. Fumarate is also a point of entry for half the carbon atoms of tyrosine and phenylalanine, as will be discussed shortly.

Alpha-Ketoglutarate

The carbon skeletons of several five-carbon amino acids enter the citric acid cycle at χ-*ketoglutarate.* These amino acids are first converted into *gluta-mate,* which is then oxidatively deaminated by glutamate dehydrogenase to yield χ-ketoglutarate. *Histidine* is converted into 4-imidazolone 5-propionate. The amide bond in the ring of this intermediate is hydrolyzed to the *N*-formimino derivative of glutamate, which is then converted into glutamate by transfer of its formimino group to tetrahydrofolate, a carrier of activated one-carbon units. *Glutamine* is hydrolyzed to glutamate and NH_4^+ by *glutaminase Proline* and *arginine* are each converted into glutamate χ-semialdehyde, which is then oxidized to glutamate.

Succinyl Coenzyme

Succinyl CoA is a point of entry for some of the carbon atoms of methio-nine, isoleucine, and valine. Propionyl CoA and then methylmalonyl CoA are intermediates in the breakdown of these three nonpolar amino acids. The mechanism for the interconversion of propionyl CoA and methylmalonyl CoA has already presented. This pathway from propionyl CoA to succinyl CoA is also used in the oxidation of fatty acids that have an odd number of carbon atoms.

Methionine Degradation

Methionine is converted into succinyl CoA in nine steps. The first step is the adenylation of methionine to form δ-adenosylmethionine (SAM), a common methyl donor in the cell. Methyl donation and deadenylation yield homocysteine, which is eventually processed to χ-ketobutyrate. The enzyme χ-ketoacid dehydrogenase complex oxidatively decarboxylates χ-ketobutyrate to propionyl CoA, which is processed to succinyl CoA.

Acetyl CoA

The degradation of the branched-chain amino acids employs reactions that we have encountered previously in the citric acid cycle and fatty acid oxidation. Leucine is transaminated to the corresponding χ-ketoacid, χ-*ketoisocaproate.* This χ-ketoacid is *oxidatively decarboxylated to isovaleryl CoA* by the *branched-chain* χ-*ketoacid dehydrogenase complex.*

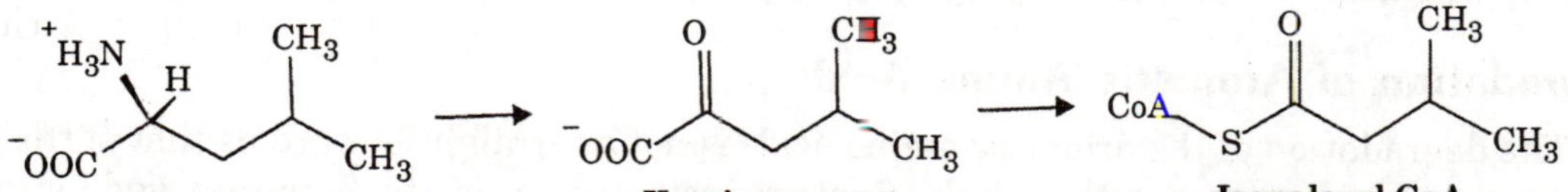

Leucine → **α-Ketoisocaproate** → **Isovaleryl CoA**

The χ-ketoacids of valine and isoleucine, the other two branched-chain aliphatic amino acids, as well as χ-ketobutyrate derived from methionine also are substrates. The oxidative decarboxylation of these χ-ketoacids is analogous to that of pyruvate to acetyl CoA and of χ-ketoglutarate to succinyl CoA. The branched-chain χ-ketoacid dehydrogenase, a multienzyme complex, is a homolog of pyruvate dehydrogenase and χ-ketoglutarate dehydrogenase.

Indeed, the E3 components of these enzymes, which regenerate the oxidized form of lipoamide, are identical. The isovaleryl CoA derived from leucine is *dehydrogenated* to yield δ-*methylcrotonyl CoA.* This oxidation is catalyzed by *isovaleryl CoA dehydrogenase.* The hydrogen acceptor is FAD, as in the analogous reaction in fatty acid oxidation that is catalyzed by acyl CoA dehydrogenase. δ-*Methylglutaconyl CoA* is then formed by the *carboxylation* of δ-methylcrotonyl CoA at the expense of the hydrolysis of a molecule of ATP. As might be expected, the carboxylation mechanism of δ-methylcrotonyl CoA carboxylase is similar to that of pyruvate carboxylase and acetyl CoA carboxylase.

Isovaleryl CoA —(FAD → $FADH_2$)→ β-Methylcrotonyl CoA —(CO_2 + ATP → ATP + P_i)→ β-Methylglutaconyl CoA

δ-Methylglutaconyl CoA is then *hydrated* to form *3-hydroxy-3-methylglutaryl CoA,* which is cleaved into *acetyl CoA* and *acetoacetate.* This reaction has already been discussed in regard to the formation of ketone bodies from fatty acids.

β-Methylglutaconyl CoA —(H_2O)→ 3-Hydroxy-3-Methylglutaryl CoA → Acetyl CoA + Aceteacetate

The degradative pathways of valine and isoleucine resemble that of leucine. After transamination and oxidative decarboxylation to yield a CoA derivative, the subsequent reactions are like those of fatty acid oxidation. Isoleucine yields acetyl CoA and propionyl CoA, whereas valine yields CO_2 and propionyl CoA. The degradation of leucine, valine, and isoleucine validate a point made earlier the number of reactions in metabolism is large, but the number *of kinds* of reactions is relatively small. The degradation of leucine, valine, and isoleucine provides a striking illustration of the underlying simplicity and elegance of metabolism.

Degradation of Aromatic Amino Acids

The degradation of the aromatic amino acids is not as straightforward as that of the amino acids previously discussed, although the final products—acetoacetate, fumarate, and pyruvate—are common intermediates. For the aromatic amino acids, *molecular oxygen is used to break an aromatic ring.*

The degradation of phenylalanine begins with its hydroxylation to tyrosine, a reaction catalyzed by *phenylalanine hydroxylase.* This enzyme is called a *monooxygenase* (or *mixed-function oxygenase)* because *one atom of* O_2 appears in the product and the other in H_2O.

Phenylalanine + O_2 + tetrahydrobiopterin —(Phenylalanine hydroxylase)→ Tyrosine + H_2O + quinonoid biopterin

The reductant here is *tetrahydrobiopterin*, an electron carrier that has not been previously discussed and is derived from the cofactor *biopterin*. Because biopterin is synthesized in the body, it is not a vitamin. Tetrahydrobiopterin is initially formed by the reduction of dihydrobiopterin by NADPH in a reaction catalyzed by *dihydrofolate reductase*. NADH reduces the quinonoid form of dihydrobiopterin produced in the hydroxylation of phenylalanine back to tetrahydrobiopterin in a reaction catalyzed by *dihydropteridine reductase.* The sum of the reactions catalyzed by phenylalanine hydroxylase and dihydropteridine reductase is

$$\text{Phenylalanine} + O_2 + \text{NADH} + H^+ \rightarrow \text{tyrosine} + \text{NAD}^+ + H_2O$$

Note that these reactions can also be used to synthesize tyrosine from phenylalanine. The next step in the degradation of phenylalanine and tyrosine is the transamination of tyrosine to *p-hydroxyphenylpyruvate*. This χ-ketoacid then reacts with O_2 to form *homogentisate*. The enzyme catalyzing this complex reaction, *p-hydroxyphenylpyruvate hydroxylase,* is called a *dioxygenase* because *both atoms of* O_2become incorporated into the product, one on the ring and one in the carboxyl group.

The aromatic ring of homogentisate is then cleaved by O_2, which yields 4-maleylacetoacetate. This reaction is catalyzed by *homogentisate oxidase,* another dioxygenase. 4-Maleylacetoacetate is then isomerized to *4-fumarylacetoacetate* by an enzyme that uses glutathione as a cofactor. Finally, 4-fumarylacetoacetate is hydrolyzed to *fumarate* and *acetoacetate*. Tryptophan degradation requires several oxygenases. Tryptophan 2,3-dioxygenase cleaves the pyrrole ring, and kynureinine 3-monooxygenase hydroxylates the remaining benzene ring, a reaction similar to the hydroxylation of phenylalanine to form tyrosine.

Alanine is removed and the 3-hydroxyanthranilic acid is cleaved with another dioxygenase and subsequently processed to acetoacetyl CoA. *Nearly all cleavages of aromatic rings in biological systems are catalyzed by dioxygenases,* a class of enzymes discovered by Osamu Hayaishi. The active sites of these enzymes contain iron that is not part of heme or an iron-sulfur cluster.

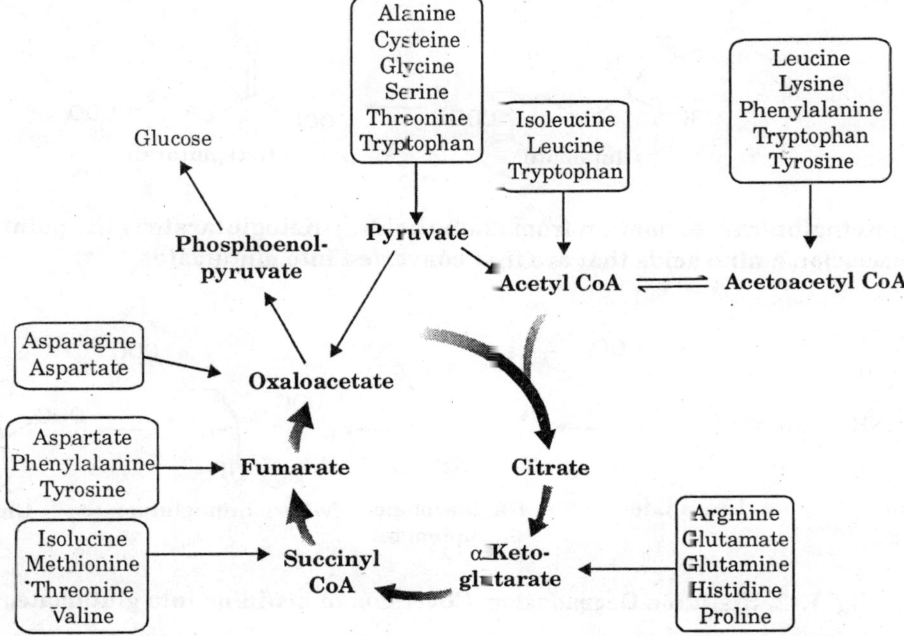

Fig. 1.39. Fates of the Carbon Skeletons of Amino Acids. Glucogenic amino acids are shaded red, and ketogenic amino acids are shaded yellow. Most amino acids are both glucogenic and ketogenic.

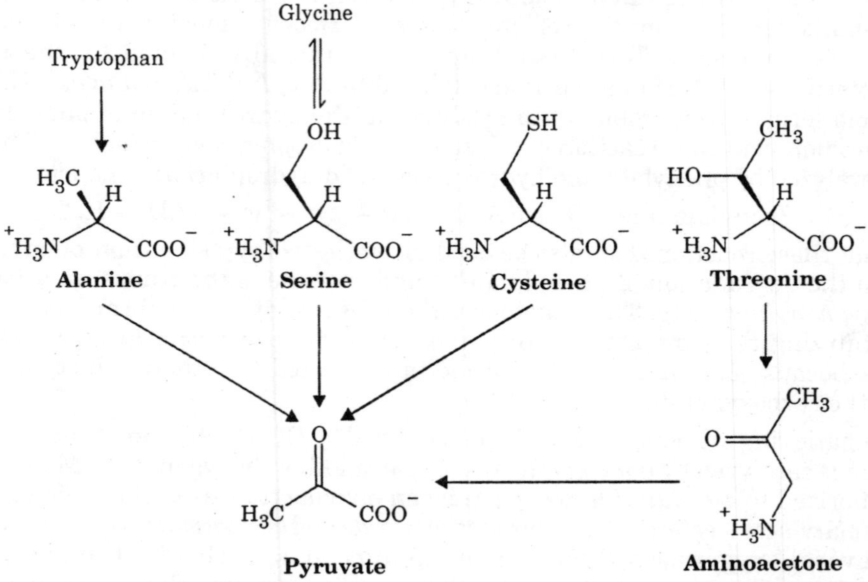

Glycine
Tryptophan
OH
SH
CH_3
H_3C H
H
H
HO— H
^+H_3N COO^-
^+H_3N COO^-
^+H_3N COO^-
^+H_3N COO^-
Alanine
Serine
Cysteine
Threonine
O
CH_3
O
H_3C COO^-
^+H_3N
Pyruvate
Aminoacetone

Fig. 1.40. Pyruvate Formation from Amino Acids. Pyruvate is the point of entry for alanine, serine, cysteine, glycine, threonine, and tryptophan.

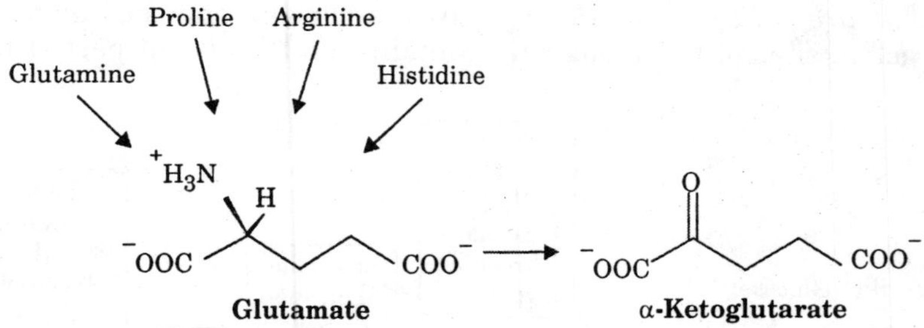

Proline Arginine
Glutamine
Histidine
^+H_3N
H
O
^-OOC COO^- ⟶ ^-OOC COO^-
Glutamate
α-Ketoglutarate

Fig. 1.41. χ-Ketoglutarate formation from amino acids. χ-Ketoglutarate is the point of entry of several fivecarbon amino acids that are first converted into glutamate.

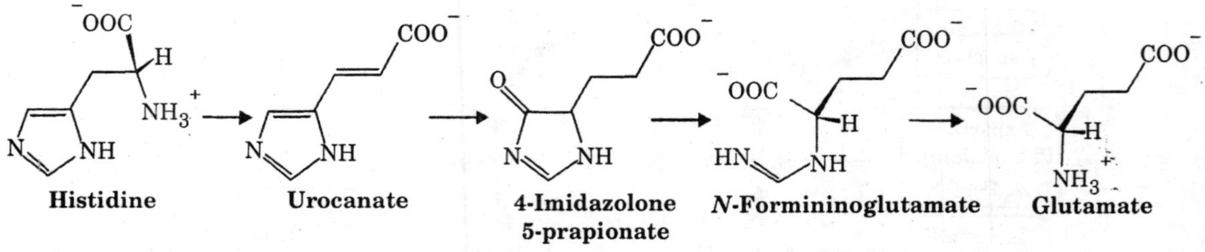

^-OOC
H
NH_3^+
N NH
Histidine
COO^-
N NH
Urocanate
COO^-
O
N NH
4-Imidazolone 5-prapionate
COO^-
^-OOC
H
HN NH
***N*-Formininoglutamate**
COO^-
^-OOC
H
NH_3^+
Glutamate

Fig. 1.42. Histidine Degradation. Coversion of histidine into glutamate.

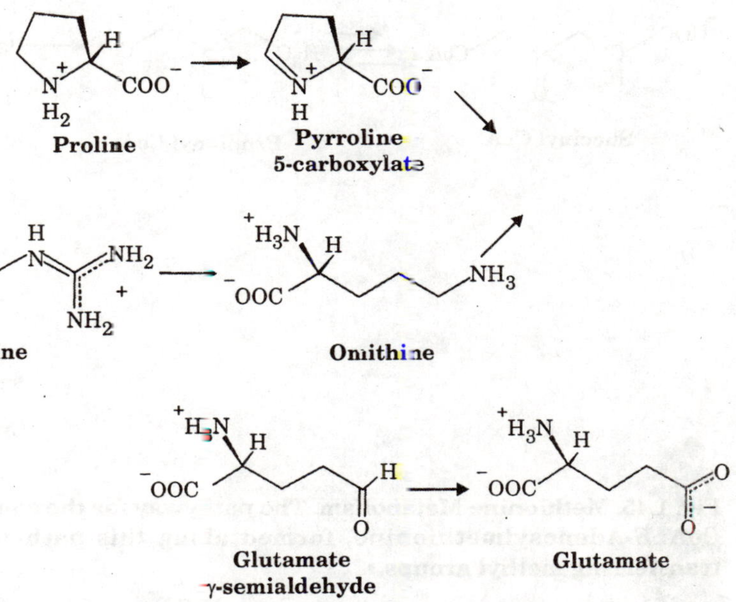

Proline → Pyrroline 5-carboxylate

Arginine → Ornithine

Glutamate γ-semialdehyde → Glutamate

Fig. 1.43. Glutamate Formation. Conversion of praline and arginine into glutamate.

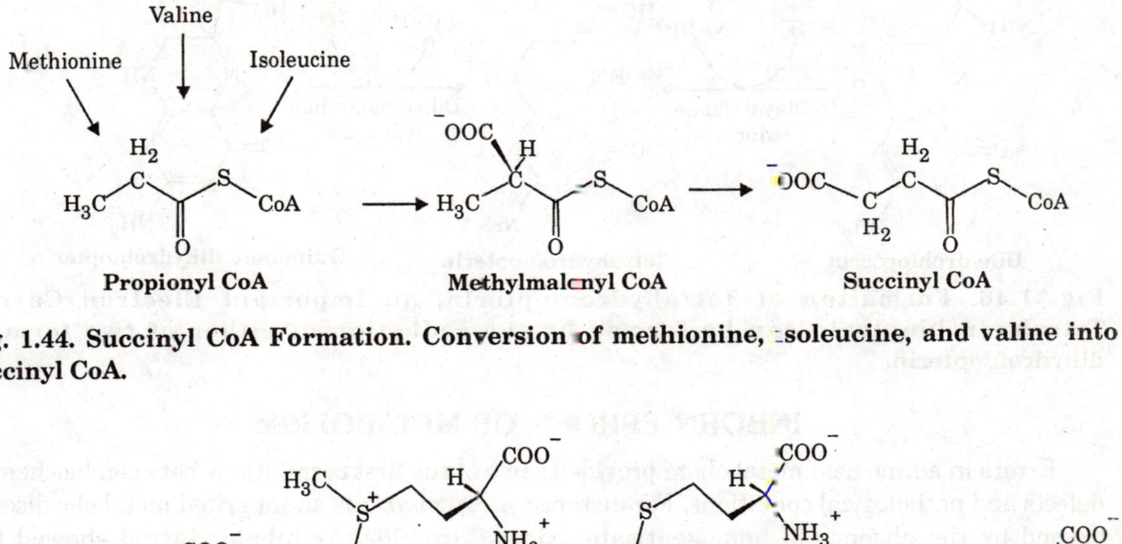

Methionine, Valine, Isoleucine → Propionyl CoA → Methylmalonyl CoA → Succinyl CoA

Fig. 1.44. Succinyl CoA Formation. Conversion of methionine, isoleucine, and valine into succinyl CoA.

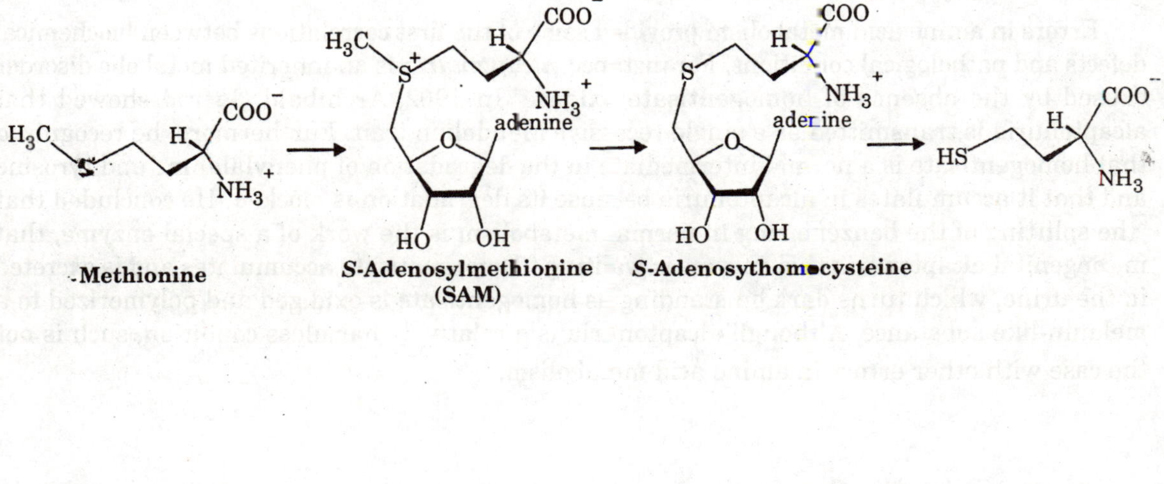

Methionine → *S*-Adenosylmethionine (SAM) → *S*-Adenosylhomocysteine →

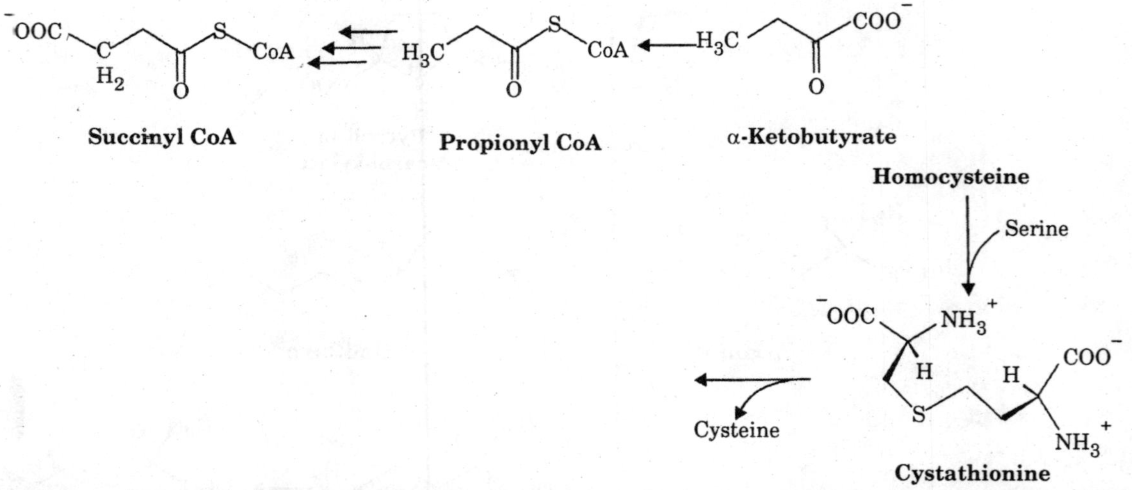

Succinyl CoA ⇌ Propionyl CoA ← α-Ketobutyrate

Homocysteine → (Serine) → Cystathionine → Cysteine

Fig. 1.45. Methionine Metabolism. The pathyway for the conversion of methionine into succinyl CoA. *S*-Adenosylmethionine, formed along this pathway, is an important molecule for transferring methyl groups.

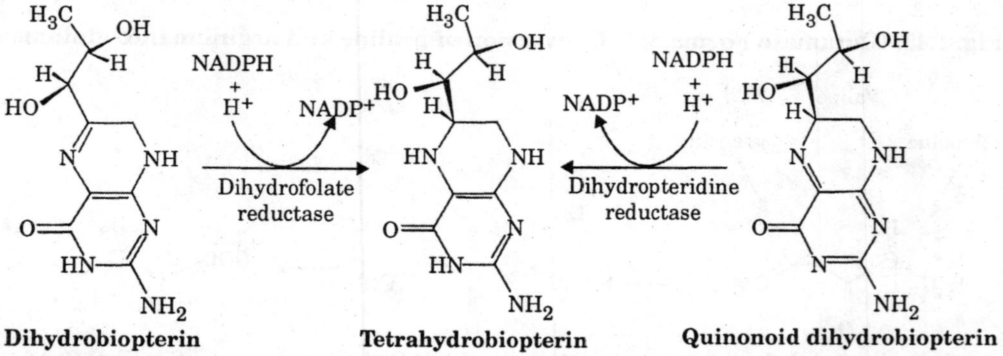

Dihydrobiopterin → (NADPH + H^+ → $NADP^+$; Dihydrofolate reductase) → Tetrahydrobiopterin ← (NADPH + H^+ → $NADP^+$; Dihydropteridine reductase) ← Quinonoid dihydrobiopterin

Fig. 1.46. Formaiton of Tetrahydrobiopterin, an Important Electron Carrier. Tetrahydrobiopterin can be formed by the reduction of either of two forms of dihydrobiopterin.

INBORN ERRORS OF METABOLISM

Errors in amino acid metabolism provided some of the first correlations between biochemical defects and pathological conditions. For instance, *alcaptonuria* is an inherited metabolic disorder caused by the absence of homogentisate oxidase. In 1902, Archibald Garrod showed that alcaptonuria is transmitted as a single recessive Mendelian trait. Furthermore, he recognized that homogentisate is a normal intermediate in the degradation of phenylalanine and tyrosine and that it accumulates in alcaptonuria because its degradation is blocked. He concluded that "the splitting of the benzene ring in normal metabolism is the work of a special enzyme, that in congenital alcaptonuria this enzyme is wanting." Homogentisate accumulates and is excreted in the urine, which turns dark on standing as homogentisate is oxidized and polymerized to a melanin-like substance. Although alcaptonuria is a relatively harmless condition, such is not the case with other errors in amino acid metabolism.

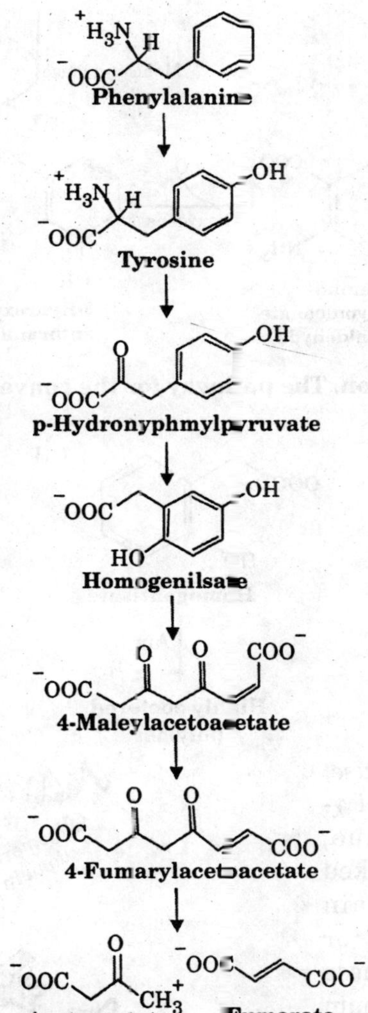

Phenylalanine

↓

Tyrosine

↓

p-Hydronyphmylpyruvate

↓

Homogenilsate

↓

4-Maleylacetoacetate

↓

4-Fumarylacetoacetate

↓

Acetoacetate + Fumarate

Fig. 1.47. Phenylalanine and Tyrosine Degradation. The pathway for the conversion of phenylalanine into acetoacetate and fumarate.

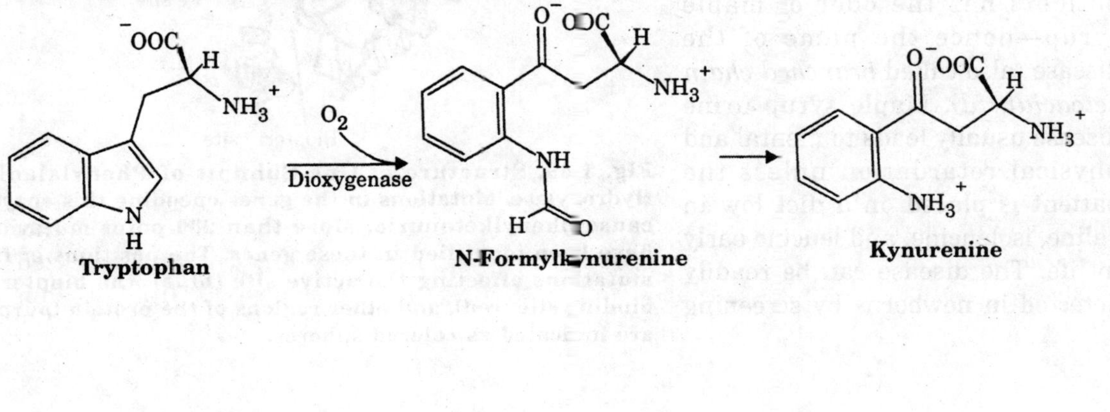

Tryptophan → (O_2, Dioxygenase) → N-Formylkynurenine → Kynurenine

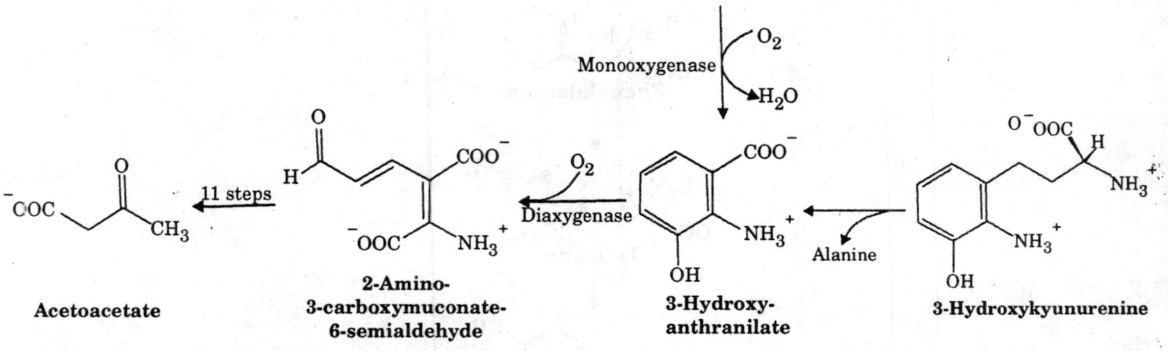

Monooxygenase
O_2
H_2O
Acetoacetate
11 steps
2-Amino-
3-carboxymuconate-
6-semialdehyde
O_2
Diaxygenase
3-Hydroxy-
anthranilate
Alanine
3-Hydroxykyunurenine

Fig. 1.48. Tryptophan Degradation. The pathway for the conversion of tryptophan into alanine and acetoacetate.

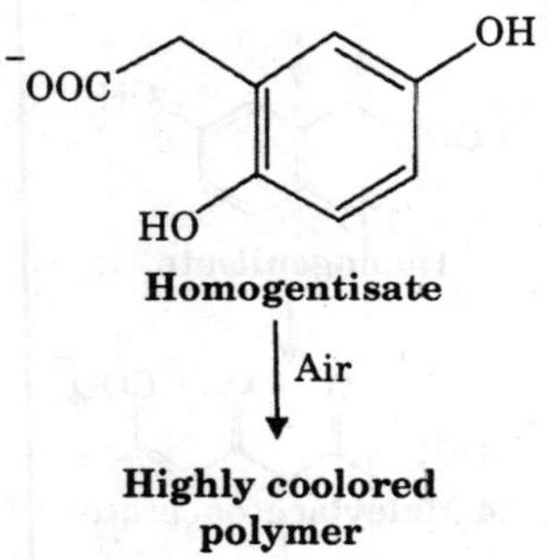

Homogentisate
Air
Highly coolored
polymer

In *maple syrup urine disease,* the oxidative decarboxylation of χ-ketoacids derived from valine, isoleucine, and leucine is blocked because the branched-chain dehydrogenase is missing or defective. Hence, the levels of these χ-ketoacids and the branched-chain amino acids that give rise to them are markedly elevated in both blood and urine. Indeed, the urine of patients has the odor of maple syrup—hence the name of the disease (also called *branched-chain ketoacidurid).* Maple syrup urine disease usually leads to mental and physical retardation unless the patient is placed on a diet low in valine, isoleucine, and leucine early in life. The disease can be readily detected in newborns by screening

Oligomerization
domain
Fe site
Biopterin site

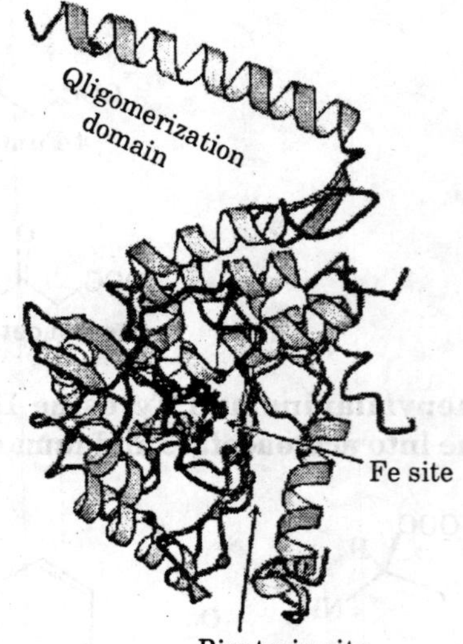

Fig. 1.49. Structure of One Suhnnit of Phenylalanine Hydroxylase. Mutations in the genes encoding this enzyme cause phenylketonuria. More than 200 point mutations have been identified in these genes. The positions of five mutations affecting the active site (blue), the biopterin-binding site (red), and other regions of the protein (purple) are indicated as colored spheres.

urine samples with 2,4-dinitrophenylhydrazine, which reacts with χ-ketoacids to form 2,4-dinitrophenylhydrazone derivatives. A definitive diagnosis can be made by mass spectrometry.

NO_2 NO_2 H_2N NH 2,4-Dinitrophenyl-hydrazine O COO^- H_3C CH_3 **α-Ketoacid** NO_2 NO_2 NH N COO^- H_3C CH_3 **2,4-Dinitrophenylhydrozone derivative**

Phenylketonuria is perhaps the best known of the diseases of amino acid metabolism. Phenylketonuria is caused by an *absence or deficiency of phenylalanine hydroxylase* or, more rarely, of its tetrahydrobiopterin cofactor. *Phenylalanine accumulates in all body fluids because it cannot be converted into tyrosine.* Normally, three-quarters of the phenylalanine is converted into tyrosine, and the other quarter becomes incorporated into proteins. Because the major outflow pathway is blocked in phenylketonuria, the blood level of phenylalanine is typically at least 20-fold as high as in normal people. Minor fates of phenylalanine in normal people, such as the formation of phenylpyruvate, become major fates in phenylketonurics.

Indeed, the initial description of phenylketonuria in 1934 was made by observing the reaction of phenylpyruvate with $FeCl_3$, which turns the urine olive green. *Almost all untreated phenylketonurics are severely mentally retarded.* In fact, about 1% of patients in mental institutions have phenylketonuria.

The brain weight of these people is below normal, myelination of their nerves is defective, and their reflexes are hyperactive. The life expectancy of untreated phenylketonurics is drastically shortened. Half are dead by age 20 and three-quarters by age 30. *The biochemical basis of their mental retardation is an enigma.* Phenylketonurics appear normal at birth, but are severely defective by age 1 if untreated. The therapy for phenylketonuria is a *low phenylalanine diet*. The aim is to provide just enough phenylalanine to meet the needs for growth and replacement.

^+H_3N H ^-OOC **Phenylalanine** α-Keto acid α-Amino acid O ^-OOC **Phenylpyruvate**

Proteins that have a low content of phenylalanine, such as casein from milk, are hydrolyzed and phenylalanine is removed by adsorption. A low phenylalanine diet must be started very soon after birth to prevent irreversible brain damage. In one study, the average IQ of phenylketonurics treated within a few weeks after birth was 93; a control group treated starting at age 1 had an average IQ of 53. Early diagnosis of phenylketonuria is essential and has been accomplished by mass screening programs. The phenylalanine level in the blood is the preferred diagnostic criterion because it is more sensitive and reliable than the $FeCl_3$ test. Prenatal diagnosis of phenylketonuria with DNA probes has become feasible because the gene has been cloned and many mutations have been pinpointed to sites in the protein.

Interestingly, whereas some mutations affect the activity of the enzyme, others do not affect the activity itself but, instead, decrease the enzyme concentration. These mutations lead to degradation of the enzyme, at least in part by the ubiquitin-proteasome pathway.

Table 1.14. Inborn errors of amino acid metabolism

Disease	*Enzyme deficiency*	*Symptoms*
Citrullinema	Agrginosuccinate lyase	Lethergy, seizures, reduced muscle tension
Tyrosinemia	Various enzymes of tyrosine degradation	Weakness, self-mutilation, liver damage, mental retardation
Albinism	Tyrosinase	Absence of pigmentation
Homocystinuria	Cystathionine δ-synthase	Scoliosis, muscle weakness, mental retardation, thin blond hair
Hyperlysinemia	χ-Aminoadipic semialdehyde dehydrogenase	Seizures, mental retardation, lack of muscle tone, ataxia

The incidence of phenylketonuria is about 1 in 20,000 newborns. The disease is inherited in an *autosomal recessive* manner. Heterozygotes, who make up about 1.5% of a typical population, appear normal.'Carriers of the phenylketonuria gene have a reduced level of phenylalanine hydroxylase, as indicated by an increased level of phenylalanine in the blood.

However, this criterion is not absolute, because the blood levels of phenylalanine in carriers and normal people overlap to some extent. The measurement of the kinetics of the disappearance of intravenously administered phenylalanine is a more definitive test for the carrier state. It should be noted that a high blood level of phenylalanine in a pregnant woman can result in abnormal development of the fetus. This is a striking example of maternal-fetal relationships at the molecular level.

THE BIOSYNTHESIS OF AMINO ACIDS

The assembly of biological molecules, including proteins and nucleic acids, requires the generation of appropriate starting materials. We have already considered the assembly of carbohydrates in regard to the Calvin cycle and the pentose phosphate pathway. The present chapter and the next two examine the assembly of the other important building blocks—namely. amino, acids. nucleotides, and lipids. The pathways for the biosynthesis of these molecules are extremely ancient, going back to the last common ancestor of all living things. Indeed, these pathways probably predate many of the pathways of energy transduction discussed in Part II and may have provided key selective advantages in early evolution.

Many of the intermediates present in energy-transduction pathways play a role in biosynthesis as well. These common intermediates allow efficient interplay between energy-transduction pathways play a role in biosynthesis as well. These common intermediates allow efficient interplay between energy-transduction (catabolic) and biosynthetic (anabolic) pathways. Thus, cells are able to balance the degradation of compounds for energy mobilization and the synthesis of starting materials for macro-molecular construction. We begin our consideration of biosynthesis with amino acids. Amino acids are the building blocks of proteins and the nitrogen source for many other important molecules, including nucleotides, neurotransmitters, and prosthetic groups such as parphyrins.

An Overview

A fundamental problem for biological systems is to obtain nitrogen in an easily usable form. This problem is solved by certain microorganisms capable of reducing the inert N ≡ N molecule (nitrogen gas) to two molecules of ammonia in one of the most remarkable reactions in biochemistry. Nitrogen in the form of ammonia is the source of nitrogen for all the amino acids. The carbon backbones come from the glycolytic pathway, the pentose phosphate pathway, or the citric acid cycle.

Anabolism

Biosynthetic processes.

Degradative processes.

Derived from the Greek *ana,* "up"; *kata,* "down"; *ballein,* "to throw."

In amino acid production, we encounter an important problem in biosynthesis—namely, stereochemical control. Because all amino acids except glycine are chiral, biosynthetic pathways must generate the correct isomer with high fidelity. In each of the 19 pathways for the generation of chiral amino acids, the stereochemistry at the χ-carbon atom is established by a transamination reaction that involves pyridoxal phosphate. Almost all the transaminases that catalyze these reactions descend from a common ancestor, illustrating once again that effective solutions to biochemical problems are retained throughout evolution. Biosynthetic pathways are often highly regulated such that building blocks are synthesized only when supplies are low.

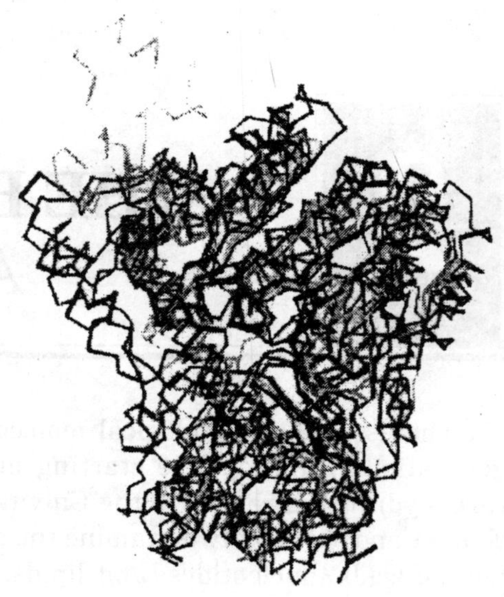

Fig. 2.1. Aminotransferase Protein Family. A superposition of the structures of four aminotransferases taking part in amino acid biosynthesis reveals a common fold, indicating that these proteins descended from a common ancestor.

Very often, a high concentration of the final product of a pathway inhibits the activity of enzymes that function early in the pathway. Often present are allosteric enzymes capable of sensing and responding to concentrations of regulatory species. These enzymes are similar in functional properties to aspartate transcarbamylase and its regulators. Feedback and allosteric mechanisms ensure that all twenty amino acids are maintained in sufficient amounts for protein synthesis and other processes.

Nitrogen is a key component of amino acids. The atmosphere is rich in nitrogen gas (N_2), a very unreactive molecule. Certain organisms such as bacteria that live in the root nodules of yellow clover (photo at left) can convert nitrogen gas into ammonia. Ammonia can then be used to synthesize first glutamate and then other amino acids. [(Left) Runk/ Schoenberger from Grant Heilman]

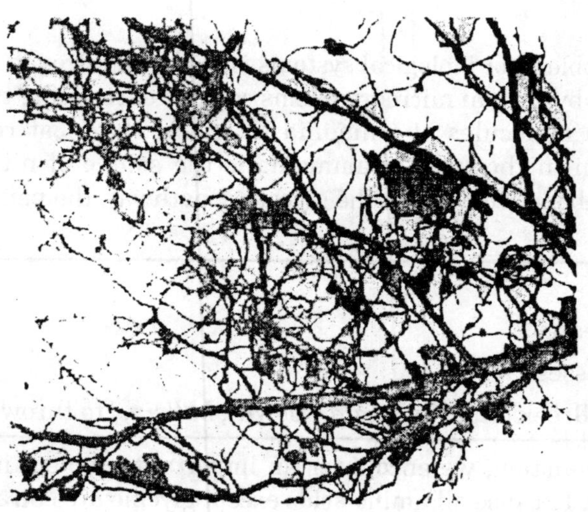

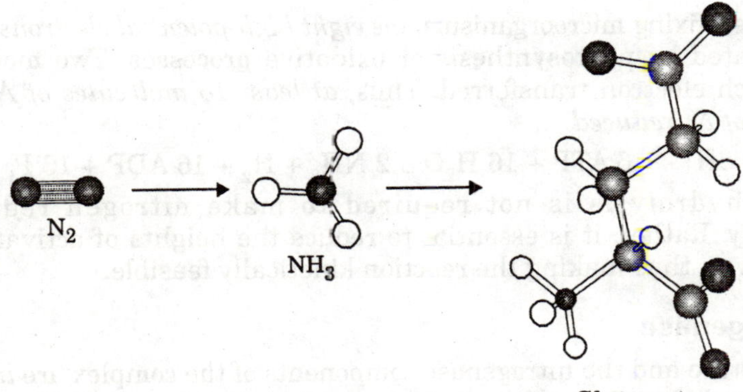

FIXATION OF ATMOSPHERIC NITROGEN TO AMMONIA

The nitrogen in amino acids, purines, pyrimidines, and other biomolecules ultimately comes from atmospheric nitrogen, N_2 The biosynthetic process starts with the reduction of N_2 to NH_3 (ammonia), a process called *nitrogen fixation*. Although higher organisms are unable to fix nitrogen, this conversion is carried out by some bacteria and archaea. Symbiotic *Rhizobium* bacteria invade the roots of leguminous plants and form root nodules in which they fix nitrogen, supplying both the bacteria and the plants. The amount of N_2 fixed by *diazotrophic (nitrogen-fixing) microorganisms* has been estimated to be 10^{11} kilograms per year, about 60% of Earth's newly fixed nitrogen. Lightning and ultraviolet radiation fix another 15%; the other 25% is fixed by industrial processes. The industrial process for nitrogen fixation devised by Fritz Haber in 1910 is still being used in fertilizer factories.

$$N_2 + 3H_2 \rightleftharpoons 2\,NH_3$$

The fixation of N_2 is typically carried out by mixing with H_2 gas over an iron catalyst at about 500°C and a pressure of 300 atmospheres. The extremely strong $N \equiv N$ bond, which has a bond energy of 225 kcal mol^{-1}, is highly resistant to chemical attack. Indeed, Lavoisier named nitrogen gas "azote," meaning "without life" because it is so unreactive.

Nevertheless, the conversion of nitrogen and hydrogen to form ammonia is thermodynamically favorable; the reaction is difficult kinetically because intermediates along the reaction pathway are unstable. To meet the kinetic challenge, the biological process of nitrogen fixation requires a complex enzyme with multiple redox centers. The *nitrogenase complex,* which carries out this fundamental transformation, consists of two proteins: a *reductase,* which provides electrons with high reducing power, and *nitrogenase,* which uses these electrons to reduce N_2 to NH_3 The transfer of electrons from the reductase to the nitrogenase component is coupled to the hydrolysis of ATP by the reductase. The nitrogenase complex is exquisitely sensitive to inactivation by O_2. Leguminous plants maintain a very low concentration of free O_2 in their root nodules by binding O_2 to *leghemoglobin,* a homolog of hemoglobin.

In principle, the reduction of N_2 to NH_3 is a six-electron process.

$$N_2 + 6e^- + 6H^+ \rightleftharpoons 2NH_3$$

However, the biological reaction always generates at least 1 mol of H_2 in addition to 2 mol of NH_3 for each mole of $N \equiv N$. Hence, an input of two additional electrons is required.

$$N_2 + 8e^- + 8H^+ \rightleftharpoons 2\,NH_3 + H_2$$

In most nitrogen-fixing microorganisms, *the eight high-potential electrons come from reduced ferredoxin,* generated by photosynthesis or oxidative processes. Two molecules of ATP are hydrolyzed for each electron transferred. Thus, *at least 16 molecules of ATP are hydrolyzed for each molecule of N_2 reduced.*

$$N_2 + 8e^- + 8H^+ + 16\ ATP + 16\ H_2O \square\ 2\ NH_3 + H_2 + 16\ ADP + 16\ P_i$$

Again, ATP hydrolysis is not required to make nitrogen reduction favorable thermodynamically. Rather, it is essential to reduce the heights of activation barriers along the reaction pathway, thus making the reaction kinetically feasible.

Effects of Nitrogenase

Both the reductase and the nitrogenase components of the complex are *iron-sulfur proteins,* in which iron is bonded to the sulfur atom of a cysteine residue and to inorganic sulfide. The *reductase* (also called the *iron protein* or the *Fe protein)* is a dimer of identical 30-kd subunits bridged by a 4Fe-4S cluster. The role of the reductase is to transfer electrons from a suitable donor, such as reduced ferredoxin, to the nitrogenase component.

The binding and hydrolysis of ATP triggers a conformational change that moves the reductase closer to the nitrogenase component from whence it is able to transfer its electron to the center of nitrogen reduction. The structure of the ATP-binding region reveals it to be a member of the P-loop NTPase family that is clearly related to ihe nucleotide-binding regions found in G proteins and related proteins. Thus, we see another example of how this domain has been recruited in evolution because of its ability to couple nucleoside triphosphate hydrolysis to conformational changes. The nitrogenase component is an $\chi_2\ \delta_2$ tetrarner (240 kd), in which the χ and δ subunits are homologous to each other and structurally quite similar. Electrons enter at the *P dusters,* which are located at the χ-δ interface. These clusters are each composed of eight iron atoms and seven sulfide ions. In the reduced form, each cluster takes the form of two 4Fe-3S partial cubes linked by a central sulfide ion. Each cluster is linked to the protein through six cysteinate residues? Electrons flow from the P cluster to the *FeMo cofactor,* a very unusual redox center. Because molybdenum is present in this cluster, the nitrogenase component is also called the *molybdenum-iron protein (MoFe protein).* The FeMo cofactor consists of two M-3Fe-3S clusters, in which molybdenum occupies the M site in one cluster and iron occupies it in the other. The two clusters are joined by three sulfide ions. The FeMo cofactor is also coordinated to a homocitrate moiety and to the χ subunit through one histidine residue and one cysteinate residue. This cofactor is distinct from the molybdenum-containing cofactor found in sulfite oxidase and apparently all other molybdenum-containing enzymes except nitrogenase.

The FeMo Cofactor is the Site of Nitrogen Fixation

Note that each of the six central iron atoms is linked to only three atoms, leaving open a binding opportunity for N_2. It seems likely that N_2 binds in the central cavity of this cofactor. The formation of multiple Fe-N interactions in this complex weakens the $N \equiv N$ bond and thereby lowers the activation barrier for reduction.

Conversion of Glutamate to Glutamine

The next step in the assimilation of nitrogen into biomolecules is the entry of NH_4^+ into amino acids. *Glutamate* and *glutamine* play pivotal roles in this regard. The χ-amino group of most amino acids comes from the χ-amino group of glutamate by transamination. Glutamine, the other major nitrogen donor, contributes its side-chain nitrogen atom in the biosynthesis of a wide range of important compounds, including the amino acids tryptophan and histidine.

Glutamate is synthesized from NH_4^+ and χ-ketoglutarate, a citric acid cycle intermediate, by the action *of glutamate dehydrogenase.* We have already encountered this enzyme in the degradation of amino acids. Recall that NAD^+ is the oxidant in catabolism, whereas NADPH is the reductant in biosyntheses. Glutamate dehydrogenase is unusual in that it does not discriminate between NADH and NADPH, at least in some species.

$$NH_4^+ + \alpha\text{-ketoglutarate} + NADPH + H^+ \rightleftharpoons \text{glutamate} + NADP^+ + H_2O_\chi$$

The reaction proceeds in two steps. First, a Schiff base forms between ammonia and x-ketoglutarate.The formation of a Schiff base between an amine and a carbonyl compound is a key reaction that takes place at many stages of amino acid biosynthesis and degradation.

Carbonyl compound + Amino donor ⇌ Schiff base + H_2O ⇌ Protonated Schiff base

Schiff bases can be easily protonated. With glutamate dehydrogenase, the protonated Schiff base is reduced by the transfer of a hydride ion from NADPH to form glutamate.

α-Ketoglutarate + NH_4^+ ⇌ (H_2O) ⇌ (H^+ + NAD(P)H → $NAD(P)^+$) Glutamate

This reaction is crucial because it establishes the stereochemistry of the χ-carbon atom *(S absolute configuration)* in glutamate. The enzyme binds the χ-ketoglutarate substrate in such a way that hydride transferred from NAD(P)H is added to form the 1 isomer of glutamate. As we shall see, this stereochemistry is established for other amino acids by transamination reactions that rely on pyridoxal phosphate.

A second ammonium ion is incorporated into glutamate to form glutamine by the action *of glutamine synthetase.* This amidation is driven by the hydrolysis of ATP. ATP participates directly in the reaction by phosphorylating the side chain of glutamate to form an acyl-phosphate intermediate, which then reacts with ammonia to form glutamine.

Glutamate → (ATP → ADP) Acyl-phosphate intermediate → (NH_3 → P_i) Glutamine

A high-affinity ammonia-binding site is formed only after the formation of the acyl-phosphate intermediate. A specific site for ammonia binding is required to prevent attack by water from hydrolyzing the intermediate and wasting a molecule of ATP. The regulation of glutamine synthetase plays a critical role in controlling nitrogen metabolism.

Glutamate dehydrogenase and glutamine synthetase are present in all organisms. Most prokaryotes also contain an evolutionary unrelated enzyme, *glutamate synthase,* which catalyzes the reductive amination of χ-ketoglutarate with the use of glutamine as the nitrogen donor.

α-Ketoglutarate + glutamine + NADPH + H^+ ⇌ 2 glutamate + $NADP^+$

The side-chain amide of glutamine is hydrolyzed to generate ammonia within the enzyme, a recurring theme throughout nitrogen metabolism. *When NH_4^+ is limiting, most of the glutamate is made by the sequential action of glutamine synthetase and glutamate synthase.* The sum of these reactions is

NH_4^+ + α-ketoglutarate + NADPH + ATP→ glutamate + $NADP^+$ + ADP + P_i

Note that this stoichiometry differs from that of the glutamate dehydrogenase reaction in that ATP is hydrolyzed. Why do prokaryotes sometimes use this more expensive pathway? The answer is that the value of K_M of glutamate dehydrogenase for NH_4^+ is high (≈ 1 mM), and so this enzyme is not saturated when NH_4^+ is limiting. In contrast, glutamine synthetase has very high affinity for NH_4^+. Thus, ATP hydrolysis is required to capture ammonia when it is scarce.

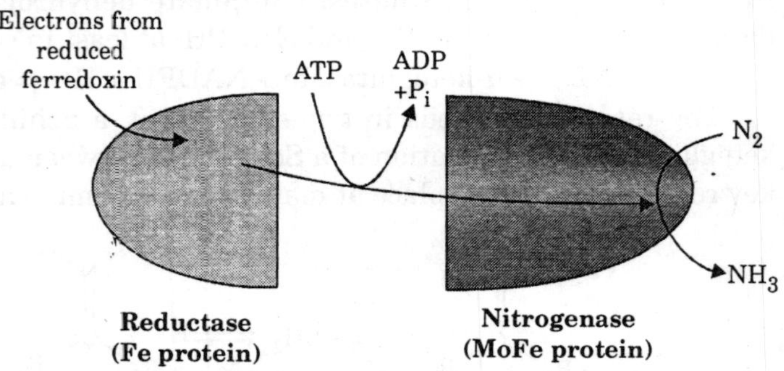

Fig. 2.2. Nitrogen Fixation. Electrons flow from ferredoxin to the reductase (iron protein, or Fe protein) to nitrogenase (molybdenum-iron protein, or MoFe protein) to reduce nitrogen to ammonia. ATP hydrolysis within the reductase drives conformational changes necessary for the efficient transfer of electrons.

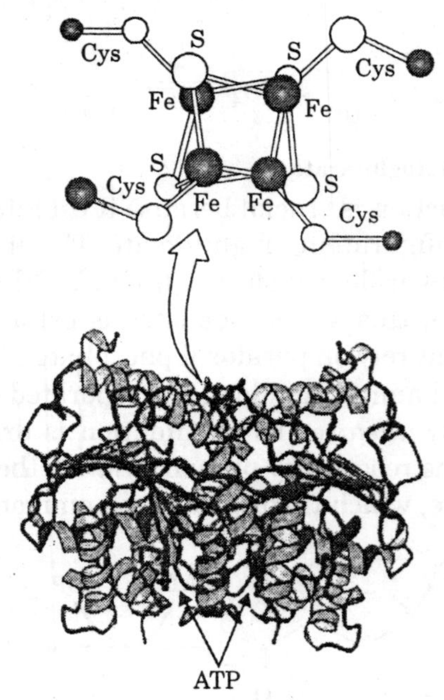

Fig. 2.3. Fe Protein. This protein is a dimmer composed of two polypeptide chains linked by a 4Fe-4S cluster. Each monomer is a member of the P-loop NTPase family and contains an ATP-binding site.

AMINO ACIDS FROM CITRIC ACID CYCLE AND OTHER MAJOR PATHWAYS

Thus far, we have considered the conversion of N_2 into NH_4^+ and the assimilation of NH_4^+ into glutamate and glutamine. We turn now to the biosynthesis of the other amino acids. The pathways for the biosynthesis of amino acids are diverse. However, they have an important common feature: *their carbon skeletons come from intermediates of glycolysis the pentose phosphate pathway, or the citric acid cycle.* On the basis of these starting materials, amino acids can be grouped into six biosynthetic families.

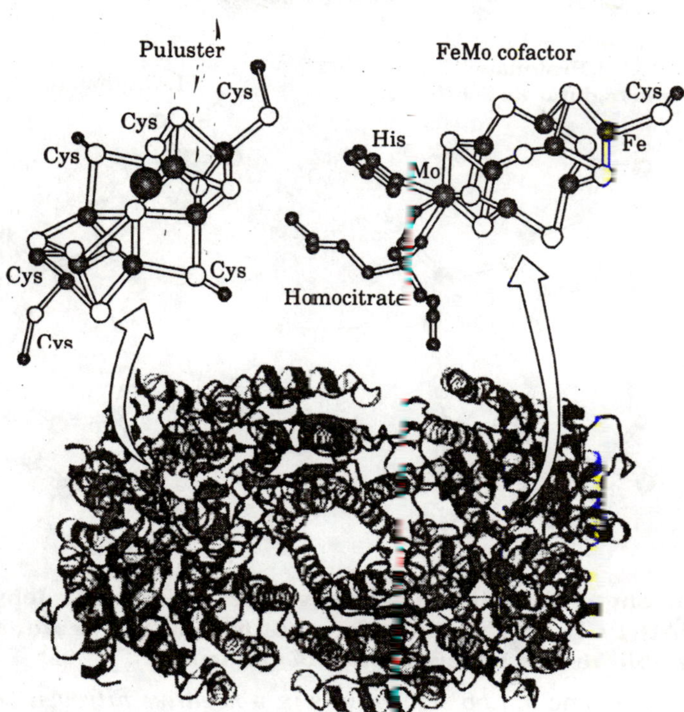

Fig. 2.4. MOFe Protein. This protein is an heterotetramer composed of two χ subunits (red) and two δ subunits (blue). The protein contains two copies each of two types of clusters: P clusters and FeMo cofactors. Each P cluster contains eight iron atoms and seven sulfldes linked to the protein by six cysteinate residues. Each FeMo cofactor contains one molybdenum atom, seven iron atoms, nine sulfldes, and a homocitrate, and is linked to the protein by one cysteinate residue and one histidine residue.

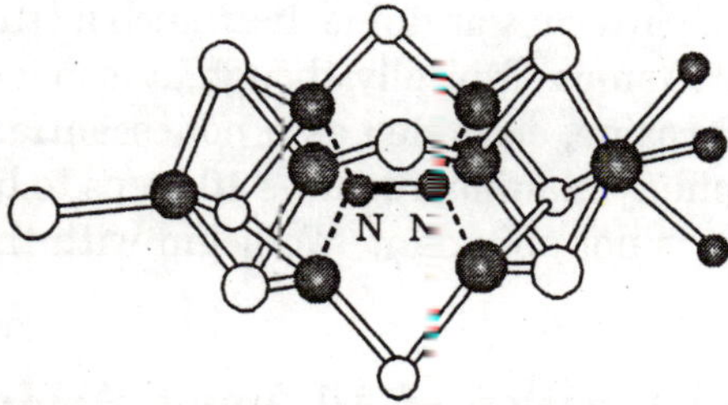

Fig. 2.5. Nitrogen-Reduction Site. The FeMo cofactor contains an open center that is the likely site of nitrogen binding and reduction.

Essential and Non Essential Amino Acids

Most microorganisms such as *E. coli* can synthesize the entire basic set of 20 amino acids, whereas human beings cannot make 9 of them. The amino acids that must be supplied in the diet are called *essential amino acids,* whereas the others are termed *nonessential amino acids*. These designations refer to the needs of an organism under a particular set of conditions. For example, enough arginine is synthesized by the urea cycle to meet the needs of an adult but perhaps not those ofagrowing child.

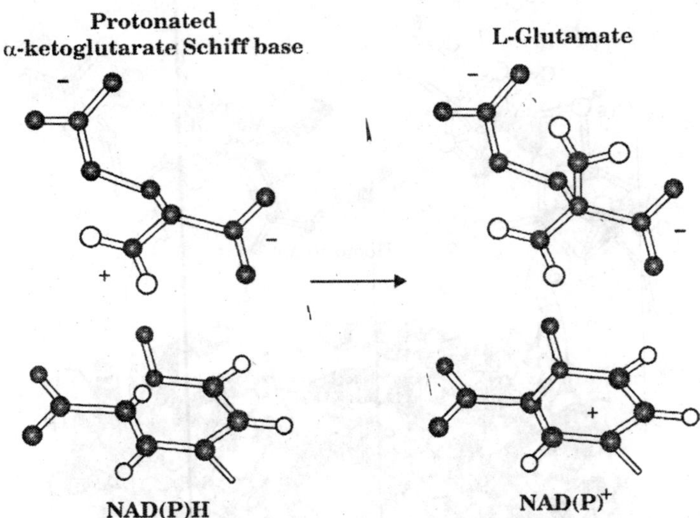

Fig. 2.6. Establishment of Chirality. In the active site of glutamate dehydrogenase, hydride transfer from NAD(P) H to a specific face of the achiral protonated Schiff base of χ-ketoglutarate establishes the 1 configuration of glutamate.

A deficiency of even one amino acid results in a *negative nitrogen balance.* In this state, more protein is degraded than is synthesized, and so more nitrogen is excreted than is ingested. The nonessential amino acids are synthesized by quite simple reactions, whereas the pathways for the formation of the essential amino acids are quite complex. For example, the nonessential amino acids *alanine* and *aspartate* are synthesized in a single step from pyruvate and oxaloacetate, respectively. In contrast, the pathways for the essential amino acids require from 5 to 16 steps.

The sole exception to this pattern is arginine, inasmuch as the synthesis of this nonessential amino acid de novo requires 10 steps. Typically, though, it is made in only 3 steps from ornithine as part of the urea cycle. Tyrosine, classified as a nonessential amino acid because it can be synthesized in 1 step from phenylalanine, requires 10 steps to be synfiesized from scratch and is essential if phenylalanine is not abundant. We begin with the biosynthesis of nonessential amino acids.

A Common Step in the Formation of All Amino Acids

Three χ-ketoacids, χ-ketoglutarate, oxaloacetate, and pyruvate—can be converted into amino acids in one step through the addition of an amino group. We have seen that χ-ketoglutarate can be converted into glutamate by reductive amination. The amino group from glutamate can be transferred to other χ-ketoacids by transamination reactions. Thus, aspartate and alanine can be made from the addition of an amino group to oxaloacetate and pyruvate, respectively.

$$\text{Oxaloacetate + glutamate} \rightleftharpoons \text{aspartate + }\alpha\text{-ketoglytarate}$$
$$\text{Pyruvate + glutamate} \rightleftharpoons \text{alanine + }\alpha\text{-ketoglutarate}$$

These reactions are carried out *by pyridoxal phosphate-dependent transaminases.* Transamination reactions participate in the synthesis of most amino acids. We shall review the transaminase mechanism as it applies to amino acid biosynthesis. The reaction pathway begins with pyridoxal phosphate in a Schiff-base linkage with lysine at the transaminase active site, forming an internal aldimine.

An external aldimine forms between PLP and the amino-group donor, glutamate, which displaces the lysine residue. The transfer of the amino group from glutamate to pyridoxal phosphate forms pyridoxamine phosphate, the actual amino donor. When the amino group has been incorporated into pyridoxamine, the reaction pathway proceeds in reverse, and the amino group is transferred to an χ-ketoacid to form an amino acid. Aspartate aminotransferase is the prototype of a large family of PLP-dependent enzymes.

Comparisons of amino acid sequences as well as several three-dimensional structures reveal that almost all transaminases having roles in amino acid biosynthesis are related to aspartate aminotransferase by divergent evolution. An examination of the aligned amino acid sequences reveals that two residues are completely conserved. These residues are the lysine residue that forms the Schiff base with the pyridoxal phosphate cofactor (lysine 258 in aspartate aminotransferase) and an arginine residue that interacts with the χ-carboxylate group of the ketoacid. An essential step in the transamination reaction is the protonation of the quinonoid intermediate to form the external aldimine.

The *chirality of the amino acid formed is determined by the direction from which this proton is added to the quinonoid form.* This protonation step determines the l configuration of the amino acids produced. The interaction between the conserved arginine residue and the χ-carboxylate group helps orient the substrate so that when the lysine residue transfers a proton to the face of the quinonoid intermediate, it generates an aldimine with an l configuration at the C_{χ} center.

Asparagine from Aspartate

The formation of asparagine from aspartate is chemically analogous to the formation of glutamine from glutamate. Both transformations are amidation reactions and both are driven by the hydrolysis of ATP. The actual reactions are different, however. In bacteria, the reaction for the asparagine synthesis is

$$\text{Aspartate} + NH_4^+ + \text{ATP} \longrightarrow \text{asparagine} + \text{AMP} + PP_i + H^+$$

Thus, the products of ATP hydrolysis are AMP and PP_i rather than ADP and P_i. Aspartate is activated by adenylation rather than by phosphorylation.

$$\text{Aspartate} \xrightarrow{\text{ATP} \;\; PP_i} \text{Acyl-adenylate Intermediate} \xrightarrow{NH_3 \;\; \text{AMP}} \text{Asparagine}$$

Aspartate **Acyl-adenylate Intermediate** **Asparagine**

We have encountered this mode of activation in fatty acid degradation and will see it again in lipid and protein synthesis.

In mammals, the nitrogen donor for asparagine is glutamine rather than ammonia as in bacteria. Ammonia is generated by hydrolysis of the side chain of glutamine and directly transferred to activated aspartate, bound in the active site. An advantage is that the cell is not directly exposed to NH_4^+, which is toxic at high levels to human beings and other mammals. *The use of glutamine hydrolysis as a mechanism for generating ammonia for use within the same enzyme is a motif common throughout biosynthetic pathways.*

Precursor of Glutamine, Proline, and Arginine

The synthesis of glutamate by the reductive amination of χ-ketoglutarate has already been discussed, as has the conversion of glutamate into glutamine. Glutamate is the precursor of two other nonessential amino acids: *proline* and *arginine*. First, the χ-carboxyl group of glutamate reacts with ATP to form an acyl phosphate. This mixed anhydride is then reduced by NADPH to an aldehyde.

Glutamate —(ATP → ADP)→ Acyl-phosphate intermediate —(H^+ + NADPH → P_i + NADP)→ Glutamic γ-semialdehyde

Glutamic χ-semialdehyde cyclizes with a loss of H_2O in a nonenzymatic process to give Φ[1]-pyrroline-5-carboxylate, which is reduced by NADPH to proline. Alternatively, the semialdehyde can be transaminated to ornithine, which is converted in several steps into arginine.

Glutamate —(ATP + NADPH → ADP + P_i + $NADP^+$)→ Glutamic γ-semialdehyde —(Glutamate → α-Keto-glutarate)→ Ornithine

Glutamic γ-semialdehyde —(H_2O)→ Δ'-Pyrroline-5-carboxylate —(H^+ + NADPH → $NADP^+$)→ Proline

Serine, Cysteine, and Glycine Are Formed from 3-Phosphoglycerate

Serine is synthesized from 3-phosphoglycerate, an intermediate in glycolysis. The first step is an oxidation to 3-phosphohydroxypyruvate. This χ-ketoacid is transaminated to 3-phosphoserine, which is then hydrolyzed to serine.

NAD^+ H^+ + NADP; Glutamate α-Ketoglutarate; H_2O P_i

3-Phosphoglycerate → 3-Phosphohydroxypyruvate → 3-Phosphoserine → Serine

Serine is the precursor of *glycine* and *cysteine*. In the formation of glycine, the side-chain methylene group of serine is transferred to *tetrahydrofolate,* a carrier of one-carbon units that will be discussed shortly.

$$\text{Serine} + \text{tetrahydrofolate} \longrightarrow \text{Glycine} + \text{methylenetetrahydrofolate} + H_2O$$

This interconversion is catalyzed by *serine transhydroxymethylase,* another PLP enzyme that is homologous to aspartate aminotransferase. The bond between the χ and δ-carbon atoms of serine is labilized by the formation of a Schiff base between serine and PLP. The side-chain methylene group of serine is then transferred to tetrahydrofolate. The conversion of serine into cysteine requires the substitution of a sulfur atom derived from methionine for the side-chain oxygen atom.

Tetrahydrofolate Carries Activated One-Carbon Units at Several Oxidation Levels

Tetrahydrofolate (also called *tetrahydropteroylglutamate*), a highly versatile carrier of activated one-carbon units, consists of three groups: a substituted pteridine, *p*-aminobenzoate, and a chain of one or more glutamate residues. Mammals can synthesize the pteridine ring, but they are unable to conjugate it to the other two units. They obtain tetrahydrofolate from their diets or from microorganisms in their intestinal tracts. The one-carbon group carried by tetrahydrofolate is bonded to its N-5 or N-10 nitrogen atom (denoted as N^5 and N^{10}) or to both. This unit can exist in three oxidation states.

The most-reduced form carries a *methyl* group, whereas the intermediate form carries a *methylene* group, whereas the intermediate form carries a methylene group. More-oxidized forms carry a *formyl, formimino,* or *methenyl* group. The fully oxidized one-carbon unit, CO_2, is carried by biotin rather than by tetrahydrofolate. The one-carbon units carried by tetrahydrofolate are interconvertible. N^5, N^{10}-*Methylenetetrahydrofolate* can be reduced to N^5-*methyl* tetrahydrofolate or oxidized to N^5, N^{10}-*methenyl* tetrahydrofolate. N^5, N^{10}-*Methyenyl* tetrahydrofolate can be converted into N^5-*formimino* tetrahydrofolate or N^{10}-*formyl* tetrahydrofolate, both of which are at the same oxidation level. N^{10}-*Formyl* tetrahydrofolate can also be synthesized from tetrahydrofolate, formate, and ATP.

$$\text{Formate} + \text{ATP} + \text{tetrahydrofolate} \longrightarrow N^{10}\text{-formyltetrahydrofolate} + \text{ADP} + P_i$$

N^5-*Formyl* tetrahydrofolate can be reversibly isomerized to N^{10}-*formyl* tetrahydrofolate or it can be converted into N^5, N^{10}–*methenyl* tetrahydrofolate.

These tetrahydrofolate derivatives serve as donors of one-carbon units in a variety of biosyntheses. Methionine is regenerated from homocysteine by transfer of the methyl group of N^5-methyltetrahydrofolate, as will be discussed shortly.

Some of the carbon atoms of *purines* are acquired from derivatives of N^{10}-formyltetrahydrofolate. The methyl group of *thymine,* a pyrimidine, comes from N^5, N^{10}-methylenetetrahydrofolate. This tetrahydrofolate derivative can also donate a one-carbon unit in an alternative synthesis of *glycine* that starts with CO_2 and NH_4^+, a reaction catalyzed by *glycine synthase* (called the *glycine* cleavage *enzyme* when it operates in the reverse direction).

$$CO_2 + NH_4^+ + N^5, N^{10}\text{-methylenetetrahydrofolate} + NADH \rightleftharpoons \text{glycine} + \text{tetrahydrofolate} + NAD^+$$

Thus, one-carbon units at each of the three oxidation levels are utilized in biosyntheses. Furthermore, *tetrahydrofolate serves as an acceptor of one-carbon units in degradative reactions.* The major source of one-carbon units is the facile conversion of serine into glycine, which yields N^5, N^{10}-methylenetetrahydrofolate. Serine can be derived from 3-phosphoglycerate, and so *this pathway enables one-carbon units to be formed de nov from carbohydrates.*

Donor of Methyl Groups

Tetrahydrofolate can carry a methyl group on its N-5 atom, but its transfer potential is not sufficiently high for most biosynthetic methylations. Rather, the activated methyl donor is usually S-*adenosylmethionine (SAM),* which is synthesized by the transfer of an adenosyl group from ATP to the sulfur atom of methionine.

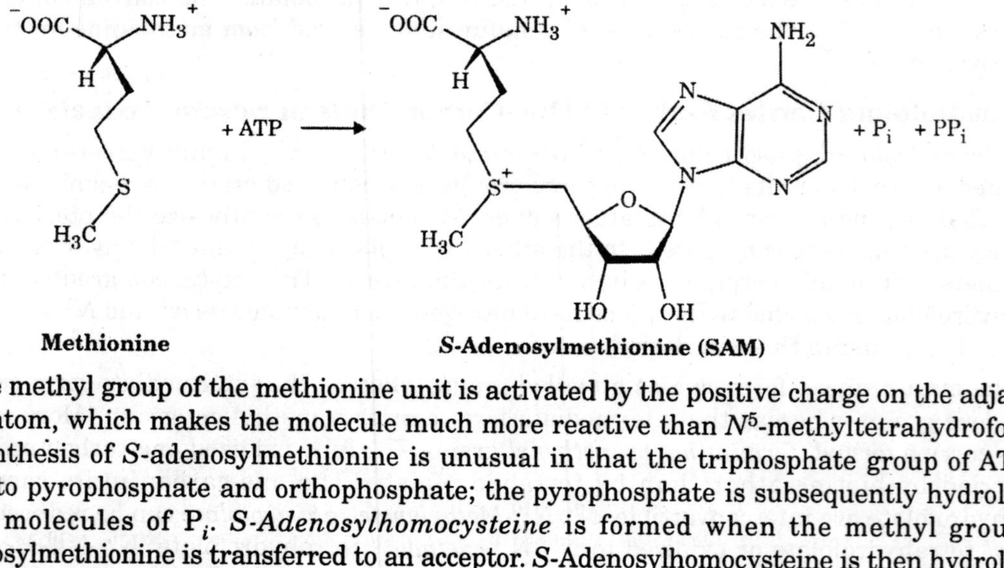

Methionine **S-Adenosylmethionine (SAM)**

The methyl group of the methionine unit is activated by the positive charge on the adjacent sulfur atom, which makes the molecule much more reactive than N^5-methyltetrahydrofolate. The synthesis of *S*-adenosylmethionine is unusual in that the triphosphate group of ATP is split into pyrophosphate and orthophosphate; the pyrophosphate is subsequently hydrolyzed to two molecules of P_i. *S-Adenosylhomocysteine* is formed when the methyl group of *S*-adenosylmethionine is transferred to an acceptor. *S*-Adenosylhomocysteine is then hydrolyzed to *homocysteine* and adenosine.

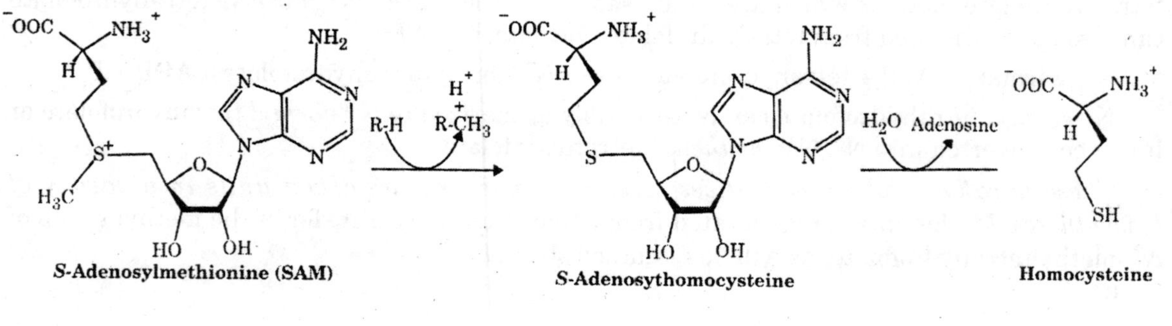

S-Adenosylmethionine (SAM) **S-Adenosythomocysteine** **Homocysteine**

Methionine can be regenerated by the transfer of a methyl group to homocysteine from N^5-methyltetrahydrofolate, a reaction catalyzed by *methionine synthase* (also known as *homocysteine methyltransferase).*

Homocysteine + N^5-Methyl-tetrahydrofolate → Methionine + Tetrahydrofolate

The coenzyme that mediates this transfer of a methyl group is *methylcobalamin,* derived from vitamin B_{12}. In fact, this reaction and the rearrangement of 1-methylmalonyl CoA to succinyl CoA, catalyzed by a homologous enzyme, are the only two B_{12}-dependent reactions known to take place in mammals. Another enzyme that converts homocysteine into methionine without vitamin B_{12} also is present in many organisms. These reactions constitute the *activated methyl cycle.*

Methyl groups enter the cycle in the conversion of homocysteine into methionine and are then made highly reactive by the addition of adenosyl groups, which make the sulfur atoms positively charged and the methyl groups much more electrophilic. The high transfer potential of the *S*-methyl group enables it to be transferred to a wide variety of accepters. Among the acceptors modified by *S*-adenosylmethionine are specific/bases in DNA. The methylation of DNA protects bacterial DNA from cleavage by restriction enzymes.

The base to be methylated is flipped out of the DNA double helix into the active site where it can accept a methyl group from *S*-adenosylmethionine. A recurring *S*-adenosylmethionine-binding domain is present in many SAM-dependent methylases. *S* Adenosylmethionine is also the precursor of *ethylene,* a gaseous plant hormone that induces the ripening of fruit. *S*-Adenosylmethionine is cyclized to a cyclopropane derivative that is then oxidized to form ethylene.

The Greek philosopher Theophrastus recognized more than 2000 years ago that sycamore figs do not ripen unless they are scraped with an iron claw. The reason is now known: *wounding triggers ethylene production, which in turn induces ripening.* Much effort is being made to understand this biosynthetic pathway because ethylene is a culprit in the spoilage of fruit.

S-Adenosylmethionine —ACC synthase→ 1-Aminocyclopropane-1-carboxylate (ACC) —ACC oxidase→ $H_2C{=}CH_2$ Ethylene

Synthesis of Cysteine

In addition to being a precursor of methionine in the activated methyl cycle, homocysteine is an intermediate in the synthesis of cysteine. Serine and homocysteine condense to form *cystathionine.* This reaction is catalyzed by *cystathionine δ-synthase.* Cystathionine is then deaminated and cleaved to cysteine and δ-ketobutyrate by *cystathionine δ-synthase*. Cystathionine is then deaminated and cleaved to cysteine and χ-ketobutyrate by *cystathionine*. Both of these enzymes utilize PLP and are homologous to aspartate aminotransferase. The net reaction is

$$\text{Homocysteine + serine} \rightleftharpoons \text{cysteine} + \alpha\text{-ketobutyrate} + NH_4^+$$

Note that the sulfur atom of cysteine is derived from homocysteine, whereas the carbon skeleton comes from serine.

Homocycteine + Serine $\xrightarrow{H_2O}$ Cystathionins $\xrightarrow{H_2O}$ NH_4^+ + α-Ketobutyrate + Cysteine

High Levels of Homocysteine

People with elevated serum levels of homocysteine or the disulfide-linked dimer homocystine have an unusually high risk for coronary heart disease and arteriosclerosis. The most common genetic cause of high homocysteine levels is a mutation within the gene encoding cystathionine δ-synthase. The molecular basis of homocysteine's action has been clearly identified, although it appears to damage cells lining blood vessels and to increase the growth of vascular smooth muscle. The amino acid raises oxidative stress as well. Vitamin treatments are effective in reducing homocysteine levels in some people.

Treatment with vitamins maximizes the activity of the two major metabolic pathways processing homocysteine. Pyridoxal phosphate, a vitamin B_6 derivative, is necessary for the activity of cystathionine δ-synthase, which converts homocysteine into cystathione; tetrahydrofolate, and vitamin B_{12}, supports the methylation of homocysteine to methionine.

Biosynthesis of Aromatic Amino Acids

We turn now to the biosynthesis of essential amino acids. These amino acids are synthesized by plants and microorganisms, and those in the human diet are ultimately derived primarily from plants. The essential amino acids are formed by much more complex routes than are the nonessential amino acids. The pathways for the synthesis of aromatic amino acids in bacteria have been selected for discussion here because they are well understood and exemplify recurring mechanistic motifs. Phenylalanine, tyrosine, and tryptophan are synthesized by a common pathway in *E. coli*.

The initial step is the condensation of phosphoenolpyruvate (a glycolytic intermediate) with erythrose 4-phosphate (a pentose phosphate pathway intermediate). The resulting seven-carbon open-chain sugar is oxidized, loses its phosphoryl group, and cyclizes to 3-dehydroquinate. Dehydration then yields 3-dehydroshikimate, which is reduced by NADPH to shikimate. Phosphorylation of shikimate by ATP gives shikimate 3-phosphate, which condenses with a second molecule of phosphoenolpyruvate. This 5-enolpyruvyl intermediate loses its phosphoryl group, yielding chorismate, the common precursor of all three aromatic amino acids. The importance of this pathway is revealed by the effectiveness of glyphosate (Roundup), a broad-spectrum herbicide. This compound inhibits the enzyme that produces 5-enolpyruvylshikimate 3-phosphate and, hence, blocks aromatic amino acid biosynthesis in plants. Because animals lack this enzyme, the herbicide is fairly nontoxic.

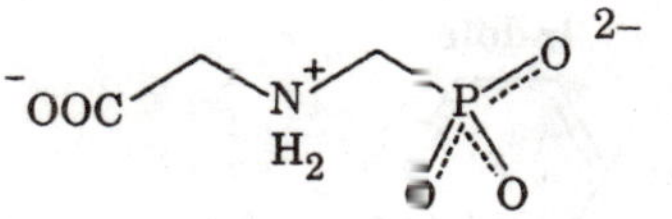

Glyphosate
(Raundup)

The pathway bifurcates at chorismate. Let us first follow the *prephenate branch.* A mutase converts chorismate into prephenate, the immediate precursor of the aromatic ring of phenylalanine and tyrosine. This fascinating conversion is a rare example of an electrocyclic reaction in biochemistry, mechanistically similar to the well-known Diels-Alder reaction from organic chemistry. Dehydration and decarboxylation yield *phenylpyruvate.* Alternatively, prephenate can be oxidatively decarboxylated to *p-hydroxyphenylpyruvate.* These χ-ketoacids are then transaminated to form *phenylalanine* and *tyrosine.* The branch starting with *anthranilate* leads to the synthesis of tryptophan.

Chorismate acquires an amino group derived from the hydrolysis of the side chain of glutamine and releases pyruvate to form anthranilate. Then anthranilate condenses with *5-phosphoribosyl-1-pyrophosphate (PRPP), an activated form of ribose phosphate.* PRPP is also an important intermediate in the synthesis of histidine, purine nucleotides, and pyrimidine nucleotides. The C-1 atom of ribose 5-phosphate becomes bonded to the nitrogen atom of anthranilate in a reaction that is driven by the release and hydrolysis of pyrophosphate. The ribose moiety of phosphoribosylanthranilate undergoes rearrangement to yield 1-(*o*-carboxyphenylamino)-1-deoxyribulose 5-phosphate. This intermediate is dehydrated and then decarboxylated to indole-3-glycerol phosphate, which is cleaved to indole. Then indole reacts with serine to form tryptophan. In these final steps, which are catalyzed by tryptophan synthetase, the side chain of indole-3-glycerol phosphate is removed as glyceraldehyde 3-phosphate and replaced by the carbon skeleton of serine.

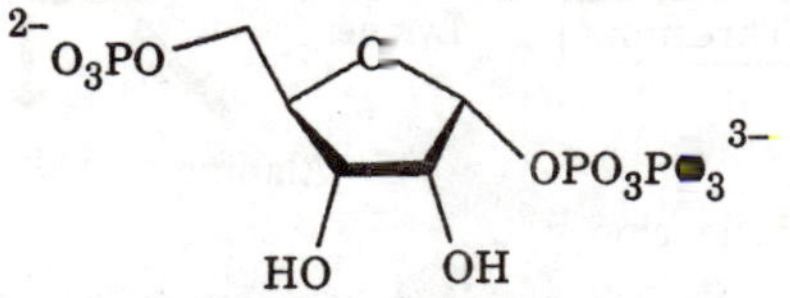

5-Phosphoribosyl-1-pyrophosphate
(PRPP)

Tryptophan Synthetase

Tryptophan synthetase of *E. coli,* an $\chi_2\ \delta_2$ tetramer, can be dissociated into two χ subunits and a δ_2 subunit. The χ subunit catalyzes the formation of indole from indole-3-glycerol phosphate, whereas each δ subunit has a PLP-containing active site that catalyzes the condensation of indole and serine to form tryptophan. The overall three-dimensional structure of this enzyme is distinct from that of aspartate aminotransferase and the other PLP enzymes already discussed. Serine forms a Schiff base with this PLP which is then dehydrated to give the *Schif base of aminoacrylate.* This reactive intermediate is attacked by indole to give tryptophan.

Indole

Schiff base of aminoacrylate
(derived from serine)

The synthesis of tryptophan poses a challenge. Indole, a hydrophobic molecule, readily traverses membranes and would be lost from the cell if it were allowed to diffuse away from the enzyme. This problem is solved in an ingenious way. A 25-Å-long channel connects the active site of the χ subunit with that of the adjacent δ subunit in the χ_2 δ_2 tetramer. Thus, indole can diffuse from one active site to the other without being released into bulk solvent.

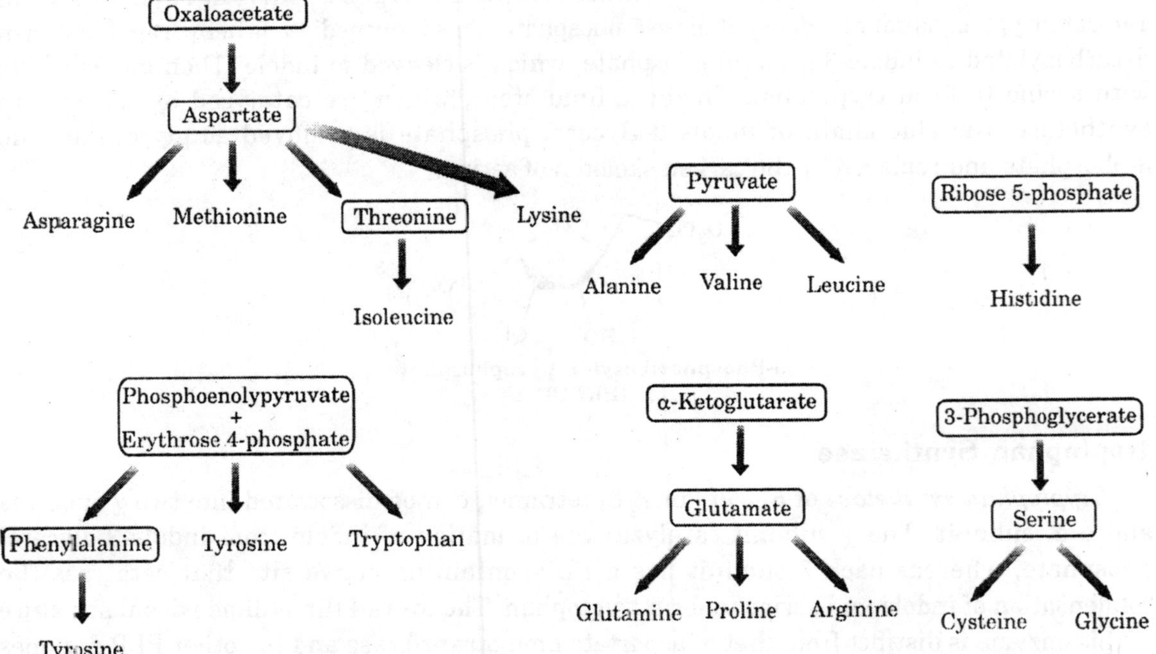

Fig. 2.7. Biosynthetic Families of Amino Acids in Bacteria and Plants. Major metabolic precursors are shaded blue. Amino acids that give rise to other amino acids are shaded yellow. Essential amino acids are in boldface type.

Indeed, the results of isotopic-labeling experiments showed that indole formed by the χ subunit does not leave the enzyme when serine is present.

Furthermore, the two partial reactions are coordinated. Indole is not formed by the χ subunit until the highly reactive aminoacrylate is ready and waiting in the δ subunit. We see here a clear-cut example of *substrate channeling* in catalysis by a multienzyme complex. Channeling substantially increases the catalytic rate. Furthermore, a deleterious side reaction—in this case, the potential loss of an intermediate—is prevented.

Table 2.1 Basic set of 20 amino acids

Nonessential	*Essential*
Alannine	Histidine
Arginine	Isoleucine
Asparagines	Leucine
Aspartate	Lysine
Cysteine	Methionine
Glutamate	Phenylalanine
Glutamine	Threonine
Glycine	Tryptophan
Proline	Valine
Serine	
Tyrosine	

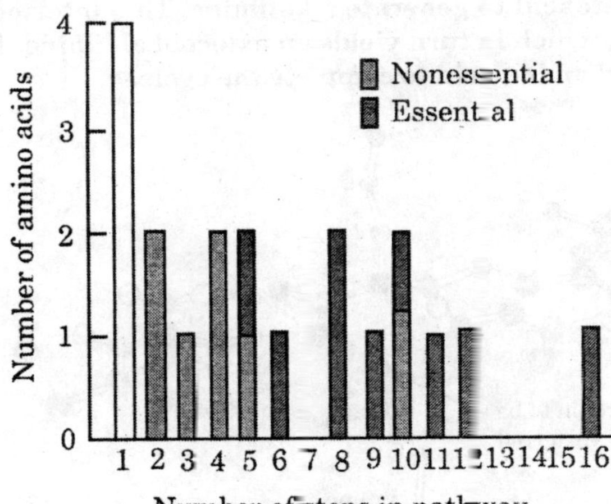

Fig. 2.8. Essential and Nonessential Amino Acids. Some amino acids are nonessential to human beings because they can be biosynthesized in a small number of steps. Those amino acids requiring a large number of steps for their synthesis are essential in the diet because some of the enzymes for these steps have been lost in the course of evolution.

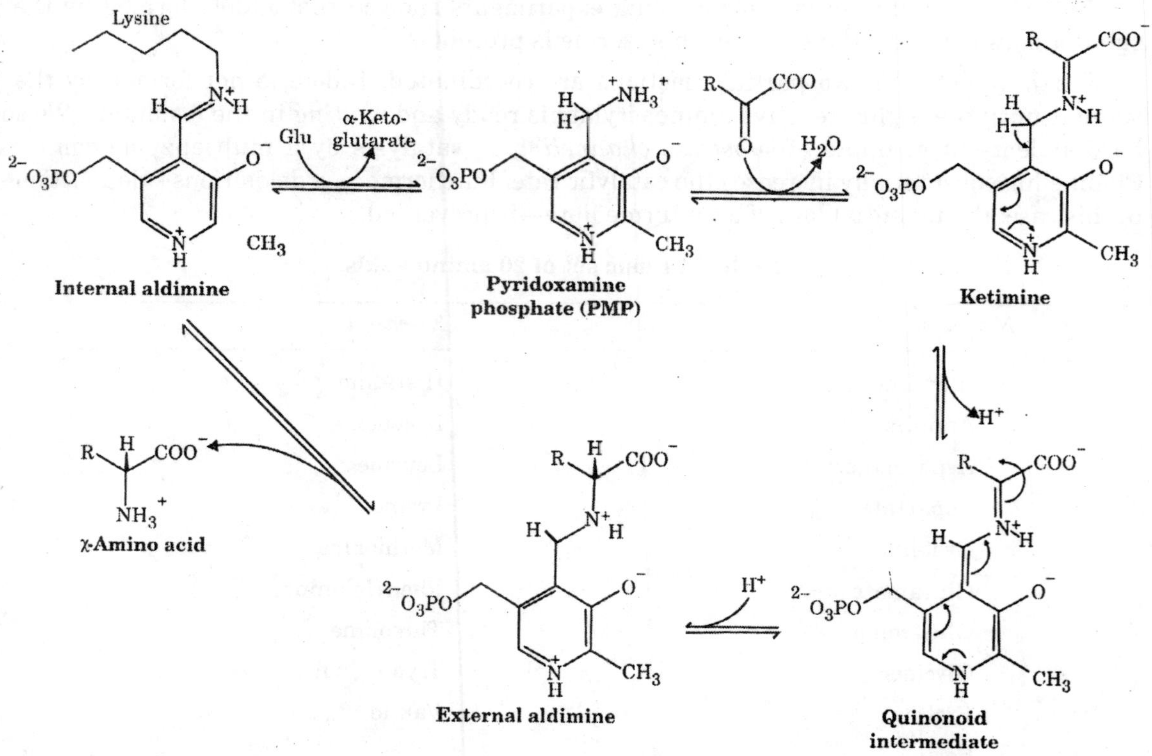

Fig. 2.9. Amino Acid Biosynthesis by Transamination. Within a transaminase, the internal aldimine is converted into pyridoxamine phosphate (PMP) by reaction with glutamate. PMP then reacts with an χ-ketoacid to generate a ketimine. This intermediate is converted into a quinonoid intermediate, which in turn yields an external aldimine. The aldimine is cleaved to release the newly formed amino acid to complete the cycle.

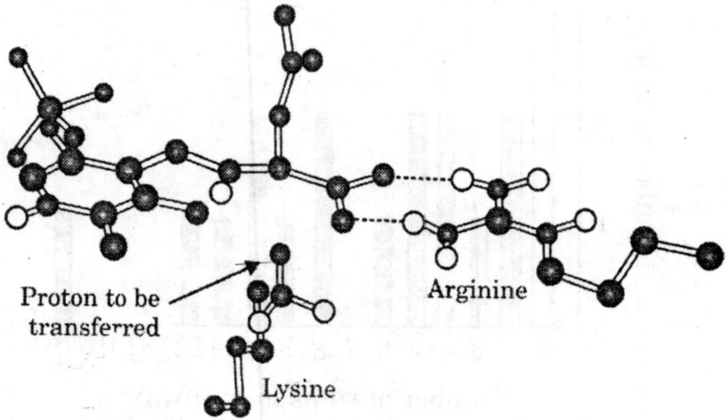

Fig. 2.10. Stereochemistry of Proton Addition. In a transaminase active site, the addition of a proton from the lysine residue to the bottom face of the quinonoid intermediate determines the l configuration of the amino acid product. The conserved arginine residue interacts with the χ-carboxylate group and helps establish the appropriate geometry of the quinonoid intermediate.

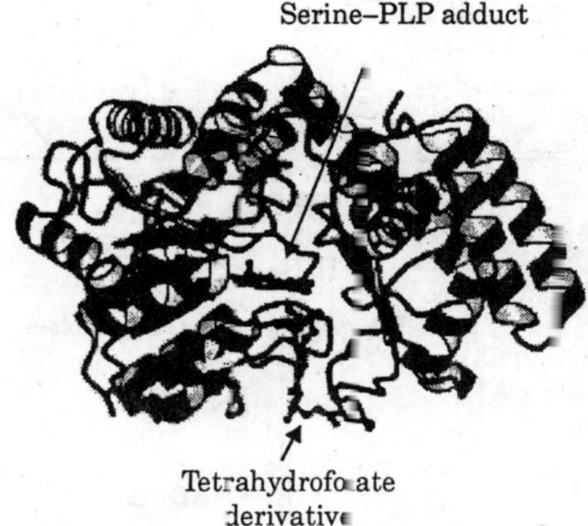

Fig. 2.11. Structure of Serine Hydroxymethyltransferase. This enzyme transfers a one-carbon unit from the side chain of serine to tetrahydrofolate. One subunit of the dimeric enzyme is shown.

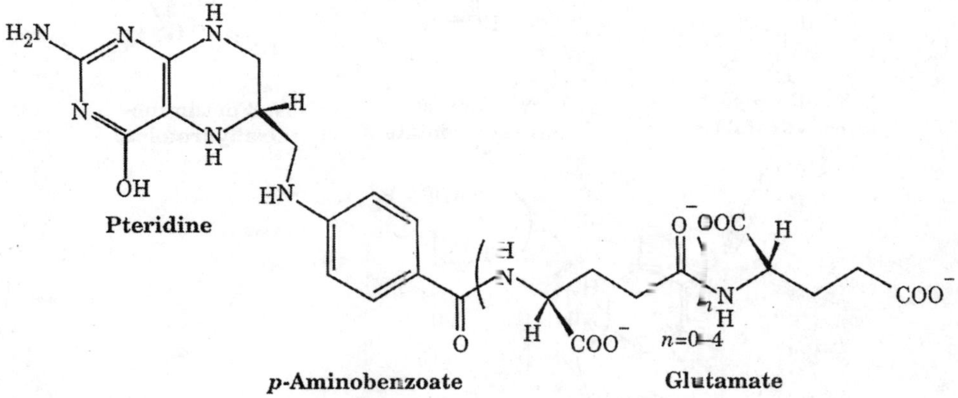

Pteridine

p-Aminobenzoate

Glutamate

$n = 0–4$

Fig. 2.12. Tetrahydrofolate. This cofactor includes three components: a pteridine ring, *p*-aminobenzoate, and one or more glutamate residues.

Table 2.2. One-carbon groups carried by tetrahydrofolate

Oxidation state	*Group*	
Most reduced (= methanol)	$—CH_3$	Methyl
Intermediate (= formaldehyde)	$—CH_2—$	Methylene
Most oxidized (= formic acid)	—CHO	Formyl
	—CHNH	Formimino
	—CH=	Methenyl

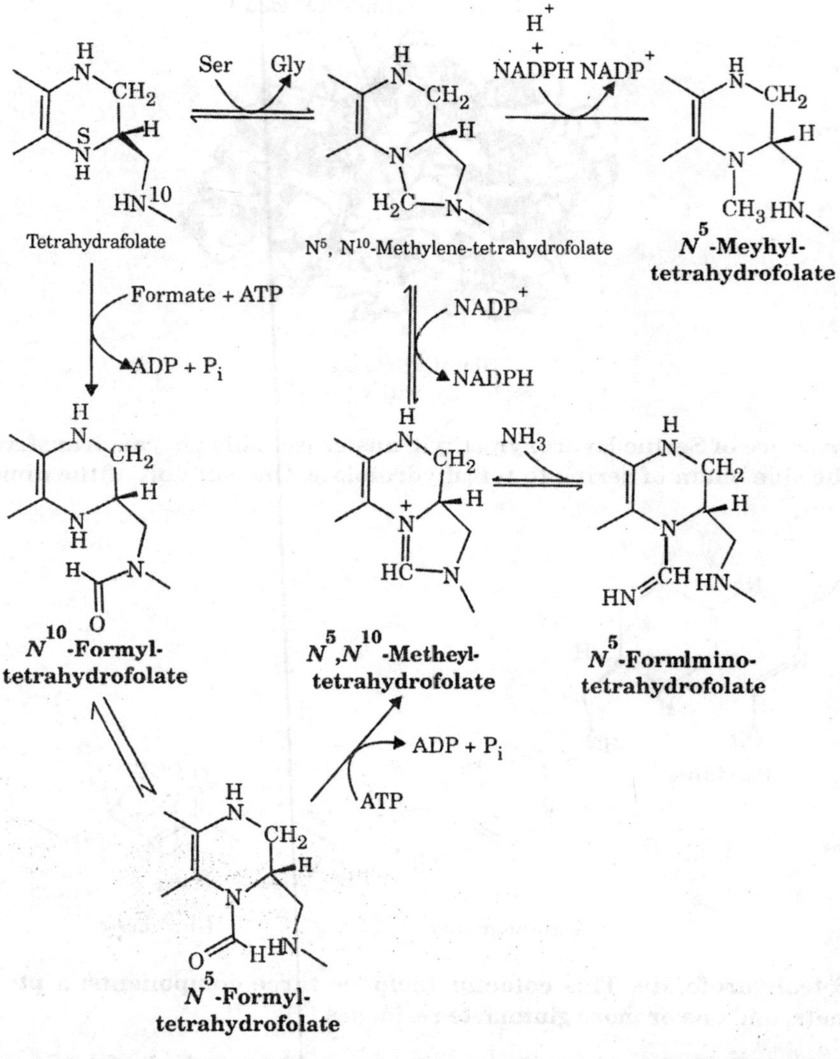

Ser Gly

NADPH NADP+

Tetrahydrafolate

N5, N10-Methylene-tetrahydrofolate

N5-Meyhyl-tetrahydrofolate

Formate + ATP

ADP + Pi

NADP+

NADPH

NH3

N10-Formyl-tetrahydrofolate

N5,N10-Metheyl-tetrahydrofolate

N5-Formlmino-tetrahydrofolate

ADP + Pi

ATP

N5-Formyl-tetrahydrofolate

Fig. 2.13. Conversions of One-Carbon Untis Attached to Tetrahydrofolate.

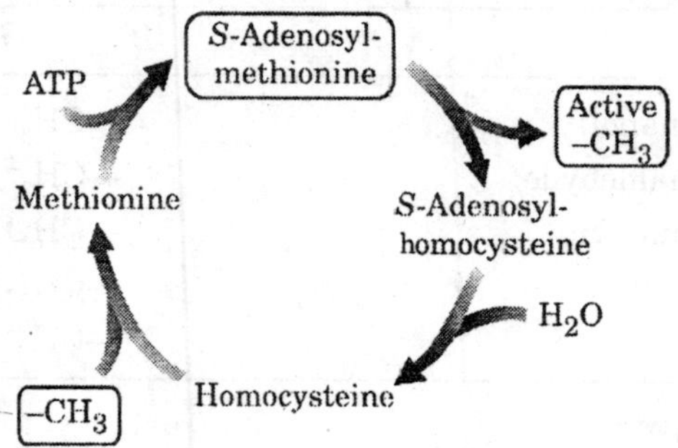

Fig. 2.14. Activated Methyl Cycle. The methyl group of methionine is activated by the formation of S-adenosylmethionine.

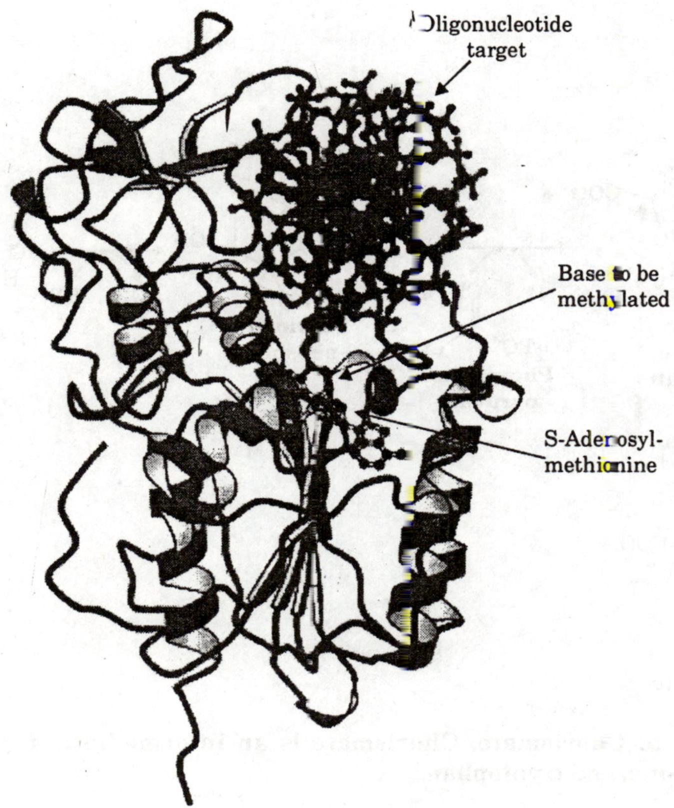

Fig. 2.15. DNA Methylation. The structure of a DNA methylase bound to an oligonucleotide target shows that the base to be methylated is flipped out of the DNA helix into the site of a SAM-dependent methylase.

$^{2-}O_3PO$ — COO^- ; CH_2

Phosphoenol-pyruvate

+

O=C–H ; H–C–OH ; H–C–OH ; $CH_2O_3PO_3^{2-}$

Erythrose 4-phosphate

H_2O → P_i

O=C–COO^- ; CH_2 ; HO–C–H ; H–C–OH ; H–C–OH ; $CH_2OPO_3^{2-}$

3-Deooxyarabino-heptullosanate 7-phosphate

NAD^+ → NADH ; P_i + H +

HO, COO^-, OH, H, O, HO H

3-Dehydro-quinate

→ H_2O

COO^-, OH, H, O, HO H

3-Dehydro-shikimate

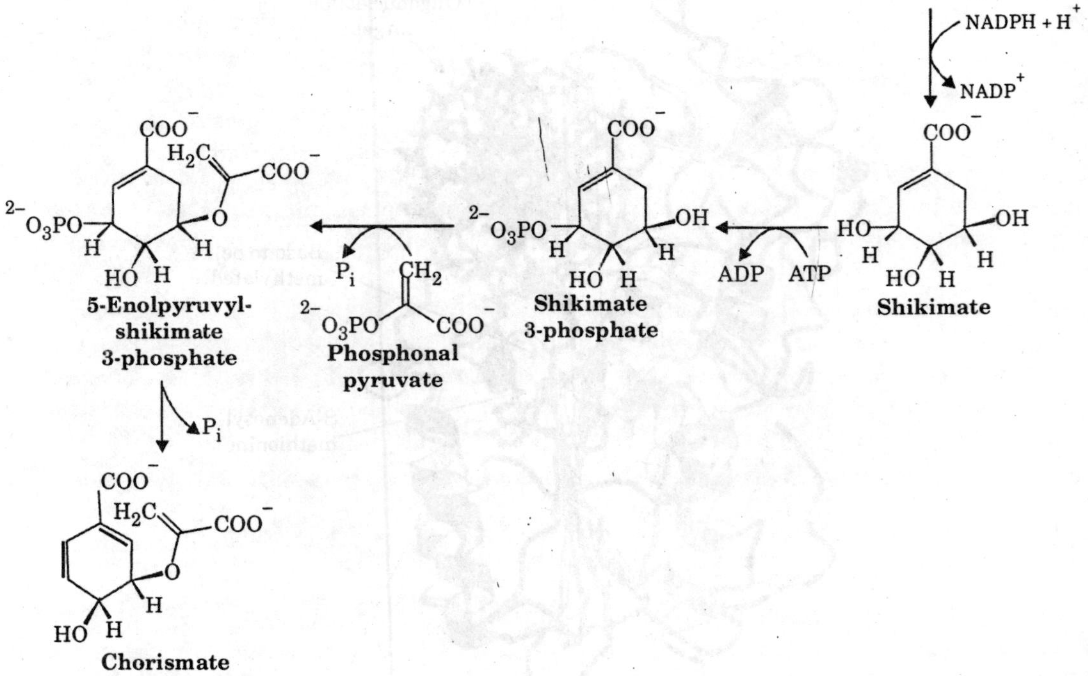

Fig. 2.16. Pathway to Chorismate. Chorismate is an intermediate in the biosynthesis of phenylalanine, tyrosine, and tryptophan.

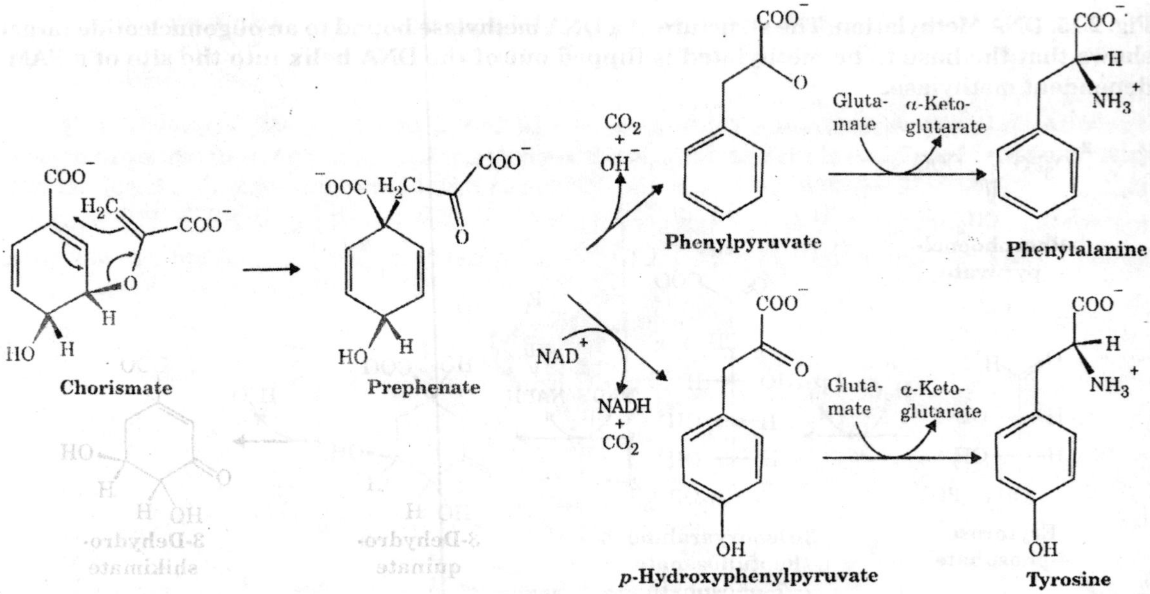

Fig. 2.17. Synthesis of Phenylalanine and Tyroshine. Chorismate can be converted into prephenate, which is subsequently converted into phenylalanine and tyrosine.

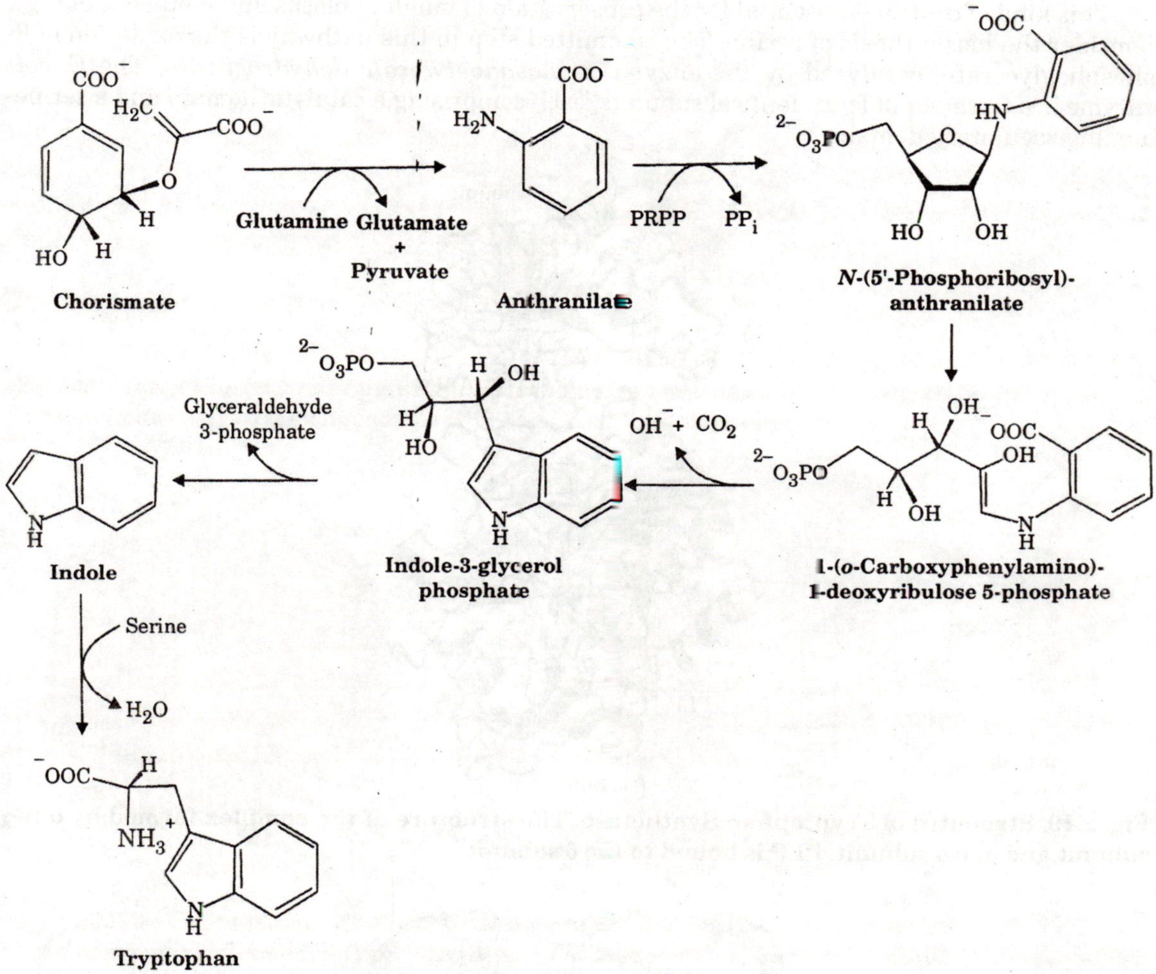

Chorismate → (Glutamine → Glutamate + Pyruvate) → Anthranilate → (PRPP → PP_i) → N-(5'-Phosphoribosyl)-anthranilate → 1-(o-Carboxyphenylamino)-1-deoxyribulose 5-phosphate → ($OH^- + CO_2$) → Indole-3-glycerol phosphate → (Glyceraldehyde 3-phosphate) → Indole → (Serine → H_2O) → Tryptophan

Fig. 2.18. Synthesis of Tryptophan. Chorismate can be converted into anthranilate, which is subsequently converted into tryptophan.

FEEDBACK INHIBITION

The rate of synthesis of amino acids depends mainly on the *amounts* of the biosynthetic enzymes and on their *activities*. We now consider the control of enzymatic activity. In a biosynthetic pathway, the first irreversible reaction, called the *committed step*, is usually an important regulatory site. *The final product of the pathway (Z) often inhibits the enzyme that catalyzes the committed step (A→B).*

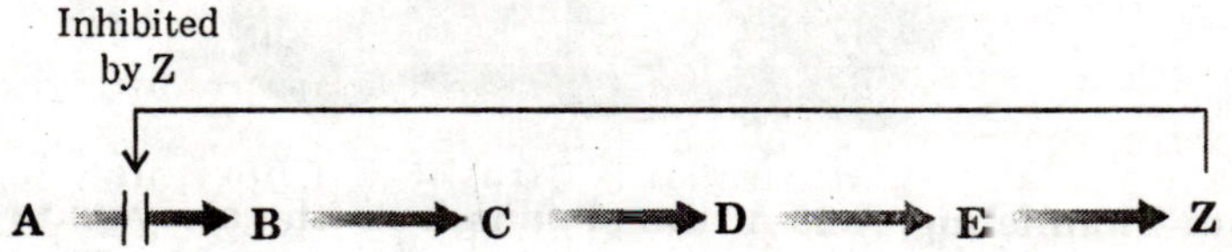

Inhibited by Z

A —|→ B → C → D → E → Z

This kind of control is essential for the conservation of building blocks and metabolic energy. Consider the biosynthesis of serine. The committed step in this pathway is the oxidation of 3-phosphoglycerate, catalyzed by the enzyme *3-phosphoglycerate dehydrogenase*. The *E. coli* enzyme is a tetramer of four identical subunits, each comprising a catalytic domain and a serine-binding regulatory domain.

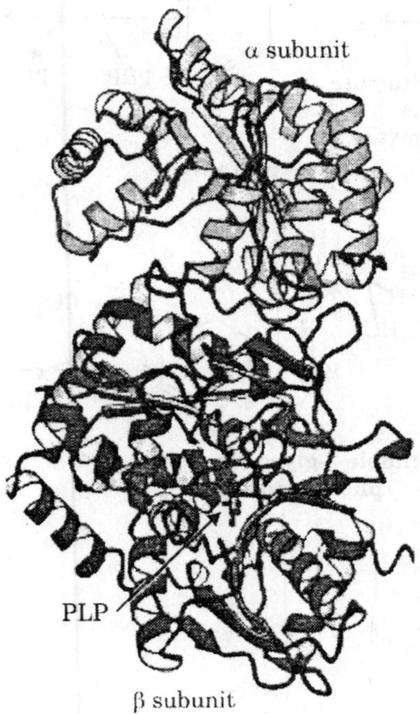

Fig. 2.19. Strucutre of Tryptophan Synthetase. The structure of the complex formed by one χ subunit and one δ subunit. PLP is bound to the δ subunit.

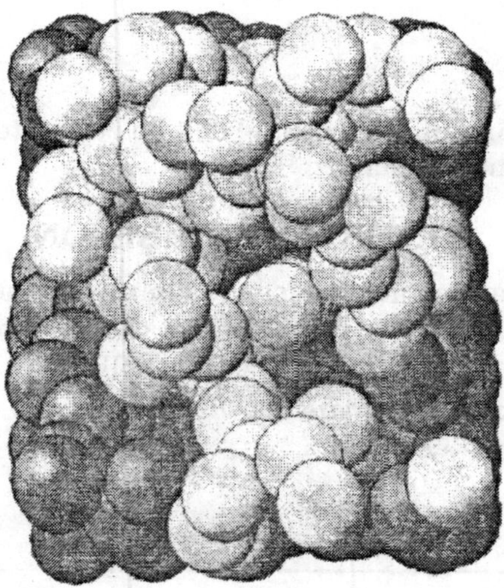

Fig. 2.20. Substrate Channeling. A 25-Å tunnel runs from the active site of the χ subunit of tryptophan synthetase to the PLP cofactor in the active site of the δ subunit.

The regulatory domains of two subunits interact to form a dimeric serine-binding regulatory unit so that the tetrameric enzyme contains two such regulatory units. Each unit is capable of binding two serine molecules. The binding of serine to a regulatory site reduces the value of V_{max} for the enzyme; an enzyme bound to four molecules of serine is essentially inactive. Thus, if serine is abundant in the cell, the enzyme activity is inhibited, and so 3-phosphoglycerate, a key building block that can be used for other processes, is not wasted.

Branched Pathways

The regulation of branched pathways is more complicated because the concentration of two products must be accounted for. In fact, several intricate feedback mechanisms have been found in branched biosynthetic pathways.

Feedback Inhibition and Activation

Consider, for example, the biosynthesis of the amino acids valine, leucine, and isoleucine. A common intermediate, hydroxyethyl thiamine pyrophosphate (hydroxyethyl-TPP; initiates the pathways leading to all three of these amino acids. Hydroxyethyl-TPP can react with χ-ketobutyrate in the initial step for the synthesis of isoleucine. Alternatively, hydroxyethyl-TPP can react with pyruvate in the committed step for the pathways leading to valine and leucine. Thus, the relative concentrations of χ-ketobutyrate and pyruvate determine how much isoleucine is produced compared with valine and leucine. *Threonine deaminase,* the PLP enzyme that catalyzes the formation of χ-ketobutyrate, is allosterically inhibited by isoleucine. This enzyme is also allosterically activated by valine. Thus, this enzyme is inhibited by the product of the pathway that it initiates and is activated by the end product of a competitive pathway.

This mechanism balances the amounts of different amino acids that are synthesized. The regulatory domain in threonine deaminase is very similar in structure to the dimeric regulatory domain in 3-phosphoglycerate dehydrogenase. In this case, the two half regulatory domains are fused into a single unit with two differentiated amino acid-binding sites, one for isoleucine and the other for valine. Sequence analysis shows that similar regulatory domains are present in other amino acid biosynthetic enzymes. *The similarities suggest that feedback-inhibition processes may have evolved by the linkage of specific regulatory domains to the catalytic domains of biosynthetic enzymes.*

Enzyme Multiplicity

Sophisticated regulation can also evolve by duplication of the genes encoding the biosynthetic enzymes. For example, the phosphorylation of aspartate is the committed step in the biosynthesis of threonine, methionine, and lysine. Three distinct aspartokinases catalyze this reaction in *E. coli,* an example of a regulatory mechanism called *enzyme multiplicity.* The catalytic domains of these enzymes show approximately 30% sequence identity. Although the mechanisms of catalysis are essentially identical, their activities are regulated differently: one enzyme is not subject to feedback inhibition, another is inhibited by threonine, and the third is inhibited by lysine.

Cumulative Feedback Inhibition

The regulation of glutamine synthetase in *E. coli* is a striking example of *cumulative feedback inhibition.* Recall that glutamine is synthesized from glutamate, NH_4^+, and ATP. *Glutamine synthetase* consists of 12 identical 50-kd subunits arranged in two hexagonal rings that face each other. Earl Stadtman showed that this enzyme regulates the flow of nitrogen and hence plays a key role in controlling bacterial metabolism. The amide group of glutamine is a source of nitrogen in the biosyntheses of a variety of compounds, such as tryptophan, histidine, carbamoyl phosphate, glucosamine 6-phosphate, cytidine triphosphate, and adenosine monophosphate.

Glutamine synthetase is cumulatively inhibited by each of these final products of glutamine metabolism, as well as by alanine and glycine. *In cumulative inhibition, each inhibitor can reduce the activity of the enzyme, even when other inhibitors are bound at saturating levels.* The enzymatic activity of glutamine synthetase is switched off almost completely when all final products are bound to the enzyme.

Enzymatic Cascade

The activity of glutamine synthetase is also controlled by *reversible covalent modification*—the attachment of an *AMP unit* by a phosphodiester bond to the hydroxyl group of a specific tyrosine residue in each subunit. *This adenylylated enzyme is less active and more susceptible to cumulative feedback inhibition than is the deadenylylated form.* The covalently attached AMP unit is removed from the adenylylated enzyme by phosphorolysis. The attachment of an AMP unit is the final step in an enzymatic cascade that is initiated several steps back by reactants and immediate products in glutamine synthesis. The adenylation and phosphorolysis reactions are catalyzed by the same enzyme, *adenylyl transferase.*

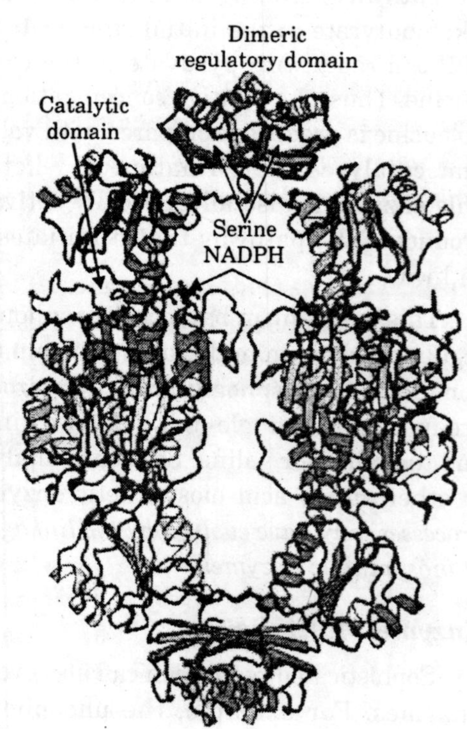

Fig. 2.21. Structure of 3-Phosphoglycerate Dehydrogenase. This enzyme, which catalyzes the committed step in the serine biosynthetic pathway, includes a serine-binding regulatory domain. Serine binding to this domain reduces the activity of the enzyme.

Sequence analysis indicates that this adenylyl transferase comprises two homologous halves, suggesting that one half catalyzes the adenylation reaction and the other half the phospholytic de-adenylation reaction. What determines whether an AMP unit is added or removed? The specificity of adenylyl transferase is controlled by a *regulatory protein* (designated P or P_{II}), a trimeric protein that can exist in two forms, P_A and P_D. The complex of P_A and adenylyl transferase catalyzes the attachment of an AMP unit to glutamine synthetase, which reduces its activity. Conversely, the complex of P_D and adenylyl transferase removes AMP from the adenylylated enzyme. This brings us to another level of reversible covalent modification. P_A is converted into P_D by the attachment of uridine monophosphate to a specific tyrosine residue.

This reaction, which is catalyzed by *uridylyl transferase,* is stimulated by ATP and χ-ketoglutarate, whereas it is inhibited by glutamine. In turn, the UMP units on P_D are removed by hydrolysis, a reaction promoted by glutamine and inhibited by χ-ketoglutarate. These opposing catalytic activities are present on a single polypeptide chain, homologous to adenylyl transferase, and are controlled so that the enzyme does not simultaneously catalyze uridylylation and hydrolysis. Why is an enzymatic cascade used to regulate glutamine synthetase? One advantage of a cascade is that it *amplifies signals,* as in blood clotting and the control of glycogen metabolism.

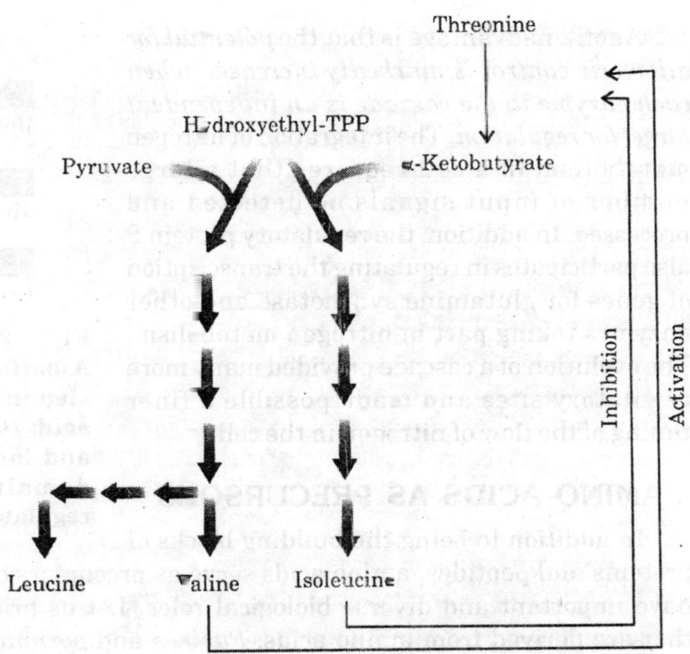

Fig. 2.22. Regulation of Threonine Deaminase. Threonine is converted into χ-ketobutyrate in the committed step leading to the synthesis of isoleucine. The enzyme that catalyzes this step, threonine deaminase, is inhibited by isoleucine and activated by valine, the product of a parallel pathway.

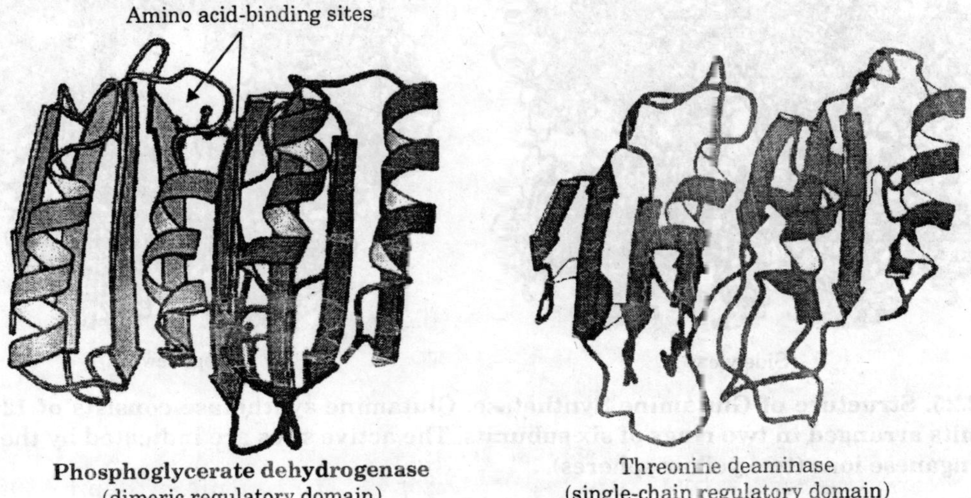

Fig. 2.23. A Recurring Regulatory Domain. The regulatory domain formed by two subunits of 3-phosphoglycerate dehydrogenase is structurally related to the single-chain regulatory domain of threonine deaminase. Sequence analyses have revealed this amino acid-binding regulatory domain to be present in other enzymes as well.

Another advantage is that the *potential for allosteric control is markedly increased when each enzyme in the cascade is an independent target for regulation.* The integration of nitrogen metabolism in a cell requires that a large number of input signals be detected and processed. In addition, the regulatory protein P also participates in regulating the transcription of genes for glutamine synthetase and other enzymes taking part in nitrogen metabolism. The evolution of a cascade provided many more regulatory sites and made possible a finer tuning of the flow of nitrogen in the cell.

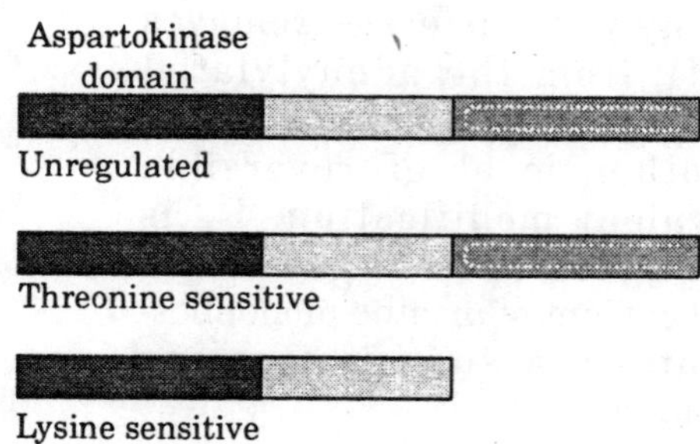

Fig. 2.24. Domain Structures to Three Aspartokinases. Each catalyzes the committed step in the biosynthesis of a different amino acid: (top) methionine, (middle) threonine, and (bottom) lysine. They have a catalytic domain in common but differ in their regulatory domains.

AMINO ACIDS AS PRECURSORS

In addition to being the building blocks of proteins and peptides, amino acids serve as precursors of many kinds of small molecules that have important and diverse biological roles. Let us briefly survey some of the biomolecules that are derived from amino acids. *Purines* and *pyrimidines* are derived largely from amino acids. The biosynthesis of these precursors of DNA, RNA, and numerous coenzymes will be discussed in detail else where.

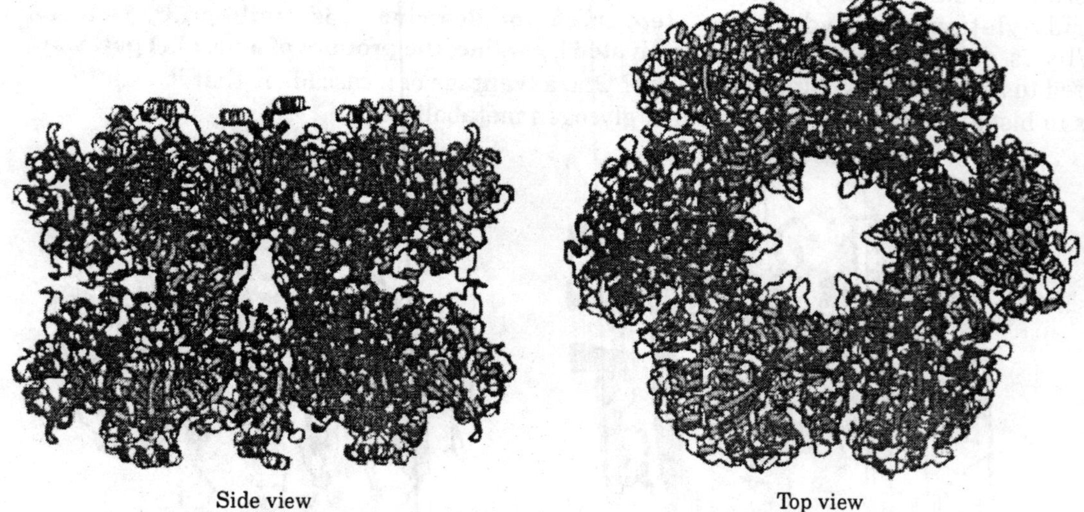

Fig. 2.25. Structure of Glutamine Synthetase. Glutamine synthetase consists of 12 identical subunits arranged in two rings of six subunits. The active sites are indicated by the presence of manganese ions (two yellow spheres).

The reactive terminus of *sphingosine,* an intermediate in the synthesis of sphingolipids, comes from serine. *Histamine,* a potent vasodilator, is derived from histidine by decarboxylation. Tyrosine is a precursor of the hormones *thyroxine* (tetraiodothyronine) and *epinephrine* and of *melanin,* a complex polymeric pigment. The neurotransmilther *serotonin* (5-hydroxytryptamine) and the *nicotinamide ring* of NAD $^+$ are synthesized from tryptophan. Let us now consider in more derail three particularly important biochemicals derived from amino acids.

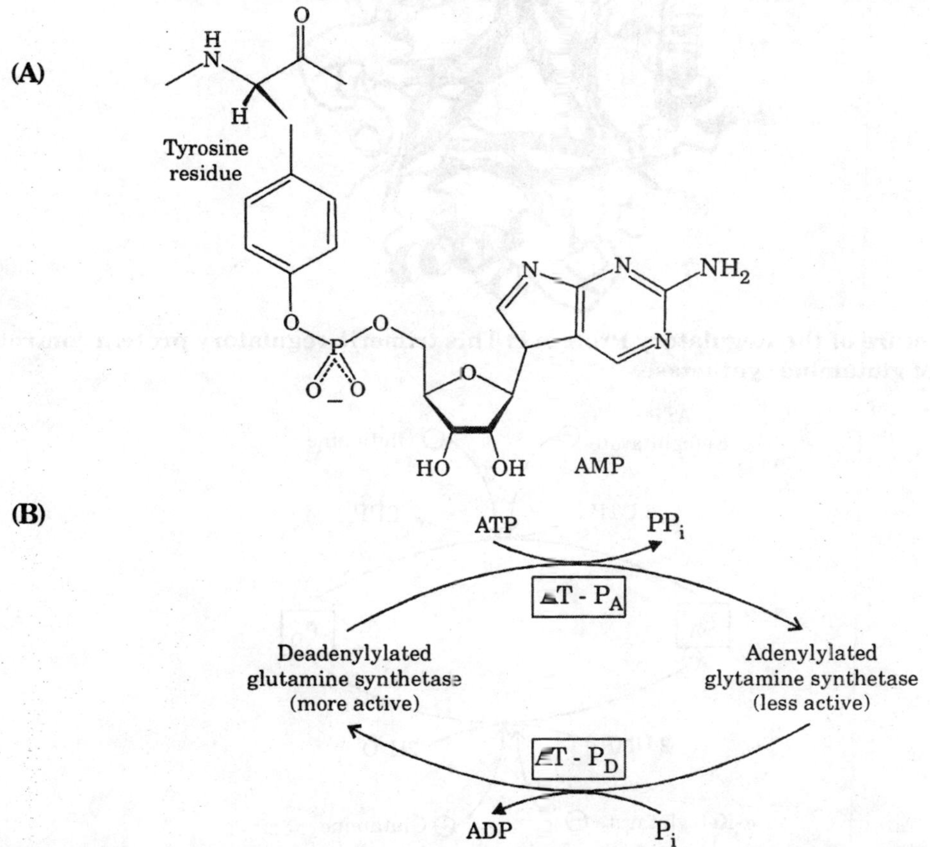

Fig. 2.26. Regulation by Adenylation. (A) specific tyrosine residue in each subunit in glutamine synthetase is modified by adenylation. (B) Adenylation of tyrosine is catalyzed by a complex of adenylyl transferase (AT) and one form of a regulatory protein (P_A). The same enzyme catalyzes deadenylation when it is complexed with the other form (P_D) of the regulatory protein.

An-antioxidant

Glutathione, a tripeptide containing a sulfhydryl group, is a highly distinctive amino acid derivative with several important roles. For example, glutathione, present at high levels ($\approx$ 5 mM) in animal cells, protects red cells from oxidative damage by serving as a sulfhydryl buffer. It cycles between a reduced thiol form (GSH) and an oxidized form (GSSG) in which two tripeptides are linked by a disulfide bond. GSSG is reduced to GSH by *glutathione reductase,* a flavoprotein that uses NADPH as the electron source. The ratio of GSH to GSSG in most cells is greater than 500. *Glutathione plays a key role in detoxification by reacting with hydrogen peroxide and organic peroxides, the harmful by-products of aerobic life.*

$$2\ GSH + RO\text{–}OH \rightleftharpoons GSSG + H_2O + ROH$$

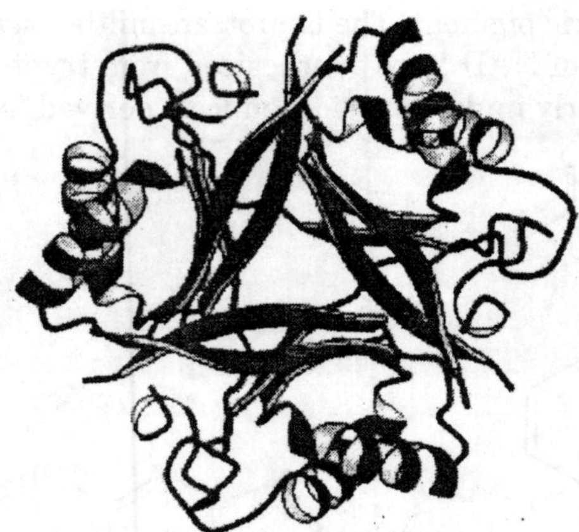

Fig. 2.27. Structure of the Regulatory Protein P. This trimeric regulatory protein controls the modification of glutamine synthetase.

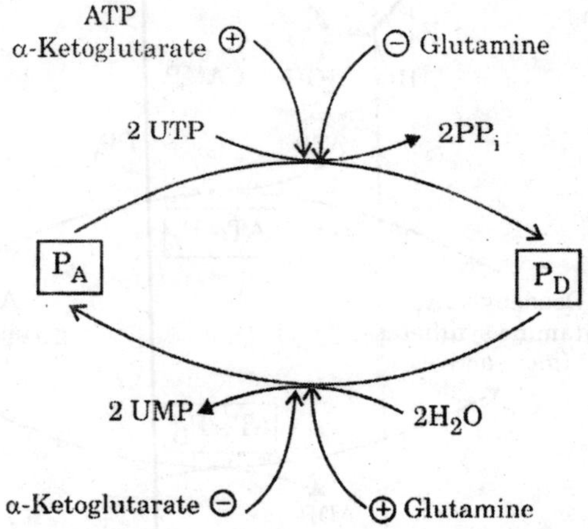

Fig. 2.28. A Higher Level in the Regulatory Cascade of Glutamine Synthetase. P_A and P_D, the regulatory proteins that control the specificity of adenylyl transferase, are interconvertible. P_A is converted into P_D by uridylylation, which is reversed by hydrolysis. The enzymes catalyzing these reactions are regulated by the concentrations of metabolic intermediates.

Glutathione peroxidase, the enzyme catalyzing this reaction, is remarkable in having a modified amino acid containing a *selenium* (Se) atom. Specifically, its active site contains the selenium analog of cysteine, in which selenium has replaced sulfur. The selenolate (E-Se$^-$) form of this residue reduces the peroxide substrate to an alcohol and is in turn oxidized to selenenic acid (E-SeOH). Glutathione then comes into action by forming a selenosulfide adduct (E-Se-S-G). A second molecule of glutathione then regenerates the active form of the enzyme by attacking the selenosulfide to form oxidized glutathione.

A Short-Lived Signal Molecule

Nitric oxide (NO) is an important messenger in many vertebrate signal-transduction processes. This free-radical gas is produced endogenously from *arginine* in a complex reaction that is catalyzed by *nitric oxide synthase.* NADPH and O_2 are required for the synthesis of nitric oxide. Nitric oxide acts by binding to and activating soluble guanylate cyclase, an important enzyme in signal transduction. This enzyme is homologous to adenylate cyclase but includes a heme-containing domain that binds NO.

Mammalian Porphyrins

The involvement of an amino acid in the biosynthesis of the porphyrin rings of hemes and chlorophylls was first revealed by the results of isotopiclabeling experiments carried out by David Shemin and his colleagues. In 1945, they showed that the nitrogen atoms of heme were labeled after the feeding of [^{15}N] glycine to human subjects (of whom Shemin was the first), whereas the ingestion of [^{15}N] glutamate resulted in very little labeling.

Using ^{14}C, which had just become available, they discovered that 8 of the carbon atoms of heme in nucleated duck erythrocytes are derived from the χ-carbon atom of glycine and none from the carboxyl carbon atom. The results of subsequent studies demonstrated that the other 26 carbon atoms of heme can arise from acetate. Moreover, the ^{14}C in methyl-labeled acetate emerged in 24 of these carbons, whereas the ^{14}C in carboxyl-labeled acetate appeared only in the other.

This highly distinctive labeling pattern led Shemin to propose that a heme precursor is formed by the condensation of glycine with an activated succinyl compound. In fact, *the first step in the biosynthesis of porphyrins in mammals is the condensation of glycine and succinyl CoA to form ϕ-aminolevulinate.*

$$^{-}OOC\text{-}CH_2\text{-}CH_2\text{-}C(=O)\text{-}S\text{-}CoA + {}^{-}OOC\text{-}CH_2\text{-}NH_3^{+} \xrightarrow[\;CoA + CO_2\;]{H^{+}} {}^{-}OOC\text{-}CH_2\text{-}CH_2\text{-}C(=O)\text{-}CH_2\text{-}NH_3^{+}$$

Succinyl CoA **Glycine** **δ-Aminolevulinate**

This reaction is catalyzed by *ϕ-aminolevulinate synthase*, a PLP enzyme present in mitochondria. Two molecules of ϕ-aminolevulinate condense to form *porphobilinogen*, the next intermediate. Four molecules of porphobilinogen then condense head to tail to form a linear *tetrapyrrole* in a reaction catalyzed by *porphobilinogen deaminase*. The enzyme-bound linear tetrapyrrole then cyclizes to form uroporphyrinogen III, which has an asymmetric arrangement of side chains. This reaction requires a *cosynthase*.

In the presence of synthase alone, uroporphyrinogen I, the nonphysiologic symmetric isomer, is produced. Uroporphyrinogen III is also a key intermediate in the synthesis of vitamin B_{12} by bacteria and that of chlorophyll by bacteria and plants. The porphyrin skeleton is now formed. Subsequent reactions alter the side chains and the degree of saturation of the porphyrin ring. *Coproporphyrinogen III* is formed by the decarboxylation of the acetate side chains. The desaturation of the porphyrin ring and the conversion of two of the propionate side chains into vinyl groups yield *protoporphyrin IX*.

The chelation of iron finally gives *heme*, the prosthetic group of proteins such as myoglobin, hemoglobin, catalase, peroxidase, and cytochrome *c*. The insertion of the *ferrous* form of iron is catalyzed by *ferrochelatase*. Iron is transported in the plasma by *transferrin*, a protein that binds two ferric ions, and stored in tissues inside molecules of *ferritin.* The large internal cavity (≈ 80 Å in diameter) of ferritin can hold as many as 4500 ferric ions. The normal human erythrocyte has a life span of about 120 days, as was first shown by the time course of ^{15}N in Shemin's own hemoglobin after he ingested ^{15}N-labeled glycine.

The first step in the degradation of the heme group is the cleavage of its χ-methene bridge to form the green pigment *biliverdin,* a linear tetrapyrrole. The central methene bridge of biliverdin is then reduced by *biliverdin reductase* to form *bilirubin,* a red pigment. The changing color of a bruise is a highly graphic indicator of these degradative reactions.

Porphyrins Accumulation

Porphyrias are inherited or acquired disorders caused by a deficiency of enzymes in the heme biosynthetic pathway. Porphyrin is synthesized in both the erythroblasts and the liver, and either one may be the site of a disorder. *Congenital erythropoietic porphyria,*for example, prematurely destroys eythrocytes. This disease results from insufficient cosynthase. In this porphyria, the synthesis of the required amount of uroporphyrinogen III is accompanied by the formation of very large quantities of uroporphyrinogen I, the useless symmetric isomer. Uroporphyrin I, coproporphyrin I, and other symmetric derivatives also accumulate.

The urine of patients having this disease is red because of the excretion of large amounts of uroporphyrin I. Their teeth exhibit a strong red fluorescence under ultraviolet light because of the deposition of porphyrins. Furthermore, their *skin is usually very sensitive to light* because photoexcited porphyrins are quite reactive. *Acute intermittent porphyria* is the most prevalent of the porphyrias affecting the liver. This porphyria is characterized by the overproduction of porphobilinogen and ϕ-aminolevulinate, which results in severe abdominal pain and neurological dysfunction. The "madness" of George III, King of England during the American Revolution, is believed to have been due to this porphyria.

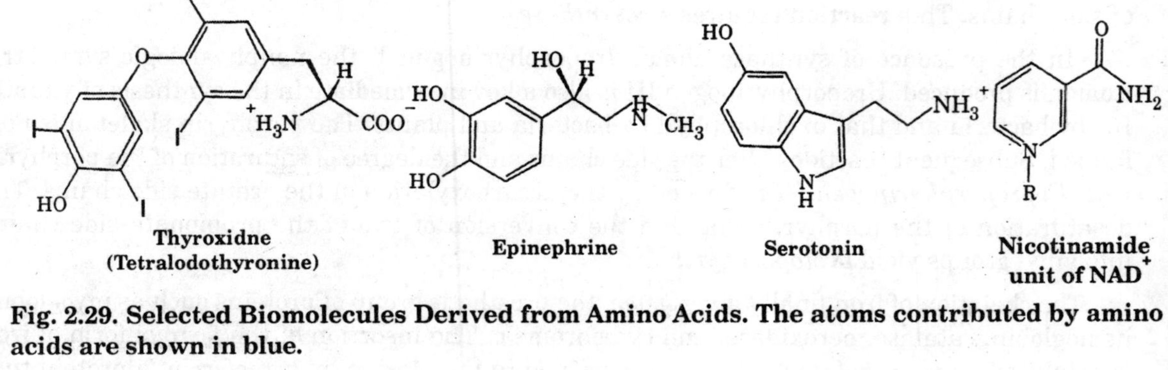

Adomine | Cytorine | Sphingosine | Histamine

Thyroxidne (Tetralodothyronine) | Epinephrine | Serotonin | Nicotinamide unit of NAD^+

Fig. 2.29. Selected Biomolecules Derived from Amino Acids. The atoms contributed by amino acids are shown in blue.

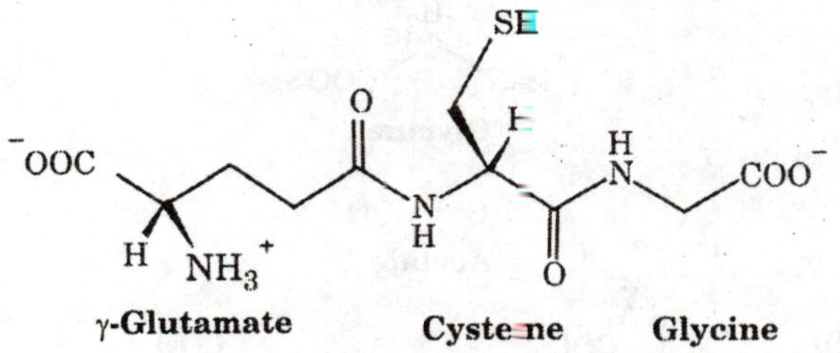

γ-Glutamate Cysteine Glycine

Fig 2.30. Glutathione. This tripeptide consists of a cysteine residue flanked by a glycine residue and a glutamate residue that is linked to cysteine by an isopeptide bond between glutamate's side-chain carboxylate group and cysteine's amino group.

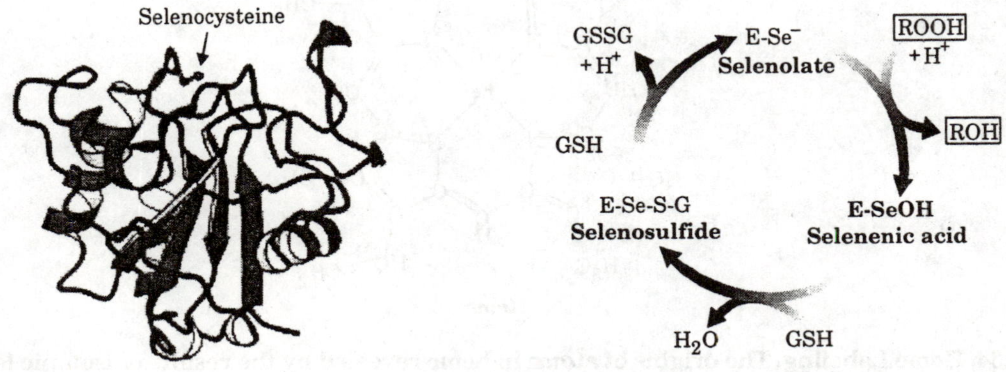

Fig. 2.31. Structure of Glutathione Peroxidase. This enzyme, which has a role in peroxide detoxification, contains a selenocysteine residue in its active site.

Fig. 2.32. Catalytic Cycle of Glutathione Peroxidase.

Arginine → (H^+ + O_2 + NADPH → H_2O + $NADP^+$) → *N*-(w)-Hydroxy-arginine → (O_2 + NADPH → H_2O + $NADP^+$) → Citrulline + ^-NO Nitric oxide

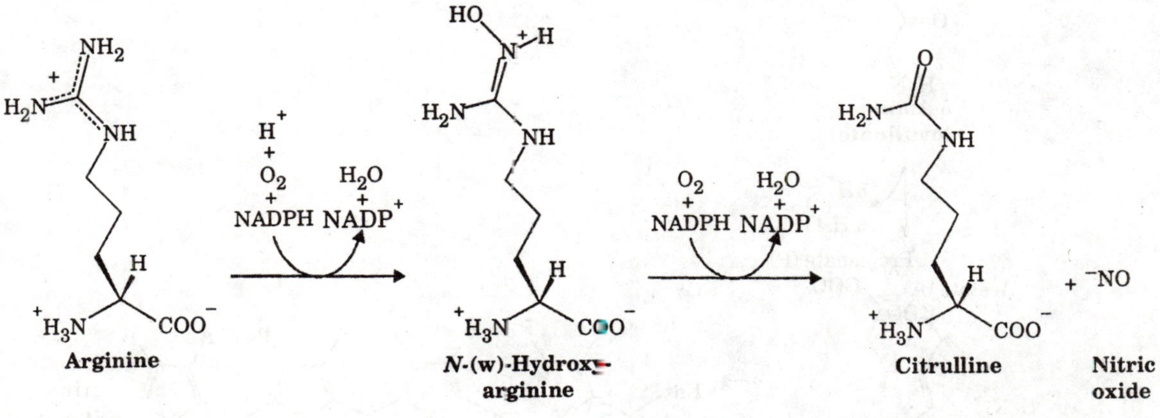

Fig. 2.33. Formation of Nitric Oxide. NO is generated by the oxidation of arginine.

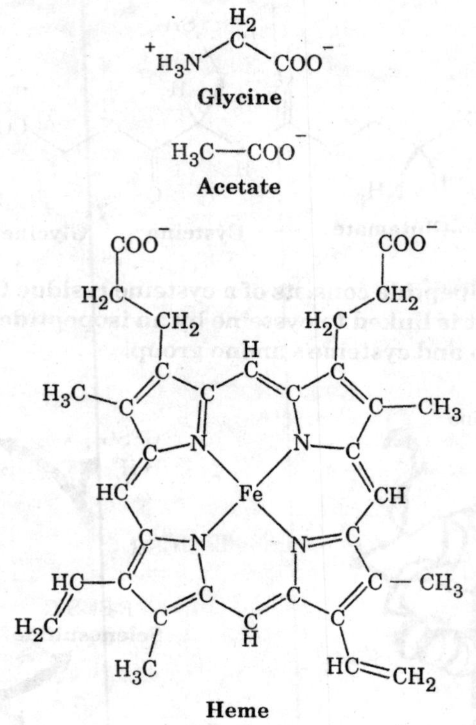

Fig. 2.34. Heme Labeling. The origins of atoms in heme revealed by the results of isotopic labeling studies.

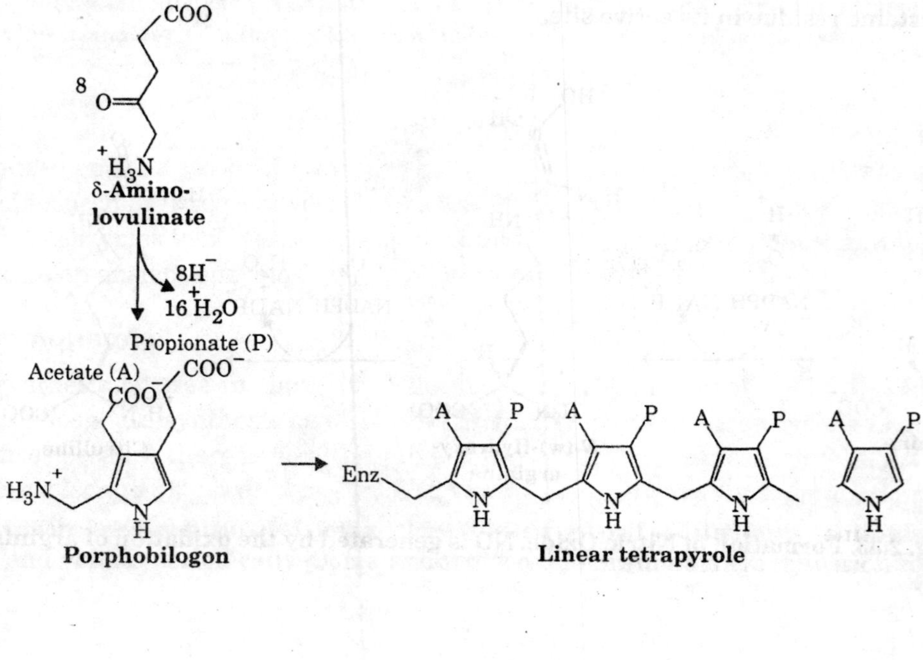

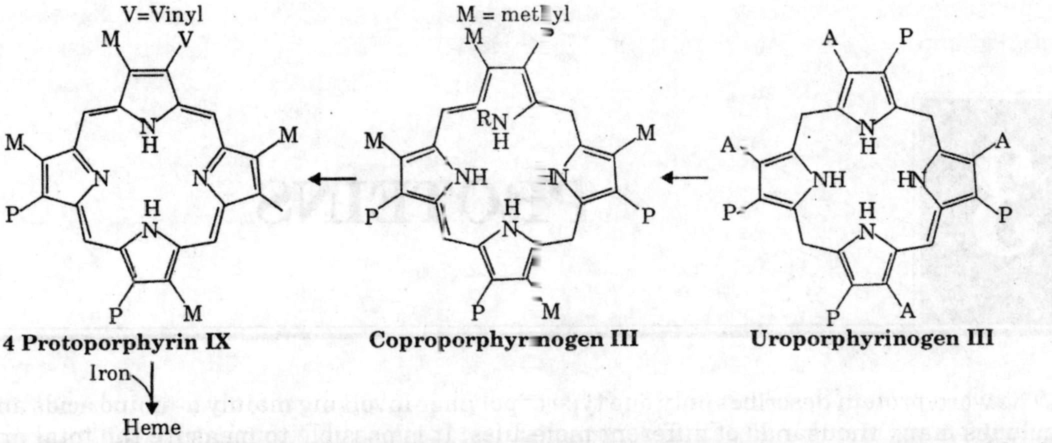

V=Vinyl   M = methyl

4 Protoporphyrin IX ← Coproporphyrinogen III ← Uroporphyrinogen III

Iron ↓

Heme

Fig. 2.35. Heme Biosynthetic Pathway. The pathway for the formation of heme starts with eight molecules of φ-aminolevulinate.

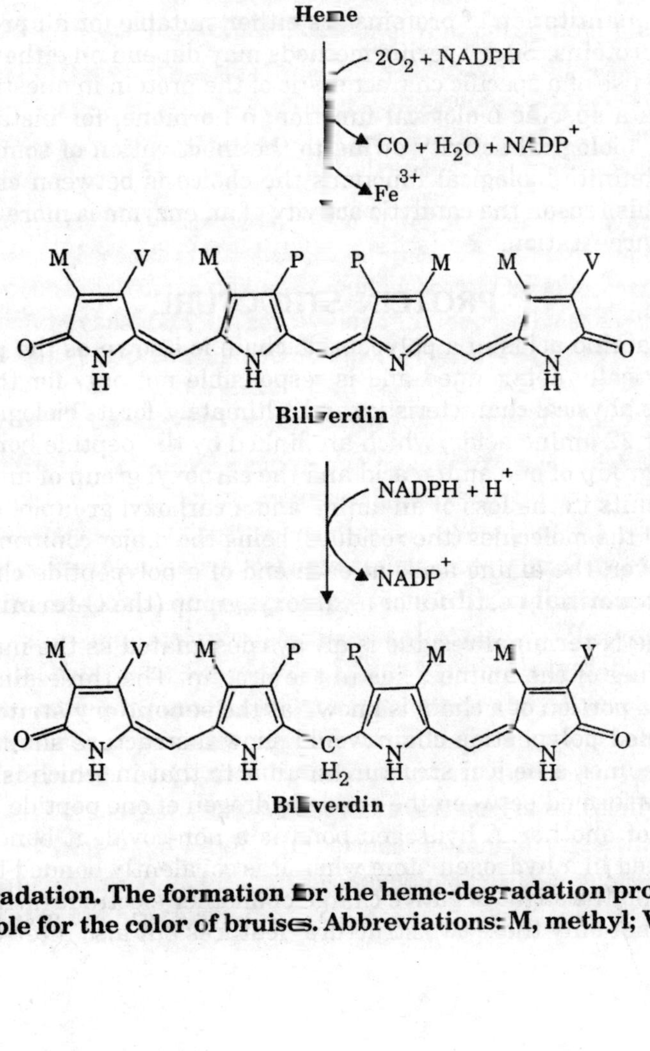

Heme

↓ $2O_2$ + NADPH → CO + H_2O + $NADP^+$, Fe^{3+}

Biliverdin

↓ NADPH + H^+ → $NADP^+$

Biliverdin

Fig. 2.36. Heme Degradation. The formation for the heme-degradation products biliverdin and bilirubin is responsible for the color of bruises. Abbreviations: M, methyl; V, vinyl.

3 PROTEINS

The word protein describes only one type of polymer involving mainly α-amino acids and yet it includes many thousands of different molecules. It is possible to measure the total protein content of a sample despite the fact that relatively simple preparative techniques may be capable of demonstrating the presence of different proteins. However, if interest lies in only one of these proteins, then a measure of the total protein content would be completely in appropriate.

Methods for the quantitation of proteins are either suitable for all proteins or designed to measure individual proteins. Such specific methods may depend on either a preparative stage in the analysis or the use of a specific characteristic of the protein in question. The quantitation of a protein that has a specific biological function, a hormone, for instance, may not give a true indication of its biological activity owing to the inactivation of some of the protein. For proteins that have definite biological functions the choice is between chemical quantitation and bio-assays. For this reason the catalytic activity of an enzyme is more frequently measured than is its protein concentration.

PROTEIN STRUCTURE

The sequence of amino acids in a polypeptide chain is known as the **primary structure.** This feature is genetically determined and is responsible not only for the final shape of the protein but also for its physical characteristics and, ultimately, for its biological function. Proteins are made up of about 22 amino acids, which are linked by the peptide bond, an amide linkage involving the amino group of one amino acid and the carboxyl group of another. The formation of a peptide bond results in the loss of an amino and a carboxyl group of each amino acid, the remaining portions of the molecules (the residues) being the major components in the structure of the protein. However, the amino acids at each end of a polypeptide chain retain either an amino group (the **N-terminal residue**) or a carboxyl group (the **C-terminal residue**).

By convention the N-terminal residue is always designated as the number one residue in a numerical sequencing of the amino acids in the protein. The three-dimensional shape of a polypeptide chain or a portion of a chain is known as the **secondary structure**. In its simplest form the fully extended polypeptide chain would show a structure similar to that indicated. However, it often assumes a helical structure similar to that in which is, stabilized by intra-chain hydrogen bonds formed between the amide hydrogen of one peptide bond and the oxygen of a carbonyl group of another. A hydrogen bond is a non-covalent bond between the slight positive charge induced in a hydrogen atom when it is covalently bonded to an electronegative atom (*e.g.* nitrogen) and the slight negative charge of another electronegative atom (*e.g.* oxygen). Many proteins show not only these two structural features but also inter-mediary forms.

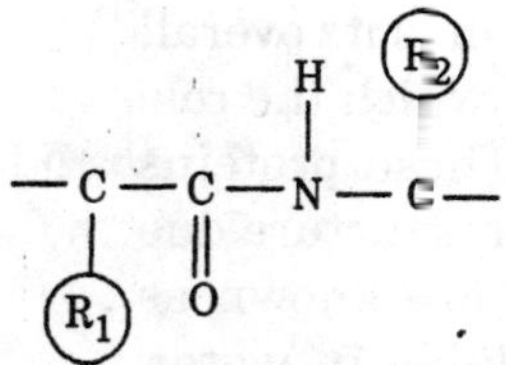

Fig. 3.1. The peptide bond.

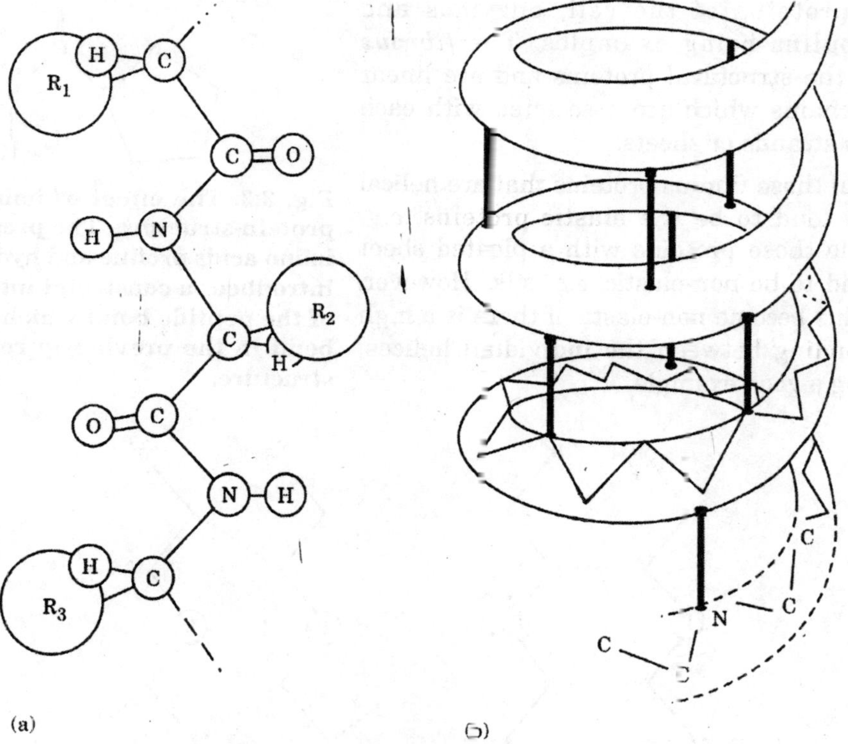

Fig. 3.2. The secondary structure of proteins. The simplest spatial arrangement of amino acids in a polypeptide chain is as a fully extended chain (*a*) which has a regular backbone structure due to the bond angles involved and from which the additional atoms, H and O, and the amino acid residues, R, project at varying angles. The helical form (*b*) is stabilized by hydrogen bonds between the —NH group of one peptide bond and the —CO group of another peptide bond. The amino acid residues project from the helix rather than internally into the helix.

The presence of the imino acids proline and hydroxyproline induces bends in the chain due to the variant of the resulting peptide bond. Glycine, which effectively has no side chain (H), permits greater flexibility at the peptide bond than do other amino acid residues. Inter-chain bonding can also occur between parallel extended chains to produce *pleated sheet structures* in which the hydrogen bonds are formed between hydrogen and oxygen atoms in different chains. Depending upon the arrangement of the chains these are known as either parallel pleated sheets or anti-parallel pleated sheets.

The **tertiary structure** of a protein is its overall shape and the level of organization at which the role of the protein becomes significant. Those proteins that have overall spherical or globular structure due to the internal folding of the chain are known as *globular proteins.* They are semi-soluble in water, forming colloidal solutions, and in the solid form often exhibit a crystalline structure. They are usually the functional proteins of the cell, enzymes and immunoglobulins being examples. The *fibrous proteins* are the structural proteins and are linear polypeptide chains which are associated with each other to form strands or sheets.

In general those fibrous proteins that are helical in structure tend to be the elastic proteins, *e.g.* keratin, while those proteins with a pleated sheet structure tend to be non-elastic, *e.g.* silk. However, helical proteins become non-elastic if there is a high degree of bonding between the individual helices, collagen being a good example.

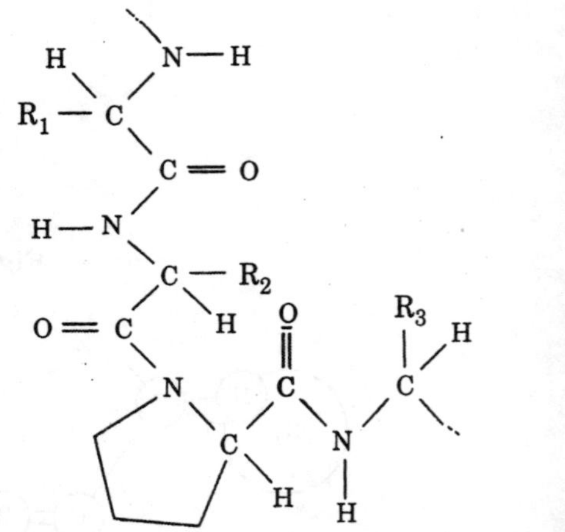

Fig. 3.3. The effect of imino acids on protein structure. The presence of the imino acids proline and hydroxyproline introduces a constraint into the angles of the peptide bond which results in a bend in the previously regular chain structure.

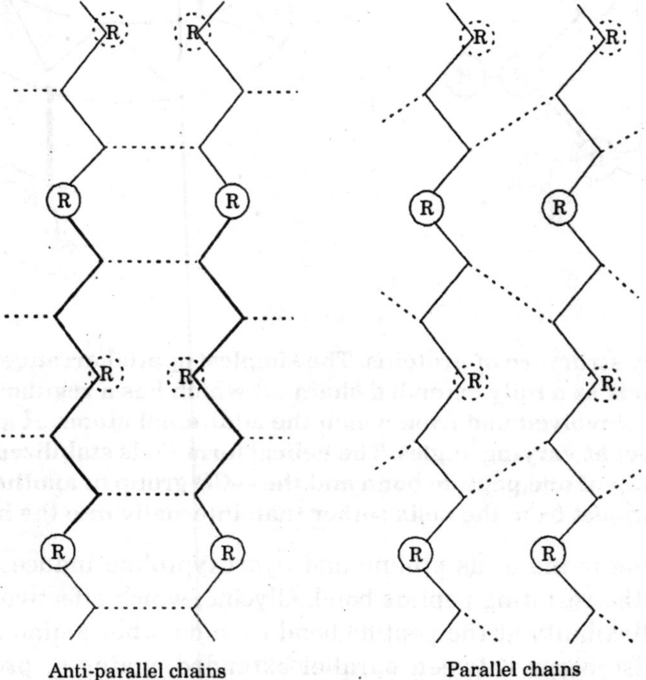

Fig. 3.4. Pleated sheets of fibrous proteins. Parallel pleated sheets are composed of polypeptide chains which all have their N-terminal amino acid at the same end whereas anti-parallel pleated sheets involve polypeptide chains which are alternately reversed in direction. Both forms of sheet show a high degree of hydrogen bonding between the chains.

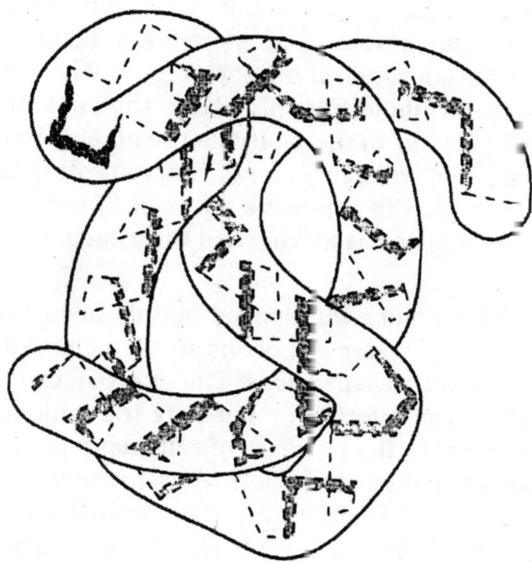

Fig. 3.5. Globular proteins. The folding of a polypeptide chain in a globular form is stabilized by hydrophobic interactions and some covalent bonding, particularly the disulphide bond between cysteine residues. The polypeptide chain shows some sections which are regular and helical in nature and other sections, particularly at bends and folds, where the conformation of the chain is distorted.

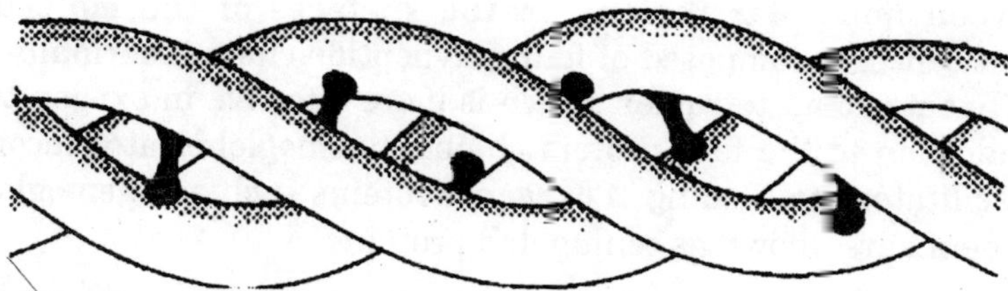

Fig. 3.6. Collagen. The collagens are major constituents of cartilage and other connective tissue and contain large amounts of the imino acids proline and hydroxy-proline. The basic structure involves three polypeptide chains with a considerable degree of inter-chain hydrogen bonding.

The nature of the amino acid residues is of prime importance in the development and maintenance of protein structure. Polypeptide chains composed of simple aliphatic amino acids tend to form helices more readily than do those involving many different amino acids. Sections of a polypeptide chain which are mainly non-polar and hydrophobic tend to be buried in the interior of the molecule away from the interface with water, whereas the polar amino acid residues usually lie on the exterior of a globular protein.

The folded polypeptide chain is further stabilized by the presence of disulphide bonds; which are produced by the oxidation of two cysteine residues. Such covalent bonds are extremely important in maintaining protein structure, both internally in the globular proteins and externally in the bonding between adjacent chains in the fibrous proteins. The residues of those amino acids that are classed as either acidic or basic are capable of accepting or donating a proton and will at any given pH carry a charge of characteristic sign and intensity.

The presence of all these ionizable groups results in proteins showing not only acidic and basic features but also the characteristics of an electrolyte. Such substances are known as ampholytes. At low pH values the ionization of the basic groups will be dominant and the protein will carry a net positive charge, while at high pH values the ionization of the anionic groups will be most evident. At a pH peculiar to the molecule the anionic nature will exactly balance the cationic nature and the protein will carry no net charge. This is known as the *isoionic pH* of the protein. At the iso-ionic pH the repulsive effect of like-charges, which is the major stabilizing force of colloidal suspensions, is lacking and the protein will show minimal solubility and may precipitate.

Whether or not precipitation occurs will depend mainly upon the degree of hydration, a phenomenon in which molecules of water are bound to the polar exterior of the protein and act as a stabilizing factor in colloidal suspensions. The effective charge carried by a colloid is known as the zeta potential and is affected by not only the ionic nature of the amino acid residues but also external factors. In the presence of salts adsorption of ions occurs, resulting in a reduction in the charge carried by the colloid. At some point the concentration of salts will be such that the zeta potential is reduced to zero and the protein will again tend to precipitate, a process known as 'salting out'. This effect is enhanced by competition for the water molecules by these high concentrations of salts.

Many globular proteins have a further level of organization known as the **quaternary structure,** which describes the association of protein units to produce an aggregate protein with a definite functional property. The complex catalytic functions of iso-enzymes is dependent upon their quaternary structure. The bonds involved are usually non-covalent and mainly hydrophobic between non-polar regions on the surfaces of the molecules concerned. Haemoglobin, for instance, is composed of four polypeptide chains, normally in two identical pairs, forming the haemoglobin tetramer, which is more effective in oxygen transport than is the monomer. In addition to the four protein chains, haemoglobin also incorporates an iron porphyrin which facilitates the binding of oxygen. Proteins such as haemoglobin that involve non-protein components are known as conjugated proteins.

GENERAL METHODS OF QUANTITATION

The nature of the sample and the presence of any interfering substances are major considerations in the selection of a suitable method. Fluid samples are the most convenient to handle but some methods are appropriate for the analysis of solid material. The presence of interfering substances may necessitate an initial purification of the protein components. This may be achieved by precipitating the soluble proteins and, after washing, quantitating using a suitable method. The use of heat or strong acids results in irreversible denaturation and a method such as the Kjeldahl will usually be necessary. However, the precipitate produced by salts, alcohol, etc., can be redissolved in alkali and a method such as the Biuret reaction could subsequently be used.

Alternatively, the substances causing the interference, which are often small molecules such as amino acids and salts, may be removed by dialysis against a large volume of a suitable buffer. The sample may be contained in a sealed bag made from suitable Visking tubing or one of the commercial hollow fibre units may be used in which a long length of narrow bore dialysis tubing gives a very large surface area for rapid dialysis to take place.

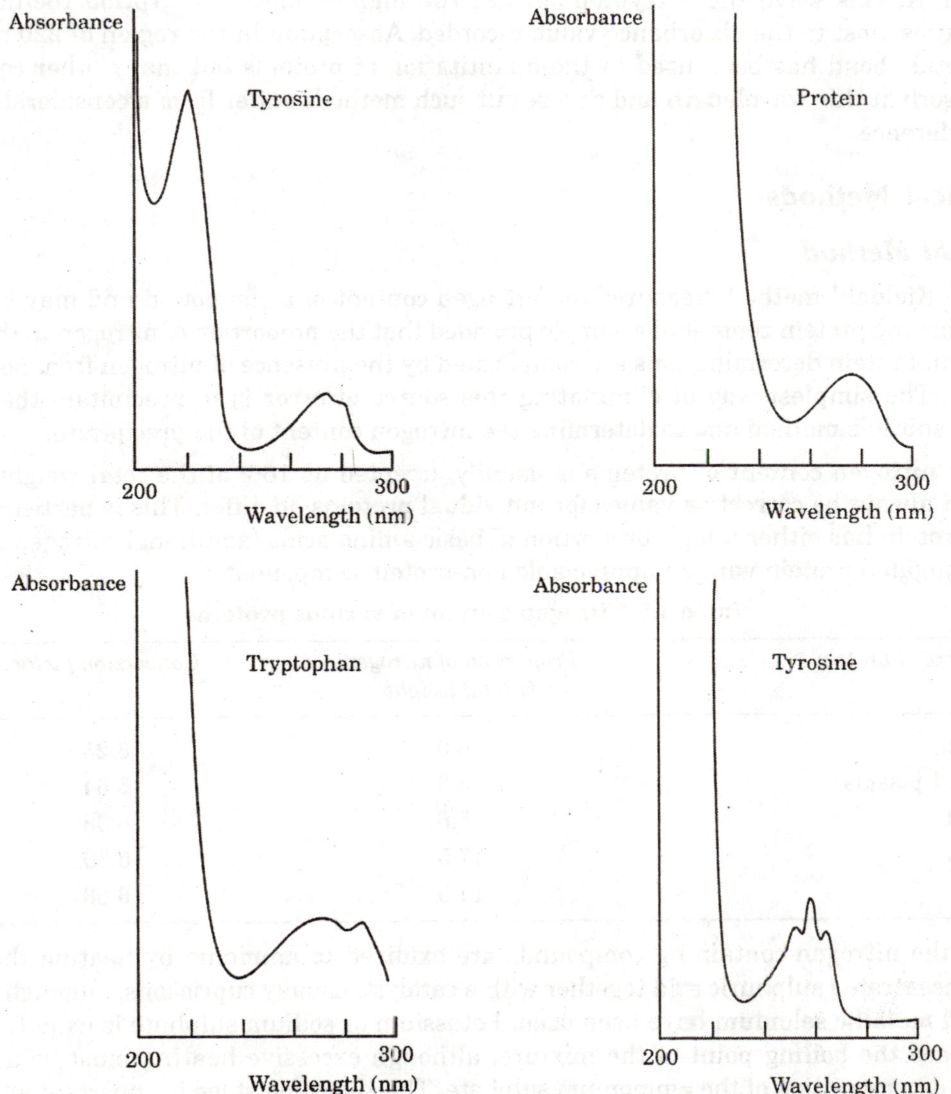

Fig. 3.7. Absorption spectra of amino acids and protein.

Spectroscopic Methods

The presence of aromatic amino acid residues results in proteins showing an absorption maximum at 280 nm and comparison of the absorbance value of a test solution with that of a standard solution provides a sensitive method of quantitation. This cannot be considered to be an absolute method of quantitation because variations in the amino acid content mean that values for absorption coefficients vary from one protein to another. Shows the absorption spectra of several amino acids and a protein and although the absorption maxima of the amino acids vary it is the cumulative effect that results in all proteins showing an absorption maximum at

280 nm. At this wavelength tryptophan has the highest molar absorption coefficient and contributes most to the absorbance value recorded. Absorption in the region of 220 nm due to the peptide bond has been used in the quantitation of proteins but many other compounds also absorb at this wavelength and as a result such methods suffer from a considerable degree of interference.

Chemical Methods

Kjeldahl Method

The Kjeldahl method measures the nitrogen content of a compound and may be used to determine the protein content of a sample provided that the proportion of nitrogen in the protein is known. Protein determinations are complicated by the presence of nitrogen from non-protein sources. The simplest way of eliminating this source of error is to precipitate the proteins using a suitable method and to determine the nitrogen content of the precipitate.

The nitrogen content of proteins is usually accepted as 16% of the total weight but this may not always be correct as values for individual proteins do differ. This is particularly true if the protein has either a high proportion of basic amino acids (additional nitrogen atoms) or is a conjugated protein with an appreciable non-protein component.

Table 3.1. Nitrogen content of various proteins

Source of protein	*Proportion of nitrogen % total weight*	*Conversion factor*
Meat	16.0	6.25
Blood plasma	15.3	6.54
Milk	15.6	6.38
Flour	17.5	5.70
Egg	14.9	6.68

All the nitrogen-containing compounds are oxidized to ammonia by heating the sample with concentrated sulphuric acid together with a catalyst, usually cupric ions, although mercuric ions and metallic selenium have been used. Potassium or sodium sulphate is usually included to increase the boiling point of the mixture, although excessive heating must be avoided to prevent decomposition of the ammonium sulphate. This digestion stage is important and usually takes several hours during which the sample initially turns brown or black and sulphur dibxide fumes are given off.

The solution eventually clears and heating should be continued for about a further 2 h. the cooled mixture is transferred to a steam distillation flask, and after making the mixture alkaline with excess sodium hydroxide the ammonia is distilled into a receiver flask. The ammonia is trapped in a boric acid solution and the ammonium ions are titrated directly with a standard hydrochloric acid solution. Although boric acid is the most convenient, it is possible to trap the ammonia in a known volume of standard hydrochloric acid and back-titrate the residual acid with a sodium hydroxide solution. The advantage of the latter method is that the detection of the end-point of the titration (a strong acid and a strong base) is easier than that for the ammonium borate and hydrochloric acid titration.

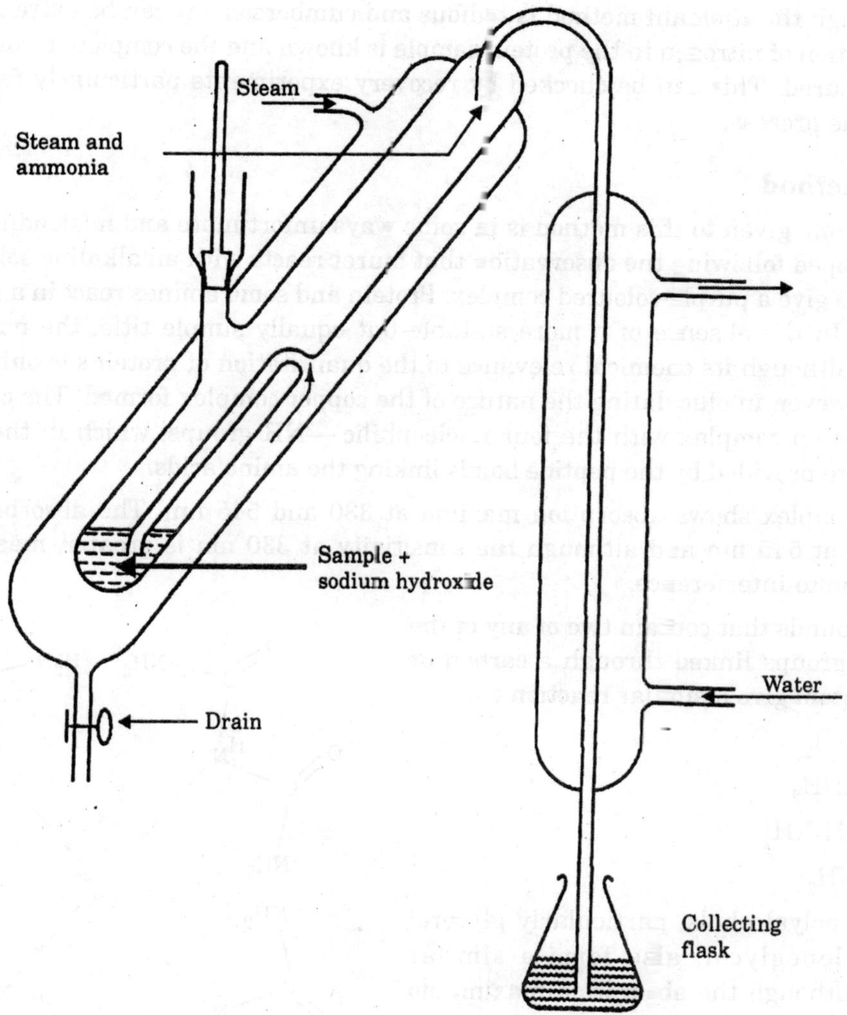

Fig. 3.8. Markham still.

The calculation is based on the fact that 1 atom of nitrogen will result in the formation of 1 mol of ammonia, which will subsequently require 1 mol of acid. Hence :

$$1.0 \text{ mol HCl} \equiv 14 \text{ g nitrogen}$$

If A ml of acid solution containing B mol l^{-1} is required to neutralize the ammonia formed from a given amount of sample, the amount of nitrogen present in the sample is :

$$\frac{14}{1000} \times A \times B \text{ grams}$$

and the amount of protein from which this nitrogen was derived:

$$\frac{14}{1000} \times A \times E \times \frac{100}{16} \text{ grams}$$

Although the Kjeldahl method is tedious and cumbersome it can be extremely accurate if the proportion of nitrogen in the protein sample is known and the complete recovery of nitrogen can be assured. This can be checked by recovery experiments particularly for the digestion stage of the process.

Biuret Method

The name given to this method is in some ways unfortunate and misleading. The method was developed following the observation that biuret reacts with an alkaline solution of copper sulphate to give a purple-coloured complex. Protein and some amines react in a similar manner to biuret. In the absence of a more suitable but equally simple title, the name biuret was retained, although its chemical relevance to the quantitation of proteins is only vague. It was useful, however, in elucidating the nature of the copper complex formed. The cupric ions form a coordination complex with the four nucleophilic —NH groups, which in the reaction with proteins are provided by the peptide bonds linking the amino acids.

The complex shows absorption maxima at 330 and 545 nm. The absorbance is usually measured at 545 nm and although the sensitivity at 330 nm is greater, measurements are more prone to interference.

Compounds that contain two of any of the following groups linked through a carbon or nitrogen atom give a similar reaction :

—$CONH_2$

—CH_2NH_2

—$C(NH)NH_2$

—$CSNH_2$

Some polyalcohols, particularly glycerol and ethyleneglycol, also form a similar complex although the absorption maxima do vary slightly from those of the protein-copper complex. The biuret reagent consists of an alkaline solution of copper sulphate with either sodium potassium tartrate or sodium citrate added to prevent the precipitation of cupric ions as the hydroxide. The method is extremely robust and reliable and obeys the Beer-Lambert relationship to a final protein concentration of about 2 g l^{-1} (a sample concentration of about 20 g l^{-1}) with a lower limit of about 100 μg, of protein (sample concentration of 1.0 g l^{-1}).

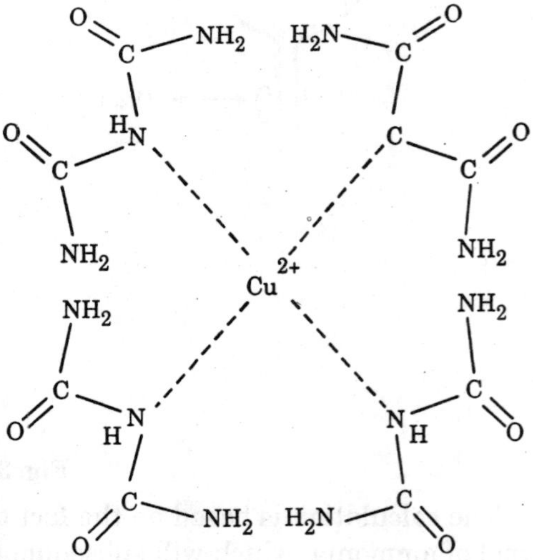

Fig. 3.9. Biuret reaction. The coordination complex formed in alkaline solution between cupric ions and the nucleophilic nitrogen atoms in four molecules of biuret.

The colour reaches maximum intensity in about 15 min and is stable for at least several hours. The method is simple and reliable and readily lends itself to automation but its lack of sensitivity is its greatest drawback. All proteins react in a similar manner and results show very little difference for different proteins.

Lowry Method

A reagent for the detection of phenolic groups known as the Folin and Ciocalteu reagent was used in the quantitation of proteins by Lowry (1951). In its simplest form the reagent detects tyrosine residues due to their phenolic nature but the sensitivity of the method was considerably improved by the incorporation of cupric ions. A copper-protein complex produced using a dilute version of the biuret reagent causes the reduction of the phosphotungstic and phosphomolybdic acids, the main constituents of the Folin and Ciocalteu reagent, to tungsten blue and molybdenum blue.

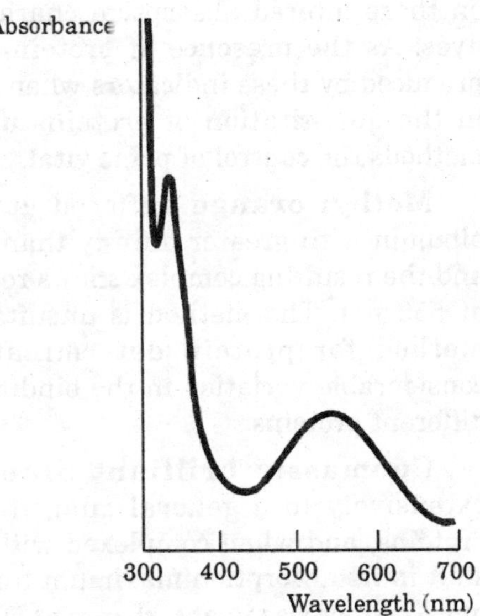

Fig. 3.10. Absorption spectrum of the protein-copper complex of the biuret reaction.

The precise composition of the blue reaction products is not known but they show broad absorption peaks in the red portion of the visible spectrum (600-800 nm). Approximately 75% of the reduction that occurs is due to the copper-protein complex, and tyrosine (and to a lesser extent, tryptophan) residues are responsible for the remainder. Folin and Ciocalteu's reagent is of complex composition and is normally purchased ready for use. It is prepared by the reflux heating of sodium tungstate and sodium molybdate with orthophosphoric acid. The reagent is normally pale yellow and of limited shelf life. The method is more sensitive than the/biuret method and has an analytical range from 10 μg to 1.0 mg of protein.

Using the method outlined below this is equivalent to sample concentrations of between 20 mg l^{-1} and 2.0 g l^{-1}. The relationship between absorbance and protein concentration deviates from a straight line and a calibration curve is necessary. The method is also subject to interference from simple ions, such as potassium and magnesium, as well as by various organic compounds, such as Tris buffer and EDTA (ethylenediamine-tetraacetic acid). Phenolic compounds present in the sample will also react and this may be of particular significance in the analysis of plant extracts.

Bicinchoninic Acid Method

This is a modification of the Lowry method involving a dye-binding step. The copper-protein complex that forms the basis of the biuret and the Lowry methods can be chelated by bicinchoninic acid to produce a very stable complex with a strong absorption maximum at 562 nm.

It is said to suffer from less interference effects than the Lowry method and is capable of detecting protein levels as low as 10 μg. The presence of lipids does interfere with the assay and modification of the technique is required if detergents are present. The reagent is only stable for a short time but if the copper sulphate is only added prior to use the stock reagents keep indefinitely.

Dye-binding Methods

Observations that the presence of protein affects the colour change of some indicators used in acid-base titrations led to the development of methods for the quantitation of proteins based on these altered absorption characteristics of such dyes. As the presence of protein alters the colour produced by these indicators when measuring pH, so in the quantitation of proteins using dye-binding methods the control of pH is vital.

Methyl orange buffered at pH 3.5 binds to albumin with greater affinity than to other proteins and the resulting complex shows reduced absorbance at 550 nm. The method is unsuitable as a general method for protein determination because of considerable variation in the binding of the dye with different proteins.

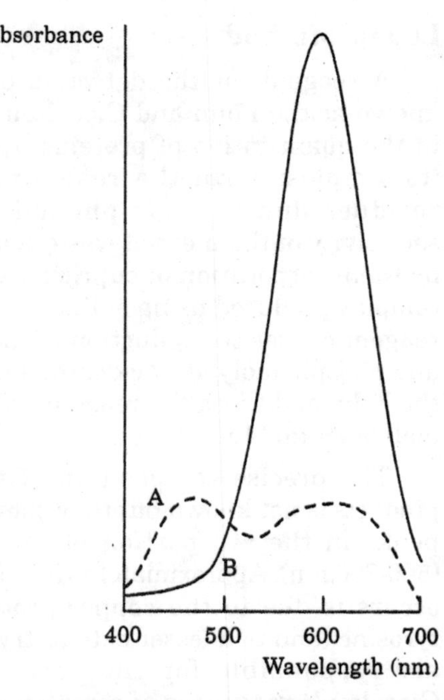

Fig. 3.11. Absorption spectrum of the protein—Coomassie brilliant blue G 250 complex: A, Coomassie brilliant blue only; B, protein—dye complex.

Coomassie brilliant blue has been used extensively in a general quantitative method for proteins, and when complexed with protein shows a shift in its absorption maximum from 464 nm to 595 nm. The increase in absorbance at 595 nm can be used as a measure of the protein concentration. The maximum absorbance is developed very rapidly (2-5 min) and is stable for at least an hour. The method is suitable for all proteins although the amount of dye bound does vary from one protein to another in a manner that seems to relate to the proportion of basic amino acid residues in the protein.

Bovine serum albumin, for instance, gives absorbance values that are 60% greater than the same concentration of egg albumin. Consequently it is important that the standard protein solutions used should be of the same composition as the test protein. The control of pH is important and although the reagent is heavily buffered, any samples that are very alkaline may alter the pH and so affect the result. Some detergents when present show significant interference resulting in increased absorbance values. The method is capable of detecting as little as 5 μg protein and a calibration curve is necessary because of the variations between different proteins and the non-linearity of the absorbance-concentration relationship.

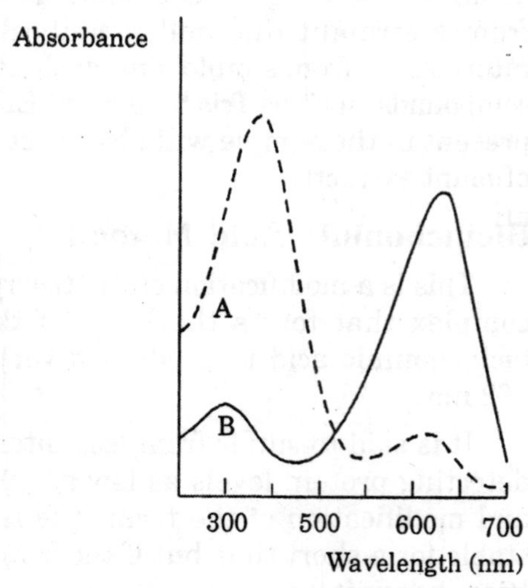

Fig. 3.12. Absorption spectrum of the albumin-bromcresol green complex : A, Bromcresol green only; B, albumin-dye complex.

Bromcresol green is frequently used for the quantitation of albumin, to which it binds selectively at pH 4.2 with a resulting increase in absorbance at 630 nm. The dye is initially a yellow colour but the resulting protein-dye complex is an intense blue colour. The method is relatively specific for albumin, bromcresol green being able to displace most substances that may be initially bound to the protein molecule

The method is sensitive, showing a lower limit of detection of about 50 μg protein and a linear relationship between absorbance and protein concentration up to about 0.2 g.

The use of bromcresol green for the measurement of albumin has been criticized on several counts. There is a tendency for the protein-dye complexes to precipitate at pH 4.2, which is very near the iso-ionic pH of albumin. It is claimed that the method is not absolutely specific for albumin and particularly with serum samples shows a positive bias in results. There is also some variability in the intensity of the colour produced with albumins from different sources, a fact which makes the choice of the standard material important.

Bromcresol purple has been suggested as showing improved specificity for albumin and less variation in colour intensity compared with bromcresol green. The dye-albumin complex shows an absorption maximum at 603 nm and the use of a reagent buffered at pH 5.2 appreciably reduces the tendency of the complex to precipitate.

Immunological Methods

Antibodies provide a very convenient and specific analytical tool for protein determinations, a role considerably enhanced by the availability of a wide range of monoclonal antibodies. Many antibodies are capable of precipitating the target protein and the resulting turbidity can be measured by either turbidometric or nephelometric methods. The various types of alternative immunoassay offer increased sensitivity over the turbidometric methods.

SEPARATION OF PROTEINS

Many of the methods described earlier do not differentiate between different proteins but it is often necessary to determine the amount of one particular protein in the presence of others. Although proteins are composed of amino acids, the problems involved in the separation of individual proteins are considerably increased compared with the separation of amino acids, and the large relative molecular mass means that some of the simpler separation techniques such as thin-layer chromatography are inappropriate. The fact that the tertiary and quaternary structure of a protein can be seriously and often permanently altered by even fairly mild conditions presents an additional problem.

Precipitation

High concentrations of a variety of salts including sulphates, sulphites and phosphates can be used to precipitate proteins but each fraction produced still consists of a mixture of proteins and usually requires further purification. Excessive denaturation of the protein is avoided by the use of low temperatures. Salt fractionation techniques prepare protein fractions by successively increased concentrations of the salt.

Sufficient salt is added to the sample to give the lowest selected concentration and the resulting precipitate is removed, usually by centrifugation or filtration. More salt is then added to the sample to increase its concentration to the next selected level and the precipitate is again removed. The process can be repeated at increasing salt concentrations and a series of precipitates obtained which can be redissolved in a suitable buffer. The salt can subsequently be removed from the protein prggaration by techniques such as dialysis or gel permeation chromatography. Various alcohols may also be used and the classical Cohn fractions of serum proteins are separated using specific concentrations of ethanol under carefully controlled conditions of temperature and pH.

Electrophoresis

Electrophoresis, in all its forms, has a major application in the separation of proteins because charge and molecular size are important to both electrophoretic separation and to protein structure. The conventional techniques which use a flat supporting medium, *e.g.* cellulose acetate strip or open gel plate, are supplemented and often replaced by capillary techniques, which offer additional analytical features.

In both techniques, the key factors in the separation of proteins are the choice of the operating pH and the electrophoretic conditions to be used. Capillary electrophoresis is an instrumental technique and is very dependent upon the availability of commercially produced equipment and reagents. The technical decisions are largely made by the manufacturer rather than the analyst. It is essential, however, that the analyst can identify the molecular feature of the analytical problem and then select an appropriate technique.

Technical Aspects

Various factors must be considered when selecting a supporting medium. Filter paper has significant disadvantages, the most serious being the adsorption of proteins. Modified cellulose media such as cellulose acetate show significantly less adsorptive effects which, together with a very uniform pore structure, result in a greatly improved resolution, although the quality of cellulose acetate membrane does vary appreciably from one manufacturer to another. The various other types or supporting media such as starch and poly-acrylamide gels show improved resolution due to a molecular sieving effect. Iso-electric focusing techniques probably give the best resolution and many of the resulting bands are due to specific proteins.

They are used mainly as a qualitative or a semi-quantitative technique due primarily to the large number of bands that develop and are of particular value when successive samples from the same source need to be compared for the presence or absence of a particular protein or for investigation of physical properties, *e.g.* pI. It is important to realize that the use of these different media for the same sample will result in separation patterns that cannot be easily compared with one another. The separation of serum proteins on cellulose acetate will result in 5–7 bands, while the use of polyacrylamide gel will give 17 bands. The pH of the buffer used affects the charge carried by the protein and although in theory any pH may be used, in practice pH values greater than the iso-electric pH of the protein (resulting in the protein being negatively charged) give better separations than other pH values.

In selecting the conditions for the separation of a particular protein mixture, a buffer pH that gives the greatest difference in the charge carried by each individual protein results in the greatest difference in velocities and hence in the final distances moved. In practice, buffers of pH 8.6 are most frequently used. In addition to its pH, the concentration of a buffer also affects the mobility of proteins. At high concentrations the zeta potential of the protein is reduced resulting in a shorter distance of migration. However, because higher concentrations of buffer give improved resolution, a compromise concentration has to be found and buffers with ionic strength (μ) varying from 0.025 to 0.075 are frequently used. The various types of capillary electrophoresis are performed either in free solution or in gels.

The choice of method depends on the nature of the sample and the analytical objective but capillary gel electrophoresis, including iso-electric focusing and SDS electrophoresis, is particularly useful for protein applications. A major problem with the use of fused silica capillaries is the adsorptive effect with proteins, which results in band broadening during the separation process. A number of techniques are available to minimize this effect, ranging from the use of extreme pHs to reduce the charge on the capillary, the use of capillaries coated with various hydrophilic groups to mask the ionized Si-OH groups in the silica and the addition of a range of substances to the buffer which are designed to compete effectively with the protein for the ionic sites.

Quantitative Aspects

In conventional electrophoresis, which uses a solid supporting medium, the sample is applied as a streak. The zones or bands of protein that develop during electrophoresis can be precipitated in the pores of the supporting medium by trichloroacetic acid and stained using a suitable dye (Ponceau S, nigrosin, etc.). The amount of dye bound is often directly correlated to the amount of protein but this quantitative relationship is open to criticism for reasons discussed under dye-binding methods. However, it provides a convenient semi-quantitative method with the various fractions usually being expressed as a percentage of the total rather than in absolute amounts.

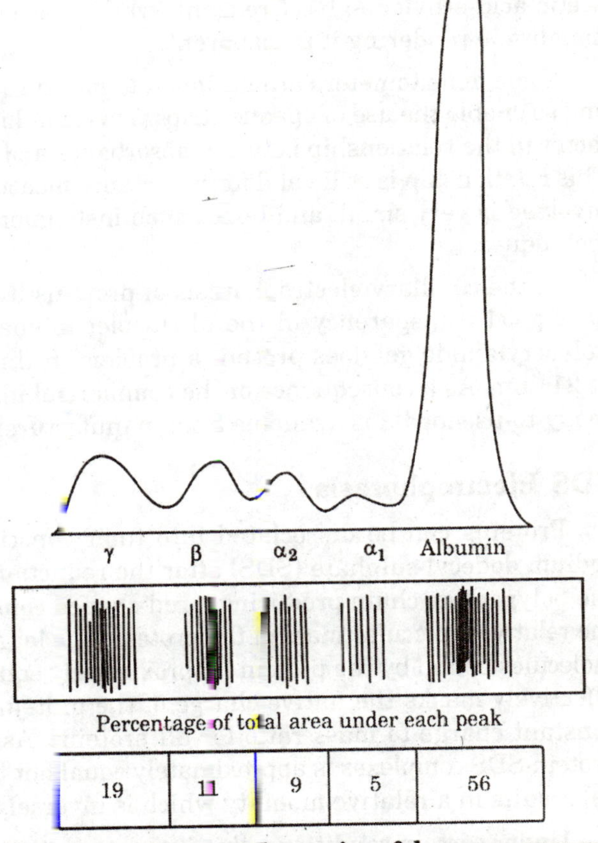

Fig. 3.13. Electrophoresis of human serum proteins. The electrophoretogram, after staining with a suitable dye, can be scanned using a densitometer, which gives a trace of the absorbance pattern of the strip.

If the total protein content of the original sample is determined using one of the general methods described earlier it is possible to calculate the amount of protein in each fraction. The amount of dye bound by each fraction can be simply determined by cutting out the stained bands and eluting the dye into a fixed volume of a suitable solvent.

The absorbance of each solution is measured and the sum. of the absorbance values assumed to be proportional to the total amount of protein. Hence the amount of protein in each fraction can be calculated as a percentage of the total. An alternative and more satisfactory method of quantitation is to scan the stained electrophoretic strip using a densitometer, which is a modified photometer in which the electrophoretogram replaces the usual glass cuvette.

The strip is slowly moved across the light path and the signal from the photoelectric detector is plotted by a pen recorder, the chart speed of which is synchronized with the movement of the strip. The result is a trace that plots the absorbance value against the distance along the electrophoretogram and the area under the trace is proportional to the total protein content. From the area under each peak, the proportion of protein associated with that peak can be calculated. A very narrow ligh path must be used in order to ensure that closely adjacent bands are resolved. If the instrument measures transmitted light it is necessary to make the supporting membrane translucent. This can be done by either impregnating the strip with an oil with a high refractive index or, for some types of cellulose acetate material, using an ethanol-acetic acid-ethylene-glycol reagent, which causes the collapse of the porous structure of the membrane rendering it transparent.

Some densitometers are designed to measure reflected light rather than transmitted light and so enable the use of opaque strips. This simplifies the technique but does introduce a further factor in the relationship between absorbance and the amount of protein present in the sample. This relationship is still valid for reflectance measurements provided that the amount of protein involved is very small, and hence such instruments are usually designed for micro-analytical techniques.

In the capillary electrophoresis of proteins it is essential that the buffer and the medium have good transparency in the ultraviolet to enable effective detection after the separation. Polyacrylamide gel does present a problem in this respect, interfering with protein detection at 214 nm. As a consequence of the commercial impact on capillary electrophoresis, a range of ready-to-use media is available from manufacturers.

SDS Electrophoresis

Proteins can be dissociated into their constituent polypeptide chains by the detergent sodium dodecyl sulphate (SDS) after the reduction of any disulphide bonds. The SDS binds to the polypeptide chain producing a rod-shaped complex, the length of which is dependent upon the relative molecular mass of the protein. The large number of these strongly anionic detergent molecules bound by the protein (approximately equal to half the number of amino acid residues) effectively masks the native charge of the protein and at a neutral pH results in a relatively constant charge to mass ratio for all proteins. As a result, the electrophoretic mobility of all protein-SDS complexes is approximately equal but the molecular sieving effect of polyacrylamide gel results in a relative mobility which is inversely related to the size of the complex.

Under certain conditions, this inverse relationship can be demonstrated by a linear plot of the relative mobility of the protein against the logarithm of its relative molecular mass. It is necessary to use a series of known proteins in order to produce a calibration curve and kits are available commercially for this purpose. Prior to electrophoresis the sample is diluted in buffer containing SDS (10-25 g l^{-1}) and β-mercaptoethanol (10-50 ml l^{-1}), which reduces any disulphide bonds stabilizing the protein. It is then heated at 100 °C for 2–5 min in order to denature the protein and expose the total length of the polypeptide chain to the detergent.

After cooling, electrophoresis is performed on polyacrylamide gel and the bands subsequently visualized using an appropriate dye.

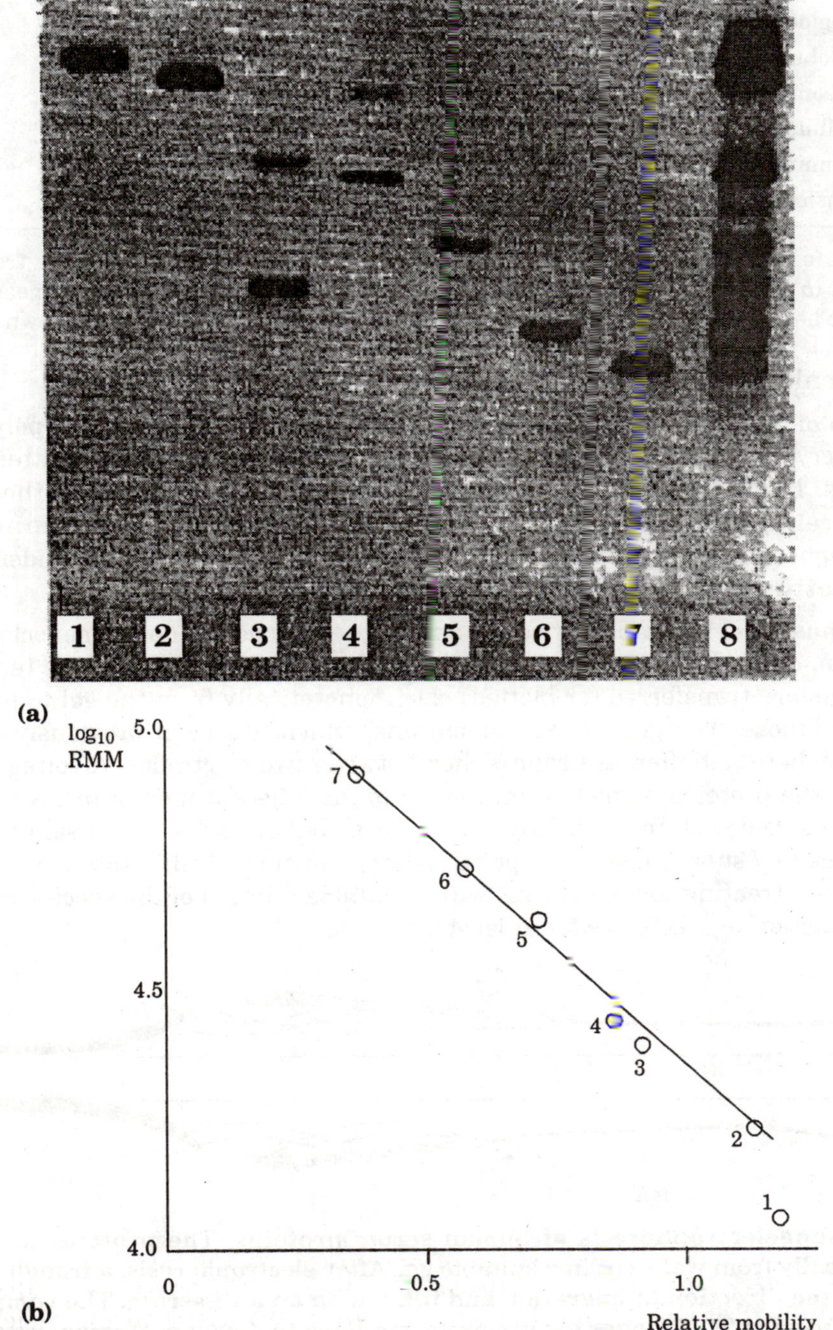

Fig. 3.14. Determination of the relative molecular mass (RMM) of a protein by SDS electrophoresis.

Number	*Protein*	*RMM*	log_{10} *RMM*
1.	Cytochrome *c* (muscle)	11 700	4.068
2.	Myoglobin (equine skeletal muscle)	17 200	4.236
3.	γ–Globulin (L-chain)	23 500	4.371
4.	Carbonic anhydrase (bovine)	29 000	4.462
5.	Ovalbumin	43 000	4.634
6.	Albumin (human)	68 000	4.832
7.	Transferrin (human)	77 000	4.886

The photograph (*a*) shows these proteins separated on a 5% polyacrylamide gel after treatment with 0.1% SDS. A plot (*b*) of $\log_{10}$ RMM against the relative mobility of each protein shows a linear relationship and provides the basis for the determination of the relative molecular mass of an unknown protein.

Immunological Methods

Separation of a mixture of proteins by electrophoretic techniques such as polyacrylamide gel, SDS polyacrylamide or iso-electric focusing usually results in a complex pattern of protein bands or zones. Interpretation of the results often involves a comparison of the patterns of test and reference mixtures and identification of an individual protein, even using immunoelectrophoresis is very difficult. However, specific proteins can often be identified using an **immunoblotting** technique known as Western blotting.

The prerequisite is the availability of an antibody, either polyclonal or monoclonal, against the test protein. After the initial separation by a conventional electrophoretic technique in a gel, the proteins are transferred (or blotted) electrophoretically from the gel to a membrane, usually nitrocellulose. The gel and the membrane, which has been previously soaked in a suitable electrophoretic buffer, are sandwiched between two electrodes. A voltage is applied, *e.g.* 100 V, and the proteins migrate from the gel to the adjacent membrane. After about 1 h the membrane is removed and carefully washed with buffer and a dilute solution of bovine albumin to block any subsequent, non-specific adsorption of antibodies to the membrane. The next step involves treating the membrane with a suitable dilution of the specific antibody and allowing the reaction to take place for at least 1 h.

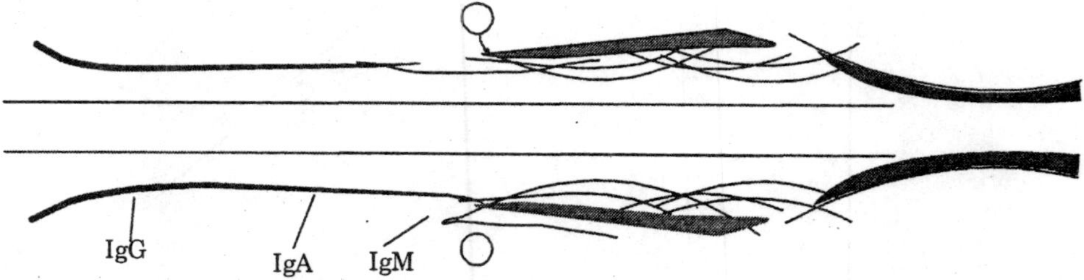

Fig. 3.15. Immunoelectrophoresis of human serum proteins. The proteins are separated electrophoretically from wells cut in a suitable gel. After electrophoresis, a trough is cut in the gel parallel to the direction of migration and filled with an anti-serum. The components are allowed to diffuse for 24-48 hours for precipitation lines to develop. Human serum contains many proteins, among which the immunoglobulins can be identified.

Excess antibodies are then washed from the membrane and the bound anti-body which remains is detected using a second antibody against the first, *e.g.* anti-rabbit immunoglobulin. This second antibody has been previously labelled with, for instance, the enzyme horseradish peroxidase but other labels such as conjugated gold or an isotope such as ^{125}I may be used. The bands can then be visualized using an appropriate method.

The resulting pattern of zones is then compared with the electrophoretogram prior to immunoblotting and the specific proteins pinpointed for any further investigation or separation. The technique may also be done in a single stage if the initial antibody is labelled but this is less convenient and often more expensive than having a single labelled antibody preparation that can be used to detect all antibodies from a single species of donor.

Immunoaffinity Purification

Proteins are frequently powerful immunogens and the availability of specific antibodies, particularly monoclonal antibodies, makes the technique of affinity chromatography very useful in the separation and purification of individual proteins. The technique has been used to purify a wide range of proteins such as hormones, membrane receptors and complement proteins. However, it is not restricted to proteins and is potentially applicable to any immunogenic substance.

The availability of suitable antibodies is essential and these may be raiea by whole animal polyclonal techniques or by monoclonal cell culture. The former antibodies may need some prior purification before being mmobilized. The most frequently used immobilization technique involves the use of cyanogen bromide-activated sepharose media.

The sepharose gel is activated by treatment with a cyanogen bromide solution for several minutes. It is then washed with ice-cold distilled water before being mixed with a dilution of the antibody to effect linking. After being allowed to react overnight, the gel is washed with Tris-HCl buffer to neutralize any remaining active groups. The column is packed with the prepared gel and equilibrated with a suitable buffer, usually with a pH in the region of 8, before the sample is slowly passed through it.

During this stage the test substance is bound by the immobilized antibody and held in the column. Large volumes of sample, typically about five times the column volume, can normally be accommodated. The column is again washed with the buffer and finally the bound antigen is eluted. There is no general rule to enable the selection of a suitable eluting solution but buffers with relative extremes of pH, often pH 2.5, are frequently used. Alternatively, high concentrations of various solutes may be employed, *e.g.* urea, guanidinium salts and SDS. The antigen is displaced quickly from the column medium and collected in as small a volume as possible. After use, the medium can be washed and regenerated for further use.

Chromatographic Methods

Various chromatographic techniques may be applied to the study of protein mixtures. Column techniques have the advantage that the resulting fractions are amenable to quantitation using the general method described earlier. Gel permeation chromatography is frequently used to separate protein mixtures but it is necessary to have some prior knowledge regarding the size of the proteins present in order to select the most suitable gel. Ion-exchange chromatography using the substituted cellulose ion-exchangers, diethylaminoethyl cellulose (DEAE) and carboxymethyl cellulose (CM), is frequently used but as with gel permeation chromatography the major applications are in the preparative aspects of protein analysis.

Affinity chromatographic techniques, including those that employ antibodies as ligands, permit highly specific separation of proteins. Reverse-phase HPLC can be used for the separation of peptides and proteins. Smaller peptides (less than 50 amino acid residues) may be satisfactorily separated on octadecylsilane (C-18) bonded phases whereas for adequate recovery of larger molecules, tetrylsilane (C-4) or octylsilane (C-8) is recommended.

Porous column packing with gel permeation and reverse phase properties is usually required for proteins with relative molecular masses greater than 50000. Hydrophobic interaction chromatography (HIC) is a variation of reverse-phase chromatography which is particularly useful for protein separations. Instead of the strongly hydrophobic stationary phases normally used such as octadecylsilane (C-18), smaller and less hydrophobic phases, such as methyl, butyl and phenyl groups, are used.

The mobile phases are aqueous and hydrophobic interactions between the proteins and the stationary phase are increased in high salt concentrations in the region of 1-2 mol l^{-1}. Typical separations involve equilibrating the column and applying the sample in a buffer with a high salt concentration and then sequentially eluting the proteins with a decreasing gradient of salt concentration. As with most separation techniques involving proteins, the pH and temperature of the eluting buffer are important.

4 SINGLE-CELL PROTEIN

The term single-cell protein (SCP) is used to describe protein derived from cells of micro-organisms such as yeast, fungi, algae and bacteria which are grown on various carbon sources for synthesis. The dried cells of microorganisms or the whole organism is harvested and consumed. This is a protein source for human food supplements and animal feeds. SCP production may have potential for feeding the ever-increasing world population. Massive quantities of SCP can be produced in a single day. As a source of protein it is very promising, with potential to satisfy the world shortage of food while population increases. There are several carbon sources that are used as energy sources for microorganisms for growing and producing CSP. In some cases, raw material requires pretreatment or hydrolysis before use.

Waste sources of carbon are customary and cheap to use. SCP technology is a suitable process for converting waste materials to useful biomass containing protein. The broth is concentrated protein which is then dried with limited moisture; it is stored for use as food or feed for humans and animals. Waste recycling has been advanced as a method for preventing environmental decay and increasing food supplies. It may be possible to convert waste streams into valuable products by separation and recovery or by biological conversion. Many products are produced biologically from food process waste. The potential benefits of successful recycling of agricultural wastes are enormous. It may be the only method for large-scale protein production that does not require a concomitant increase in energy consumption.

In addition, it may be the most effective method for producing animal and human food from lignocellulose materials that are otherwise of little nutritive value and are therefore used for fuel production. One advantage of the biological process is the flexibility of the microorganisms to adapt to different feedstocks. Therefore, when combined with the treatment of process waste streams, the biological conversion of these wastes to products can be both environmentally and economically favourable. In the production of antibiotics, sufficient growth of fungi in submerged cultures has created potential sources of biomass as SCP and as flavour additives to replace mushrooms : the biomass contains 50-65% protein. Production of mushroom from lignocellulosic waste seems to be a suitable and economical process since the raw material is inexpensive and available in most countries.

SEPARATION OF MICROBIAL BIOMASS

Bacteria, yeast and algae are produced in massive quantities of protein sources as food for animals and humans. SCP is considered a major source of feed for animals. The production of valuable biological products from industrial and agricultural wastes is considered through the bioconversion of solid wastes to added-value fermented product, which is easily marketable as animal feedstock. The waste streams that otherwise would cause pollution and threaten the environment can be considered raw material for CSP production using suitable strains of microorganisms.

BACKGROUND

It is evident that the conversion of photosynthetically produced organic compounds into human and animal food is the limiting process in human food production. The worldwide annual production of organic material by photosynthesis has been estimated to be between 25 and 50 tons. Any practical method capable of converting a small fraction of this yield into human food should find wide application and go a long way to reducing chronic food shortages. The growth of microorganisms, more rapid than that of the higher plants, makes them very attractive as high-protein crops; whereas only one or two grain crops can be grown per year, a crop of yeasts or moulds may be harvested weekly, and bacteria may be harvested daily.

The use of microorganisms as a source of protein for human and animal food is not a new development. Traditional foods and feeds such as cheese, sauerkraut, miso and silage have a high content of microorganisms to which their nutritional properties are due in part. The high-quality proteins synthesised during the growth of these microorganisms compare favourably with those derived from the better grains. There are many convincing reports on the availability of essential amino acids and the protein quality of SCP. Although there are few data on animal feeding trials using SCP produced from lignocellulose wastes, there is a large and growing body of information about SCP from petroleum and methanol. This information should be applicable to SCP from agricultural wastes with proper allowance for the undesirable contaminants in the sources.

In petroleum, there has been concern about accumulation of carcinogenic hydrocarbons. In agricultural wastes, there is concern about accumulation of pesticides and herbicides. The guidelines for testing SCP as a major supplement in animal diets and should be consulted for further details on feeding trials. Many more feeding trials will be needed before SCP from lignocellulose wastes is accepted for routine feeding. The main carbon source for production of SCP is petroleum. It has been practised in many companies around the world. Other potential substrates for SCP include bagasse, citrus wastes, sulphite waste liquor from pulp and paper, molasses, animal manure, whey, starch, sewage and agricultural wastes.

PRODUCTION METHODS

Some of the propesed methods for conversion of agricultural wastes into animal feed are presented. These methods will be briefly evaluated in the following. First, a distinction should be made between the production of SCP from the lignocellulose parts of the plant and the production of SCP from the soluble carbohydrates of many agricultural wastes. At present, there are many plants around the world that operate for the production of SCP. The Ceres Ecology Corporation of Chino, California, in the USA will process waste from over 100,000 dairy cattle for feed recycling and for control of salt in ground-water. Other plants in Toulouse (France), Zacantecos (Mexico) and Sterling (Colorado, USA) use processes that depend upon an anaerobic fermentation in a silo or covered ditch.

The manure undergoes a lactic acid fermentation due to the action of anaerobic bacteria (chiefly streptococci and lactobacilli), and a typical silage odour results in place of the odour of manure. The processing of poultry for production of prepared food generates waste containing fats and starchy materials. The waste is used in a biological process to be converted to SCP using several microorganisms. These short-time anaerobic fermentation processes do not utilise the fibre, and are therefore a partial solution to the waste problem. The fibrous residue may be used as a soil conditioner before or after composting.

Table 4.1. Methods for conversion of cellulosic agricultural wastes into animal feed

Treatment	*Microorganism*	*Substrate*	*Protein produced*	*Fibre utilized*
Dilute alkali	None	Straw	No	Yes
Aerobic mesophiles, 25 °C	*Cellulmonas*	Bagasse	Yes	Yes
Mould growth, 25 °C	*Trichoderma viride*	Waste paper	Yes	Yes
Aerobic thermophiles 55°C	*Thermoactinomyces*	Fermented livestock wastes	Yes	Yes

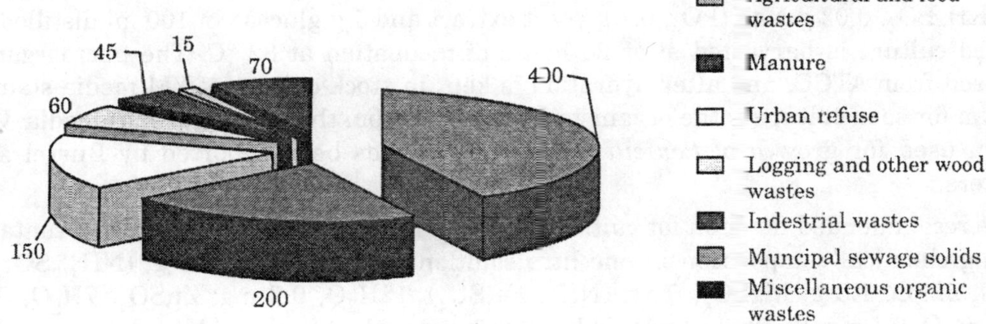

Fig. 4.1. Solid wastes in units of tons per year.

Rates of soluble sugar utilisation of 10-30 grams per litre per hour have been reported for SCP production by yeast. Rates of 5-15 grams per litre per hour have been claimed for utilisation of selected hydrocarbons. For the process of SCP production under present market conditions to be an economical, the rate of utilisation of cellulose must be at least 1-5 grams per litre per hour. As no pilot plants have been operated, it is not possible to report commercial rates, but laboratory-scale fermenters have been run at 1 gram per litre hour on pretreated wastes.

It has been reported that Gram-negative aerobic bacteria, *cellulomonas* sp., grow rapidly on cellulose at 25-30 °C, but cannot utilise lignin or ligno-cellulose. Therefore extensive pretreatment of lignocellulosic material such as rice husks with hot alkali is required. The microorganism must be harvested by centrifugation. Amino-acid analysis and animal feeding trials have shown that a high-quality SCP can be produced. Enzymes produced by the mould *Trichoderma viride* are used for production of soluble sugars from waste paper cellulose. Yeasts or bacteria for SCP production can then ferment these sugars. The enzyme reaction takes place in four steps :

1. Pretreatment of waste by ball milling or hot alkali.
2. Enzyme production by growth of *T. viride* on pretreated cellulose.
3. Depolymerization of cellulose by *T. viride* enzymes.
4. SCP production by yeast or bacteria.

Several potential and mutant strains of *T. viride* have been identified in SCP production. Their capacity for amyloletic enzyme production was enhanced severalfold in SCP from lingnocellulosic resources. The process of bioconversion of agricultural wastes to SCP appeared to be too complex to find an economic application for agricultural waste.

MEDIA PREPARATION FOR SCP PRODUCTION

Sago starch in Malaysia is abundant, inexpensive and common as raw material for SCP production. Fifty grams of sago starch is dissolved in one litre of 0.1 M NaOH solution. The mixture is heated treated until it is absolutely dissolved in deionised water with 4 g of NaOH, then the pH is adjusted to 7. Supplementary nutrients are added: 3.3g KH_2PO_4, 0.3 g Na_2HPO_4 and 1 g yeast extract; autoclave the media, and use it as feed for a fermenter. One hundred millilitres of seed culture are prepared a day in advance for inoculation of fermentation.

Saccharomycopsis fibuligera ATCC 9947 or ATCC 9266 is grown in a media comprising of 0.33 g KH_2PO_4, 0.03 g Na_2HPO_4, 0.1 g yeast extract and 1 g glucose in 100 ml distilled water. The seed culture is harvested after 24 hours of incubation at 32 °C. The microorganism is purchased from ATCC, and after hydration is kept in stock culture of YM media slants. The inoculum for seed culture is the organism transferred from the prepared slant media. Culture medium used for growth of *Penicillium javanicum* has been reported by Burrel and his coworkers.

The recommended medium for cultivation of the fungi without any alteration contains the following chemical composition in one litre solution: Fumaric acid, 2.0 g, $(NH_4)_2SO_4$, 2.5 g; $KH_2PO_4.2H_2O$, 1.0 g; $MgSO_4$, 0.5g; (NH_4) $Fe(SO_4)_2.12H_2O$, 0.2 mg; $ZnSO_4$. $7H_2O$, 0.2 mg; $MnSO_4$-H_2O, 0.1 mg; thiamine hydro-chloride 0.1 mg, and add a suitable carbon source.

ANALYTICAL METHODS

Protein concentration can be determined by using method of Bradford, which utilises Pierce reagent 23200 (Pierce Chemical Company, Rockford, IL, USA) in combination with an acidic Coomassie Brilliant Blue G-20 solution to absorb at 595 nm when reagent binds to the protein. A 20mg/l bovine serum albumin (Pierce Chemical) solution was used as the standard. Starch concentration was measured by the orcinol method using synthetic starch as the reference. A yellow to orange colour is obtained and measured at 420 nm when orcinol reacts with carbohydrates. Absorbance is determined by spectrometry.

Coomassie-Protein Reaction Scheme

This protein assay works by forming a complex between the protein and the Coomassie dye. When bound to the protein, the absorbance of the dye shifts from a wavelength of 465 nm to 595 nm (λ_{595}). The reagent generates a stronger blue colour which is detected at the specified wavelength. You will first generate a standard curve using the protein bovine serum albumin (BSA) by measuring the absorbance at 595 nm of a series of standards of known concentration. Next, you will measure the absorbance at wavelength of λ_{595} for all of your samples and determine its concentration by comparison with the standard curve.

Protein + Coomassie G-250 in acidic medium → protein-dye complex (blue; measured at 595 nm)

Preparation of Diluted BSA Standards

Prepare a fresh set of protein standards by diluting the 2.0 mg per ml BSA stock standard (stock solution). There will be sufficient volume for three replications of each diluted BSA standard if necessary.

Table 4.2. Preparation of BSA concentration for standard calibration curve

Volume of BSA to add	*Volume of diluents (buffer) to add*	*Final BSA concentration*
300 μL of Stock	0 μL	**Stock** – 2000 μg/mL
375 μL of Stock	125 μL	**A** – 1500 μg/mL
325 μL of Stock	325 μL	**B** – 1000 μg/mL
175 μL of A	175 μL	**C** – 750 μg/mL
325 μL of B	325 μL	**D** – 500 μg/mL
325 μL of D	325 μL	**E** – 250 μg/mL
325 μL of E	325 μL	**F** – 125 μg/mL
100 μL of F	400 μL	**G** – 25 μg/mL

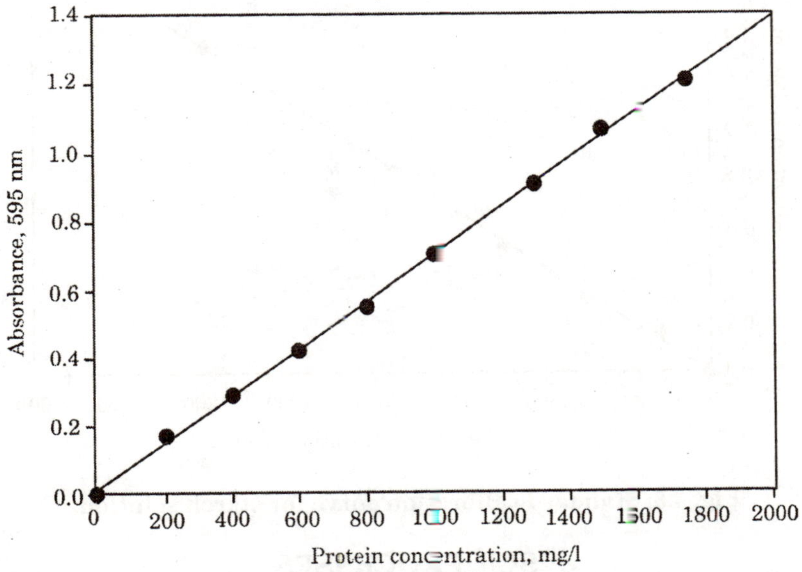

Fig. 4.2. Calibration curve for BSA standard solution.

Mixing of the Coomassie Plus Protein Assay Reagent

Allow the Coomassie Plus reagent to come to room temperature. Mix the Coomassie Plus reagent solution just before use by gently inverting the bottle several times. Do not shake.

Standard Calibration Curve

Prepare a standard curve by plotting the average blank corrected 595 nm reading for each BSA standard versus its concentration in mg/l, using the standard curve; determine the protein concentration for each unknown sample. Calibration curve for BSA standard is prepared using standard albumin, 50ml Pierce 23210 with concentration of 2g/l, diluted with 1M NaOH solution. Add 1 ml of 1M NaOH with 0.1 ml of diluted sample plus 5 ml of reagent, protein assay 23200 Pierce, stirred with a vortex mixer. Read the absorbance with a spectrophotometer at 595 nm.

Standard Calibration Curve for Starch

A standard solution of soluble strach, 2 g/l well dissolved in an alkali solution was prepared. With heating, the powder becomes a clear solution. A diluted solution from 200 to 2000 mg/l was prepared. Dissolved 0.4 g of orcinol (3, 5 dihydroxy toluene) in 99.6 g of H_2SO_4 (66% acid). Prepare 500 ml orcinol reagent, 170 ml water with 330 ml acid. Add acid to distilled water and gradually dissolve 3.115 g orcinol in the diluted acid solution. Take 0.1 ml of the diluted starch sample with 0.9 ml of distilled water, and then add 2 ml of the prepared orcinol reagent, heated in a boiling water bath for 15 min, stop the reaction by cooling it in an ice water bath. Add 7 ml of distilled water, read the absorbance at 420 nm.

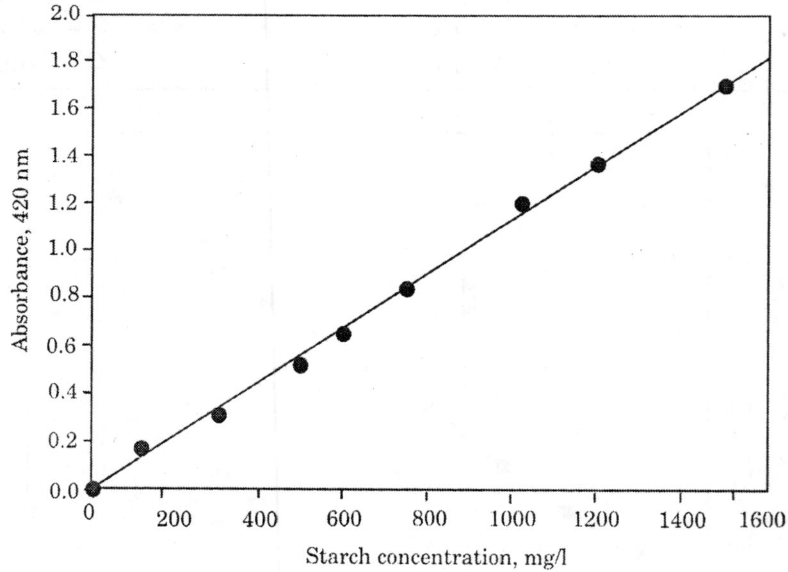

Fig. 4.3. Standard calibration curve for starch solution.

SCP PROCESSES

Using Brewer's yeast, *Saccharomyces cerevisiae,* the process was developed for the large-scale production of food. During World War I, a large-scale process for production of SCP was developed in Germany. About 60% of food was replaced by massive brewing of yeast, *S. cerevisiae.* In World War II, the yeast-based food had an important contribution in the German diet. The yeast was incorporated mainly into soup and sausages. Special strains of yeast, *Candida arborea* and *C. utilis,* were predominantly used. In the 1960s, several large-scale plants for the production of SCP were developed by British Petroleum (BP), as the organisms were able to utilise the aliphatic compounds of petroleum. *Candida lipolytica* was able to convert carbon sources originating from petroleum to protein. *C. lipolytica* was grown on alkanes, a yeast based food.

Other strains and *Candia* also are used for CSP production. For the CSP produced by BP from distilled *n*-alkanes, the cells were separated, salted, dried and used as animal feed. In the process developed by BF, the protein was named 'Toprina'. BP planned to go for large-scale production, and the protein was proved toxicologically safe. The rising price of petroleum initiated the plan, but it was unable to contribute to reducing shortages of food and feed sources for humans and animals.

Production of SCP named 'Pruteen' from methanol by bio-oxidation of methane was another a successful case in CSP process development using *Methylophilus methylotrophus.* However, the whole project of SCP has been a victim of political or economic issues in Europe and Japan. Methane is an abundant and cheap carbon source without any toxicity; it is the main constituent of natural gas and is also produced from anaerobic digestion tanks. There are many biocatalysts involved in methane bio-oxidation using stable mixed cultures. This mixed culture is one of the best examples of symbiosis. At first, methane in the gas phase is bubbled into media then biologically oxidised to methanol. The Gram-negative bacilli (rod shapes) are used, while methanol is produced in the presence of biocatalysts.

Methane is a pure carbon source; it is easily utilised by several microorganisms such as *Methylomonas* and *Methylococcus.* Another species that has been extensively studied is *Methylomonas methanica.* Supplements of nitrogen, trace metals and minerals are required for optimal growth. Mixed cultures of Gram-negative rod-shaped bacilli have the potential to oxidise methane and produce methanol, whereas other organisms present utilise methanol as a carbon source to produce SCP. This is an example of symbiosis in mixed culture, where an intermediate product is formed, which is then used by the second organism. *Acinetobacter* and *Flavobacterium* are often used in mixed culture for producting SCP from methane.

The fermentation of methanol to SCP was done in an airlift fermenter with sufficient aeration and without any mechanical agitation, using *Methylophilus methylotropha.* The SCP known as 'Pruteen' contained 72% crude protein. The product was marketed for feed as a source of energy, vitamins and minerals with sufficient protein content. The amino-acid analysis was satisfactory as the methionine and lysine content of Pruteen were compared to the protein content of white fish meal.

NUTRITIONAL VALUE OF SCP

The nutritional value of SCP depends on the composition of its amino acids, vitamins and nucleic acids. The nutrient value of SCP may have a positive and negative impact. The rigid cell wall, the high content of nucleic acids, allergies and the gastrointestinal effect should be considered as a negative impact. However, with special treatment, it is possible to eliminate these from the product. A long-term use of SCP is required to consider and remove any toxicological effects and carcinogenesis. The positive point of view for SCP is the high content of protein with sufficient enzymes, minerals and vitamins. The protein content of SCP is very high. Dried cells of *Pseudomonas* sp. grown on normal petroleum-based liquid paraffin contain 69% protein.

Algae normally possess about 40% protein. The protein content of SCP is absolutely dependent on the raw material used as a carbon source and the microorganisms grown on the media. The proteins of the microorganisms contain all the essential amino acids. Presents the average protein contents of bacteria, yeast, fungi and algae. Microorganisms such yeast and bacteria have a short doubling time, normally in the range of 5-15 minutes; mould and algae are 2-4 hours.

The fast brewing of microorganisms compared with plant cells is a promising point for food replacement and shortage in the new millennium. In terms of amino acids bacterial protein is similar to fish protein. The yeast's protein is almost identical to soya protein; fungal protein is lower than yeast protein. In addition, SCP is deficient in amino acids with a sulphur bridge, such as cystine, cysteine and methionine. SCP as a food may require supplements of cysteine and methionine; whereas they have high levels of lysine vitamins and other amino acids. The vitamins of microorganism are primarily of the B type. Vitamin B_{12} occurs mostly in bacteria, whereas algae are usually rich in vitamin A. The most common vitamins in SCP are thiamine, ribofloavin, niacin, pyridoxum pantothenic acid, choline, folic acid, inositol, B_{12} and P-aminobenzoic acid the essential amino acid analysis of SCP compared with several sources of protein.

Table 4.4. Cellular composition of SCP from various microorganisms (dry weight per cent)

	Yeast	*Bacteria*	*Fungi*	*Algae*
Protein	45-55	50-65	30-45	40-60
Nucleic acid	6-12	8-12	7-10	3-8
Fat	2-6	1.5-3	2-8	7-20
Ash	5-9.5	3-7	9-14	8-10

Table 4.5. Essential amino acid content of the cell protein in comparison with other reference proteins (weight %)

Amino acid	*Cellulo-monas cerevisiae*	*Saccharo-myces*	*Penicillium notatum*	*SCP (BP)*	*Egg*	*Cow milk*
Lysine	7.6	7.7	3.9	7.0	6.3	7.8
Threonine	5.4	4.8	—	4.9	5.0	4.6
Methionine	2.0	1.7	1.0	1.8	3.2	2.4
Cysteine	—	—	—	—	2.4	—
Tryptophane	—	1.0	1.25	—	1.6	—
Isoleucine	5.3	4.6	3.2	4.5	6.8	6.4
Leucine	7.3	7.0	5.5	7.0	9.0	9.9
Valine	7.1	5.3	3.9	5.4	7.4	6.9
Phyenyalanine	4.6	4.1	2.8	4.4	6.3	4.9
Histidine	7.8	2.7	—	2.0	—	—
Arginine	6.4	2.4	—	4.8	—	—

ADVANTAGES AND DISADVANTAGES OF SCP

Addition of SCP to the diet of a milking cow increases milk production and production efficiency by 15%. Microorganisms such as bacteria and yeasts grow rapidly and contain more uric acid than slower-growing plants and animals. Although the uric acid limits the daily intake of SCP for humans and monogastric animals such as pigs and chickens, ruminants such as cattle, sheep and goats can tolerate higher levels of uric acid or break down urea and excrete it as ammonia. About 80% ot the total cell nitrogen is amino acids whereas the remaining 20% is possibly fat, ash and nucleic acids.

The concentration of nucleic acids in SCP is higher than in conventional proteins. That is the characteristic of all fast-growing organisms. The problem occurs with the consumption of proteins, roughly 10% of nucleic acids. The high nucleic acid content of SCP results in an increase in uric acid in serum and urine. Uric acid is the final product of purine degradation in humans. Most mammals, reptiles and molluscs possess the enzyme uricase, and they are able to oxidise uric acid to allantoic acid. High uric acid in humans causes 'gout'. That disease results from the elevation of uric acid in body fluid. Its manifestation is painful inflammation of arthritic joints. In animals with urease and allantoicase, the biodegradation of uric acid is accelerated, and the end product is ammonia.

In the human body lack of such enzymes causes the catabolism of uric acid to be terminated, and uric acid has to be excreted in urine through the kidneys. The removal and reduction of the nucleic acid content of various SCPs is achieved by chemical treatment with sodium hydroxide solution or high salt solution (10%). As a result, crystals of sodium urate form and are removed from the SCP solution. The quality of SCP can be upgraded by the destruction of cell walls. That may enhance the digestibility of SCP. With chemical treatment the nucleic acid content of SCP is reduced. The presence of uricase assists the uric acid to be hydrolysed, and the end product of purine degradation is completed with the addition of uricase.

PREPARATION FOR EXPERIMENTAL RUN

1. Prepare seed culture and use it for inoculation of 21 airlift and 21 B. Braun biostat B using soluble starch or glucose.
2. Perform fermentation for 24 hours in a batch system.
3. Monitor DO level and control pH at 6.7-7 by using 0.2 M phosphate buffer solution.
4. Measure SCP based on standard methods and analysis explained above.
5. Determine yield of SCP-based carbon sources.
6. Take usual samples at intervals of 4-6 hours.
7. Measure carbon sources remaining.

5 THE BUILDING BLOCKS

In the middle of the nineteenth century, the Dutch chemist Gerardus Mulder extracted a substance common to animal tissues and the juices of plants, which he believed to be "without doubt the most important of all substances of the organic kingdom, and without it life on our planet would probably not exist." At the suggestion of the famous Swedish chemist Jons Jakob Berzelius, Mulder named this substance protein (from the Greek *proteios,* meaning "of first importance") and assigned to it a specific chemical formula ($C_{40}H_{62}N_{10}O_{12}$). Although he was wrong about the chemistry of proteins, he was right about their being indispensable to living organisms. The term "protein" endures. Proteins are the most abundant of cellular components. They include enzymes, antibodies, hormones, transport molecules, and even components for the cytoskeleton of the cell itself.

Proteins are also informational macromolecules, the ultimate heirs of the genetic information encoded in the sequence of nucleotide bases within the chromosomes. Structurally and functionally, they are the most diverse and dynamic of molecules and play key roles in nearly every biological process. Proteins are complex macromolecules with exquisite specificity; each is a specialized player in the orchestrated activity of the cell. Together they tear down and build up molecules, extract energy, repel invaders, act as delivery systems, and even synthesize the genetic apparatus itself.

In this chapter we concentrate on the structural and chemical properties of amino acids, peptides, and polypeptides—the building blocks of proteins. From our presentation you will leam the following :

1. Certain acidic and basic properties are common to all amino acids found in proteins except for the amino acid proline.
2. Side chains give amino acids their individuality. These side chains serve a variety of structural and functional roles.
3. The α-carboxyl group of one amino acid can react with the α-amino group of another amino acid to form a di-peptide.
4. Many amino acids, reacting in a similar way, can become linked to form a linear polypeptide chain.
5. The amino acid sequence in a polypeptide can be determined by a process of partial breakdown into manageable fragments, followed by stepwise analysis proceeding from one end of the chain to the other.

6. Polypeptide chains with a prespecified sequence can be synthesized by well-established chemical methods.

Amino Acids

Every protein molecule can be viewed as a polymer of amino acids. There are 20 common amino acids. The structure of a single amino acid At the center is a tetrahedral carbon atom called the alpha (α) carbon (C_α). It is covalently bonded on one side to an amino group (NH_2) and on the other side to a carboxyl group (COOH . A third bond is always to a hydrogen, and the fourth bond is to a variable side chain (R). In neutral solution (pH 7), the carboxyl group loses a proton and the amino group gains one. Thus an amino acid in solution, while neutral overall, is a doubly charged species called a zwitterion.

Almost all amino acids except proline have a protonated α-amino group ($—NH_3^+$) attached to the α carbon. In proline one of the N—H linkages is replaced by an N—C linkage forming part of a cyclic structure.

Various ways of classifying amino acids according to their R groups have been proposed. In the amino acids into three categories. The first category contains eight amino acids with relatively apolar R groups; the second category contains seven amino acids with uncharged polar R groups; and the third category contains five amino acids with R groups that normally exist in the charged state. Ammo acids are often abbreviated by three-letter symbols; when this proves to be too cumbersome (as in certain kinds of charts and figures), one-letter symbols are used. Both designations are given together with the molecular weight (M_r) of each amino acid. In addition to the 20 commonly occurring α-amino acids, a variety of other amino acids are found in minor amounts in proteins and in nonprotein compounds.

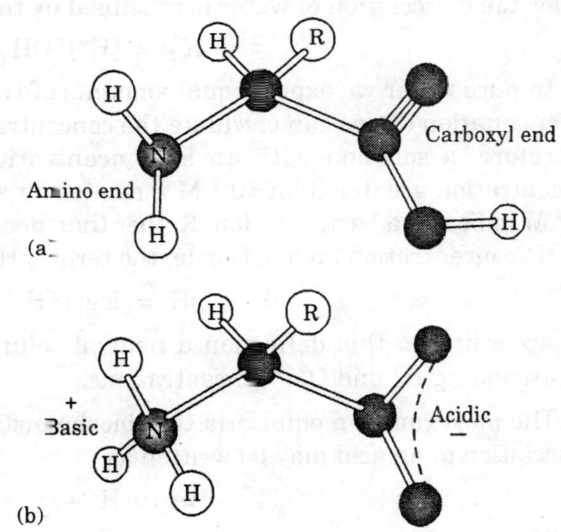

Fig. 5.1. Amino acid anatomy, (*a*) Uncharged amino acid, (*b*) Doubly charged zwitterion.

The unusual amino acids found in proteins result from modification of the common amino acids. In a few cases these amino acids are incorporated directly into the polypeptide chains during synthesis. Most frequently the amino acid is modified after incorporation. The unusual amino aicids found in nonprotein compounds are extremely varied in type and are formed by a number of different metabolic pathyways.

Properties of Amino Acids

The charge properties of amino acids are very important in determining the reactivity of certain amino acid side chains and in the properties they confer on proteins. The charge properties of amino acids in aqueous solution may best be considered under the general treatment of acid-base ionization theory. We find this treatment useful at other points in the text as well.

Recall that water can be considered a weak acid (or a weak base) because it dissociates into a proton and a hydroxide ion, according to the equilibrium

$$H_2O \rightleftharpoons H^+ + OH^- \quad ...(1)$$

The equilibrium expression for this reaction is

$$K_{eq} = \frac{[H^+][OH^-]}{[H_2O]} \quad ...(2)$$

Because water dissociates to such a small extent, the concentration of undissociated water is high and does not vary significantly for chemical reactions in aqueous solution. Therefore, the denominator in this equation is effectively constant, with a value of 55.5. The constant K_w for the dissociation of water is redefined by the expression at 25°C.

$$K_w = [H^+][OH^-] = 10^{-14} \text{ (mole/l)}^2 \quad ...(3)$$

In pure water we expect equal amounts of H^+ ("hydrogen ion") and OH^- ("hydroxide ion"). From equation (3) we can calculate the concentration of H^+ or OH^- in pure water to be 10^{-7} M Therefore, a solution with an H^+ concentration of 10^{-7} M is defined as neutral. An H^+ concentration greater than 10^{-7} M indicates an acidic solution; an H^+ concentration less than 10^{-7} M indicates a basic solution. Rather than deal with exponentials, it is convenient to express the H^+ concentration on a pH scale, the term "pH" being defined by the equation

$$pH = \log(1/[H^+]) = -\log [H^+] \quad ...(4)$$

According to this definition a neutral solution has a pH of 7. Other values of pH and corresponding H^+ and OH^- concentrations.

The most common equilibria that biochemists encounter are those of acids and bases. The dissociation of an acid may be written as

$$HA \rightleftharpoons H^+ + A^- \quad ...(5)$$

The equilibrium constant for this reaction is called the acid dissociation constant K_a written as

$$K_a = \frac{[H^+][A^-]}{[HA]} \quad ...(6)$$

Strong acids in aqueous solution dissociate completely into anions and protons. The concentration of hydrogen ion $[H^+]$ is therefore equal to the total concentration C_{HA} of the acid HA that is added to the solution. Thus the pH of the solution of a strong acid is simply $-\log C_{HA}$.

The pH of the solution of a weak acid is a function of both the C_{HA} and the acid dissociation constant. The dissociation constant of a weak acid may be written in terms of the species present in the equation for the acid dissociation constant.

First solving equation (6) for $[H^+]$ gives

$$[H^+] = \frac{K_a[HA]}{[A^-]} \quad ...(7)$$

Table 5.1. The pH Scale.

pH	*[H⁺]*	*[OH⁻]*
0	10^{0}	10^{-14}
1	10^{-1}	10^{-13}
2	10^{-2}	10^{-12}
3	10^{-3}	10^{-11}
4	10^{-4}	10^{-10}
5	10^{-5}	10^{-9}
6	10^{-6}	10^{-8}
7	10^{-7}	10^{-7}
8	10^{-8}	10^{-6}
9	10^{-9}	10^{-5}
10	10^{-10}	10^{-4}
11	10^{-11}	10^{-3}
12	10^{-12}	10^{-2}
13	10^{-13}	10^{-1}
14	10^{-14}	10^{0}

Taking the logarithm of both sides and changing signs gives us

$$-\log [H^+] = -\log K_a + \log \frac{[A^-]}{[HA]} \qquad ...(8)$$

Substituting pH for –log [H⁺] and pK_a for –log K_a in equation (8), we obtain the Henderson-Hasselbach equation :

$$pH = pK_a + \log \frac{[A^-]}{[HA]} = pK_a + \log \frac{[base]}{[acid]} \qquad ...(9)$$

The Henderson-Hasselbach equation is useful for calculating the molar ratio of base (proton acceptor) to acid (proton donor) for a given pH and pK or for calculating the pK, given the ratio of base (proton acceptor) to acid (proton donor). It can be seen that when the concentration of anion or base is equal to the concentration of undissociated acid (*i.e.,* when the acid is half neutralized), the pH of the solution is equal to the pK of the acid.

The values of pK for a particular molecule are determined by titration. A typical pH dependence curve for the titration of a weak acid by a strong base. The concentration of the anion equals the concentration of the acid when the acid is exactly half neutralized. Note that at this point on the curve, the pH is least sensitive to the quantity of added base (or acid). Under these conditions, the solution is said to be buffered. Bio-chemical reactions are typically highly dependent on the pH of the solution. Therefore, it is frequently advantageous to study reactions in buffered solutions.

The ideal buffer is one that has a pK numerically equivalent to the working pH. A simple ammo acid with a gives a complex titration curve with two inflection points. At very low pH, alanine carries a single positive charge on the α-amino group. The first inflection point occurs at a pH of 2.3. This is the pK for titration of the carboxyl group, pK_1 (—COOH→—COO⁻). At a pH of 6.0, alanine has an equal amount of positive and negative charge. This value is referred to as the isoelectric point (pI), or the isoelectric pH. As the titration continues, a second inflection point is reached at a pH of 9.7. The pK at this point, pK_2, the [$-NH_3^+$] and [$-NH_2$] are equal. Amino acids with an ionizable group show even more complex titration curves, indicative of three ionizable groups.

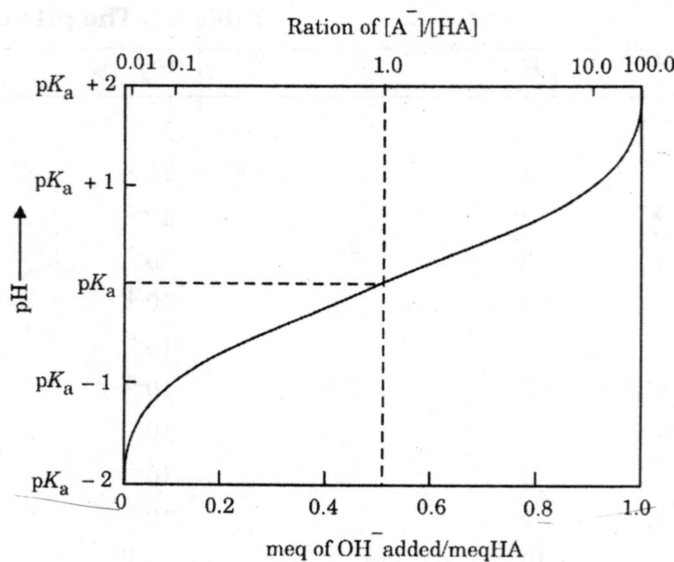

Fig. 5.2. The dependence of pH on the equivalents of base added to a typical weak acid. Note that at the pK_a [A⁻] = [HA].

The pK for the ionizable side chain, pK_R, is usually readily distinguishable from the pK values for the ionizable α-carboxyl and α-amino groups, pK_1 and pK_2, respectively, because the α-amino groups have numerical values close to the comparable pK values of alanine. Note that the only ionizable R group with a pK_R in the vicinity of 7, where most biological systems function, is that for histidine. This means that although other ionizable groups are usually fully charged under biological conditions, the side chain of histidine can be fully charged, uncharged, or partially charged, depending on the precise situation. This variability has major implications for the way the histidine side chain functions in enzyme catalysis.

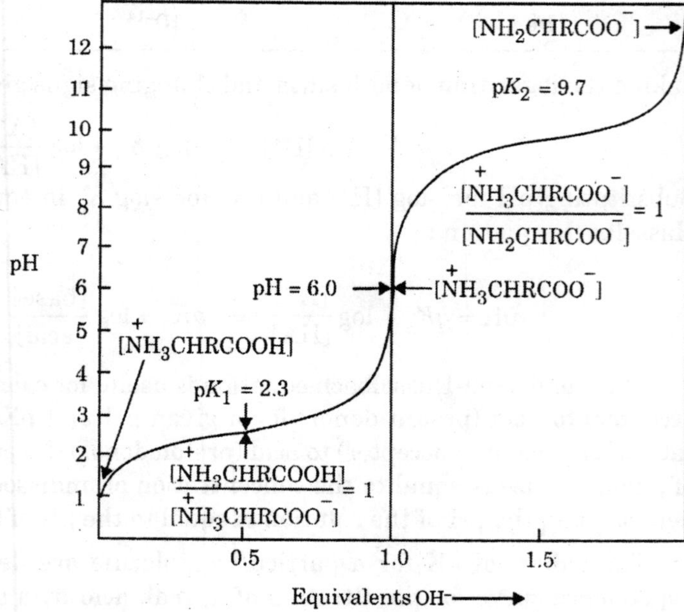

Fig. 5.3. Titration curve of alanine. The predominant ionic species at each cardinal point in the titration is indicated.

The side chain can serve as either a proton donor or a proton acceptor. An additional point, whereas the amino acid side chains (R groups) that are normally charged at physiological pH are restricted to five amino acids (aspartic acid, glutamic acid, lysine, arginine, and sometimes histidine), a number of potentially ionizable R groups are part of other arnino acids. These include cysteine, serine, threonine, and tyrosine. The ionization reactions for all of the potentially ionizable side chains are indicated.

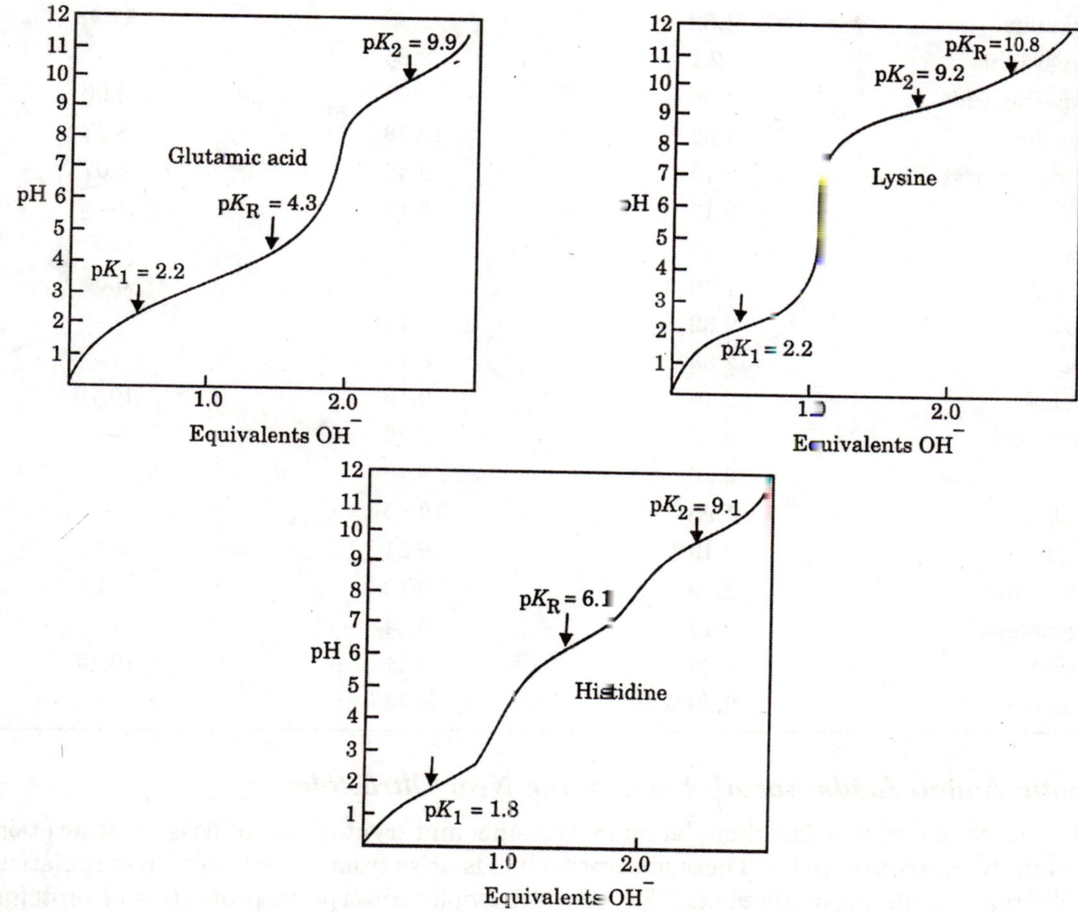

Fig. 5.4. Titration curves of glutamic acid, lysine, and histidine. In each case, the pK of the R group is designated pK_R.

The acidic and basic groups within a protein can be titrated just like free amino acids to determine their number and their pK_a values. A titration curve for β-lactoglobulin. This protein contains 94 potentially ionizable groups. The protein is positively charged at low pH and negatively charged at high pH.

At intermediate pH values a point is found where the sum of the positive side-chain charges exactly equals the sum of the negative charges, so that the net charge on the protein is zero. This value, as we have noted, is the isoelectric point (pI) of the protein; for β-lactoglobulin the pI is about 5.2. The isoelectric point is not an invariant quantity. The binding of charged species present in the solution could raise or lower the p_g, depending on their charge.

Table 5.2. Values of p*K* for the Ionizable Groups of the Twenty Amino Acids Commonly Found in Proteins

Amino Acid	pK_1 *(α—COOH)*	pK_2 *(α—NH_3^+)*	pK_R *(R Group)*
Alanine	2.35	9.87	—
Arginine.	1.82	8.99	12.48
Asparagine	2.1	8.84	—
Aspartic acid	1.99	9.90	3.90
Cysteine	1.92	10.78	8.33
Glutamic acid	2.10	9.47	4.07
Glutamine	2.17	9.13	—
Glycine	2.35	9.78	—
Histidine	1.80	9.33	6.04
Isoleucine	2.32	9.76	—
Leucine	2.33	9.74	—
Lysine	2.16	9.18	10.79
Methionie	2.13	9.28	—
Phenylalanine	2.16	9.18	—
Proline	1.95	10.65	—
Serine	2.19	9.21	≈ 13
Threonine	2.09	9.10	≈ 13
Tryptophan	2.43	9.44	—
Tyrosine	2.20	9.11	10.13
Valine	2.29	9.74	—

Aromatic Amino Acids Absorb Light in the Near-Ultraviolet

The aromatic amino acids phenylalanine, tyrosine, and tryptophan all possess absorption maxima in the near-ultraviolet. These absorption bands arise from the interaction of radiation with electrons in the aromatic rings. The near-ultraviolet absorption properties of proteins are determined solely by their content of these three aromatic amino acids. In solution, UV absorption can be quantified with the help of a conventional spectrophotometer and used as a measure of the concentration of proteins.

All Amino Acids Except Glycine Show Asymmetry

One of the most striking and significant properties of amino acids is their chirality, or handedness. The word "criral" is related to the Greek word meaning hand. Just as the right hand is related to the left hand by a mirror image, so, in general, a naturally occurring amino acid is related to a stereoisomer by its mirror image. The observation is true of 19 out of the 20 amino acids; the one exception is glycine. The chirality of amino acids stems from the chiral, or asymmetric, center, the α-carbon atom.

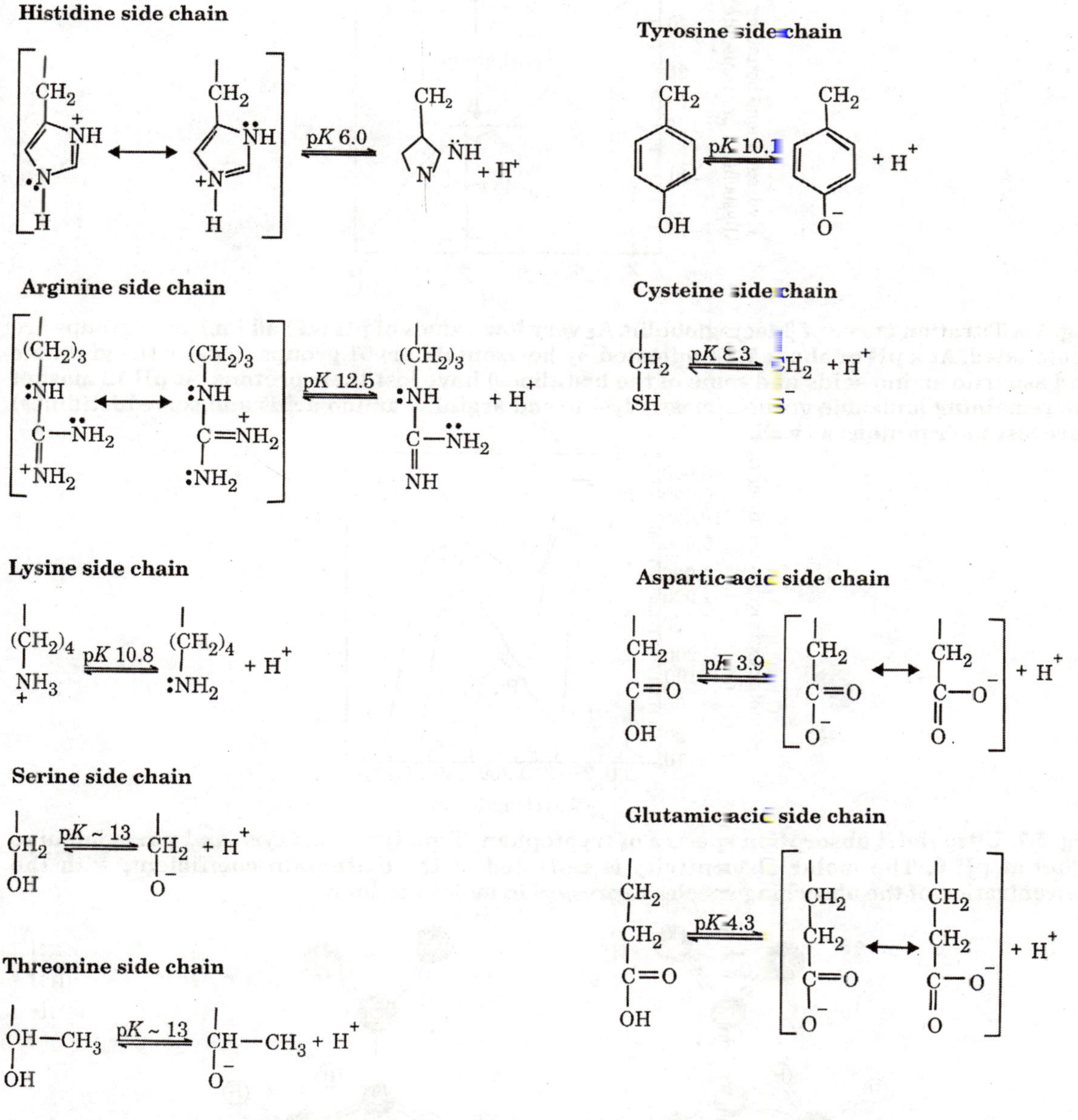

Histidine side chain

pK 6.0

Tyrosine side chain

pK 10.1

Arginine side chain

pK 12.5

Cysteine side chain

pK 8.3

Lysine side chain

pK 10.8

Aspartic acid side chain

pK 3.9

Serine side chain

pK ~ 13

Glutamic acid side chain

pK 4.3

Threonine side chain

pK ~ 13

Fig. 5.5. Equilibrium between charged and uncharged forms of amino acid side chains.

The α-carbon atom is a chiral center if it is connected to four different substituents. Thus glycine has no chiral center. Two of the amino acids, isoleucine and threonine, possess additional chiral centers because each has one additional asymmetric carbon. You should be able to locate these carbons by simple inspection. Two structures that constitute a stereoisomeric pair are referred to as enantiomers. These two isomers are called L-alanine and D-alanine, according to the way in which the substituents are arranged about the asymmetric carbon atom.

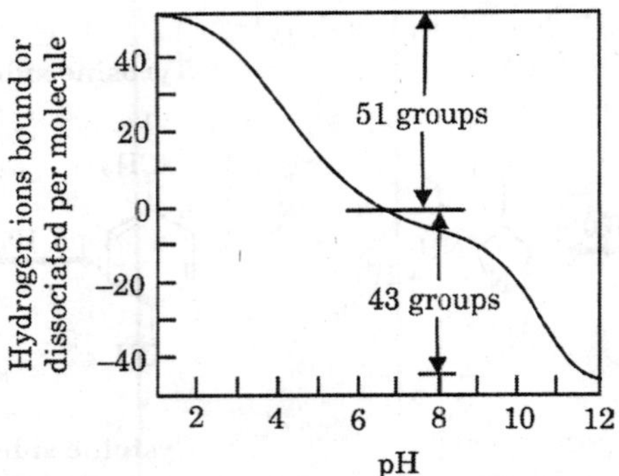

Fig. 5.6. Titration curve of β-lactoglobulin. At very low values of pH (<2) all ionizable groups are protonated. At a pH of about 7.2 (indicated by horizontal bar) 51 groups (mostly the glutamic and aspartic amino acids and some of the histidines) have lost their protons. At pH 12 most of the remaining ionizable groups (mostly lysine and arginine amino acids and some histidines) have lost their protons as well.

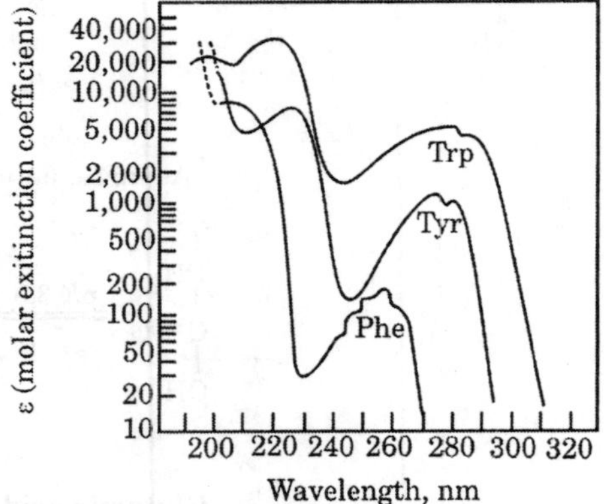

Fig. 5.7. Ultraviolet absorption spectra of tryptophan (Trp), tyrosine (Tyr), and phenylalanine (Phe) at pH 6. The molar absorptivity is reflected in the extinction coefficient, with the concentration of the absorbing species expressed in moles per liter.

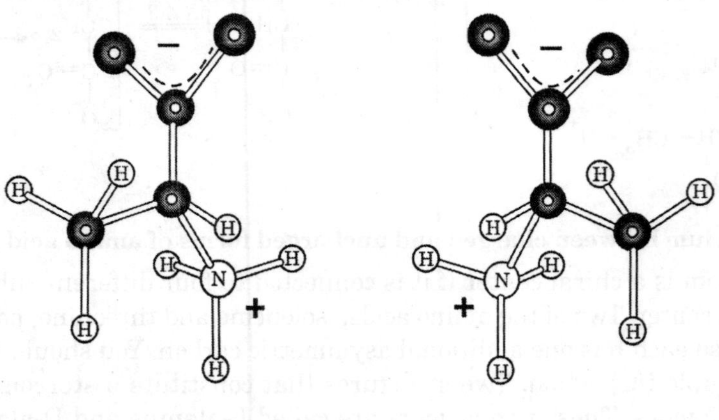

Fig. 5.8. The covalent structure of alanine, showing the three-dimensional structure of the L and D stereoisomeric forms.

The naming by D and L (for "dextrorotatory" and "levorotatory"; refers to a convention established by Emil Fischer many years ago. According to this convention all amino acids found in proteins are of the L form. Some D-amino acids are found in bacterial cell walls and certain antibiotics.

Peptides and Polypeptides

Amino acids can link together by a covalent peptide bond between the α-carboxyl end of one amino acid and the α-amino end of another. Formally, this bond is formed by the loss of a water molecule. The peptide bond has partial double-bond character owing to resonance effects; as a result, the C—N peptide linkage and all of the atoms directly connected to C and N lie in a planar configuration called the amide plane. In the following chapter we see that this amide plane, by limiting the number of orientations available to the polypeptide chain, plays a major role in determining the three-dimensional structures of proteins. Any number of amino acids can be joined by successive peptide linkages, forming a polypeptide chain.

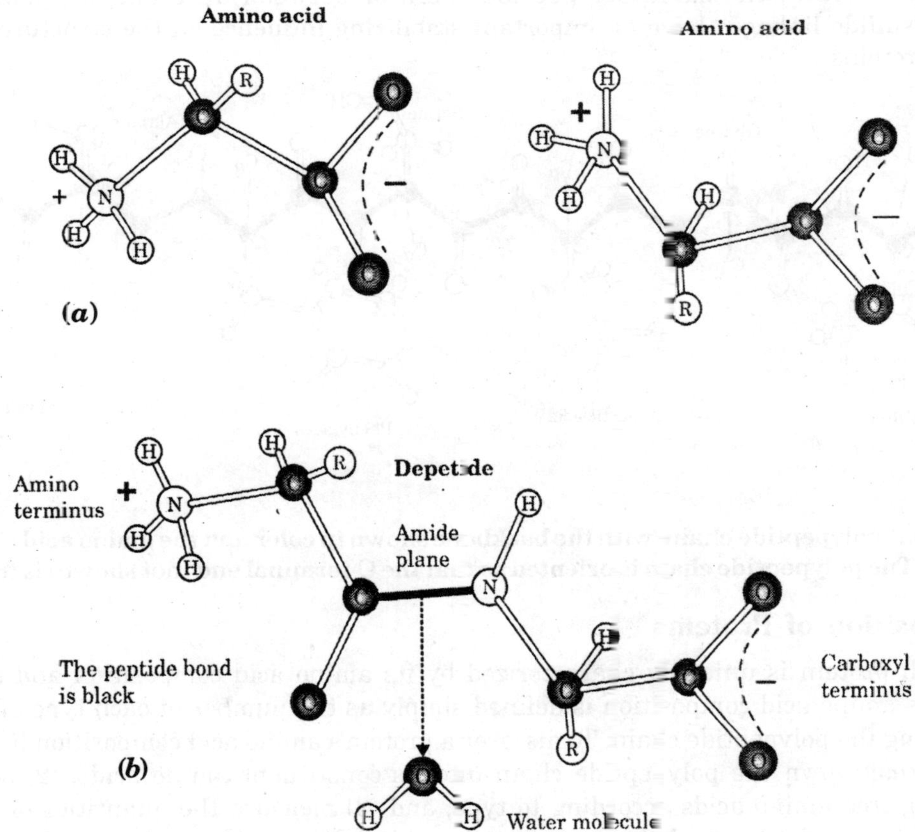

Fig. 5.9. Formation of a dipepetide from two amino acids, *(a)* Two amino acids, *(b)* A peptide bond (CO—NH) links amino acids by joining the α-carboxyl group of one with the α-amino group of another. A water molecule is lost in the reaction. It is conventional to draw dipeptides and polypeptides so that their free amino terminus is to the left and their free carboxyl terminus is to the right. The amide plane refers to six atoms that lie in the same plane.

The polypeptide chain, like the dipeptide, has a directional sense. One end, called the N-terminal, or amino-terminal, end, has a free α-amino group, whereas the other end, the C-terminal, or carboxyl-terminal, end, has a free α-carboxyl group. The sequence of main-chain atoms from the N-terminal end to the C-terminal end is N—C_α—C—NSr and so on, and in the opposite direction it is C—C_α—N—TC and so on. Short polypeptide chains, up to a length of about 20 amino acids, are called peptides or oligopeptides if they are fragments of whole polypeptide chains. A small protein molecule may contain a polypeptide chain of only 50 amino acids; a large protein may contain chains of 3,000 amino acids or more.

One of the larger single polypeptide chains is that of the muscle protein myosin, which consists of approximately 1,750 amino acid residues. Section of a polypeptide chain as a linear array with α carbons and planar amides alternating as repeating units of the main chain. Different side chains are attached to each α carbon. In addition to the covalent peptide bonds formed between adjacent amino acids within a polypeptide chain, covalent disulfide bonds can be formed within the same polypeptide chain or between different polypeptide chains. Such disulfide linkages have an important stabilizing influence on the structures formed by many proteins.

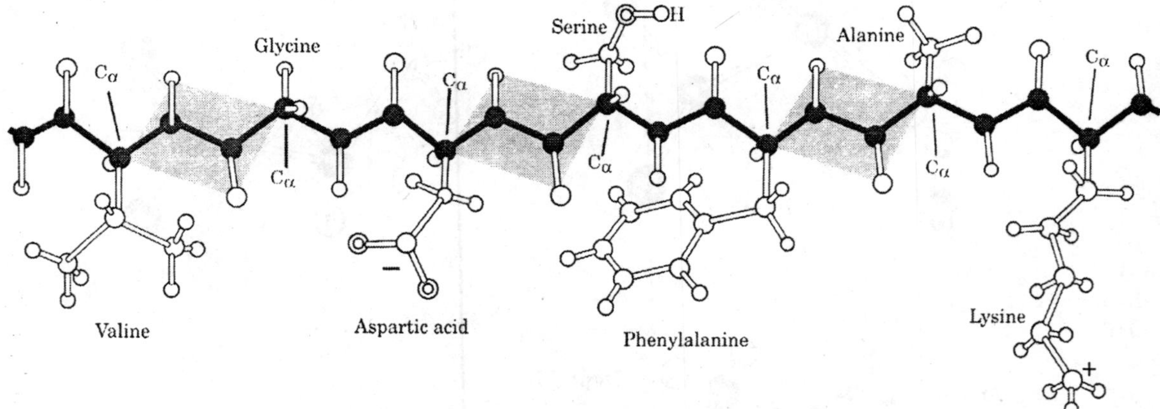

Fig. 5.10. A polypeptide chain, with the backbone shown in color and the amino acid side chains in outline. The polypeptide chain is oriented so that the C-terminal end (not shown) is to the left.

Composition of Proteins

Each protein is uniquely characterized by its amino acid composition and sequence. A protein's amino acid composition is defined simply as the number of each type of amino acid composing the polypeptide chain. To discover a protein's amino acid composition it is necessary to (1) break down the polypeptide chain into its constituent amino acids, (2) separate the resulting free amino acids according to type, and (3) measure the quantities of each amino acid. Cleavage of the peptide bonds is usually achieved by boiling the protein in 6-N HCl; this treatment causes hydrolysis of the peptide bonds and the consequent release of free amino acids.

Although acid hydrolysis is the most frequently used means of breaking a protein into its constituent amino acids, it results in the partial destruction of the indole ring of tryptophan. Consequently, the amount of tryptophan in the prolein must be estimated by an alternative method (*e.g.*, spectroscopic absorption) when using acid hydrolysis. In addition, acid hydrolysis results in the loss of ammonia from the side-chain amide groups of glutamine and asparagine, with the consequent production of glutamic and aspartic acids. Therefore, estimates of amino acid composition based on acid hydrolysis show glutamine and glutamic acid combined and measured as glutamic acid. Similarly, asparagine and aspartic acid are combined and measured as aspartic acid. Separation of amino acids for quantitative analytical purposes is usually achieved by ion-exchange chromatography.

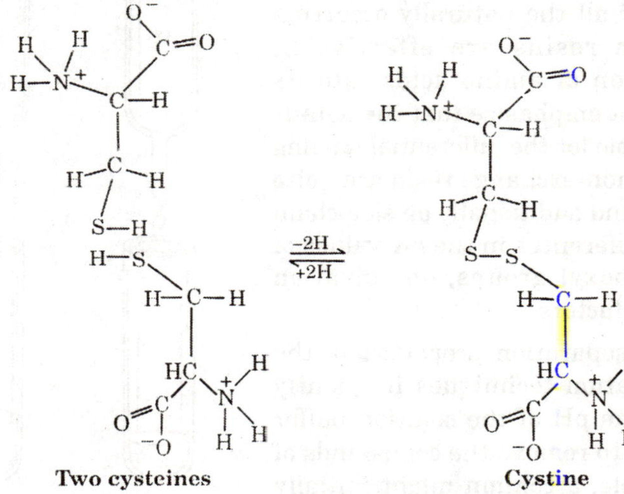

Two cysteines $\underset{+2H}{\overset{-2H}{\rightleftharpoons}}$ Cystine

Fig. 5.11. Disulfide bonds can form between two cysteines. The cysteines can exist in the cytosol as free amino acids (as shown), in which case they give rise to cystine, or they can be on polypeptide chains. In the latter instance, they can be on the same polypeptide chains or different polypeptide chains. In either case the formation of covalent disulfide bonds stabilizes structural relationships.

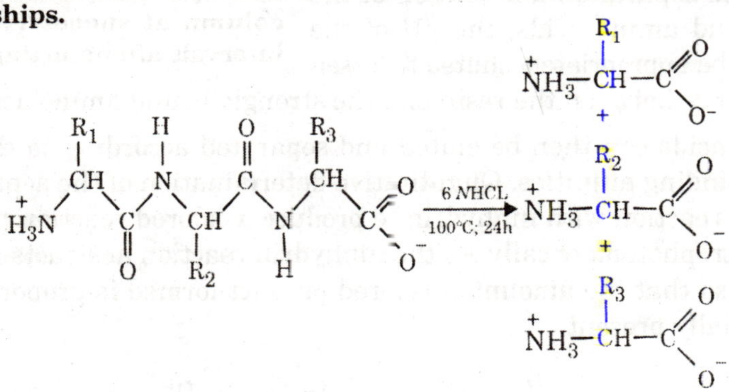

6 NHCL, 100°C; 24h

Fig. 5.12. Acid hydrolysis of a protein or polypeptide to yield amino acids.

The general efficacy of chromatographic techniques is based on a difference in affinity between each compound to be separated and an immobile phase or resin. The resin consists of some relatively chemically inert polymer, which has weakly basic side-chain constituents that are positively charged at pH 7. If we were to add some of this resin to a solution containing free aspartic acid and lysine at pH 7, the negatively charged aspartic acid would have a higher affinity for the resin than would the positively charged lysine.

If we then pump a solution of these two amino acids througn a column containing such a positively charged resin, the progress of the aspartic acid through the column would be retarded relative to the lysine, owing to the greater affinity of the aspartic acid for the resin. Ion-exchange resins exist that have differential binding affinities for all the naturally occurring amino acids. Such resins are effective in separating a solution of amino acids into its components. We must emphasize that the details of the forces responsible for the differential binding of amino acids to an ion-exchange resin are quite complicated and depend additionally on side-chain polarity, on subtle differences in the *pK* values of α-amino and α-carboxyl groups, on solvation effects, and on other factors.

To enhance the separation properties of the column, such separation techniques frequently exploit changes in the pH of the solution buffer (eluting buffer) used to remove the compounds of interest : For example, a column might initially be run with the eluting buffer at a pH that results in some amino acids being so strongly bound to the resin that they are essentially immobile. However, after the separation and elution of the less strongly bound amino acids, the pH of the eluting buffer can be appropriately shifted to lessen the charge difference between the resin and the strongly bound amino acids.

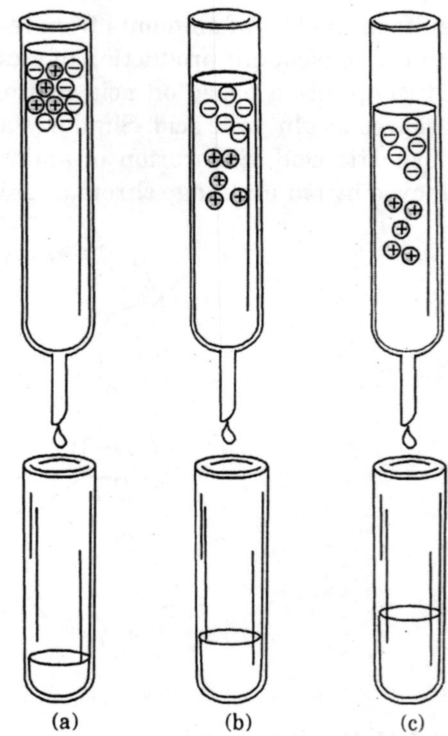

Fig. 5.13. Migration of aspartic acid ⊖ and lysine ⊕ through a column with a higher affinity for aspartic acid. Views show the column at successively increasing time intervals after starting the elution.

These amino acids can then be eluted and separated according to the newly established pattern of resin-binding affinities. Quantitative determination of the separated amino acids is achieved by their reaction with ninhydrin to produce a colored reaction product. This product is measured spectrophotometrically. As the ninhydrin reaction abstracts an amino group from each amino acid, so that the amount of colored product formed is proportional to the amount of amino acid initially present.

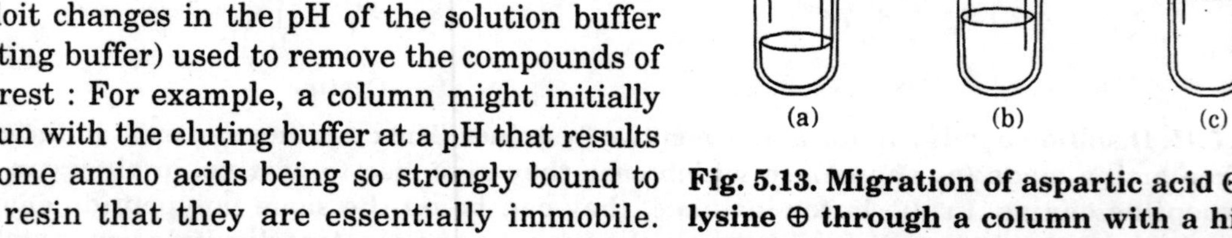

$$R-\underset{H}{\overset{NH_2}{C}}-COOH + 2\ \text{(ninhydrin)} \rightarrow \text{(indanedione)}-N=C\text{(indanedione)} + CO_2 + R-\overset{H}{C}=O + 3H_2O$$

Fig. 5.14. Reaction of ninhydrin with an amino acid yields a colored complex. The ninhydrin reaction permits qualitative location of amino acids in chromatography and quantitative assay of separated amino acids.

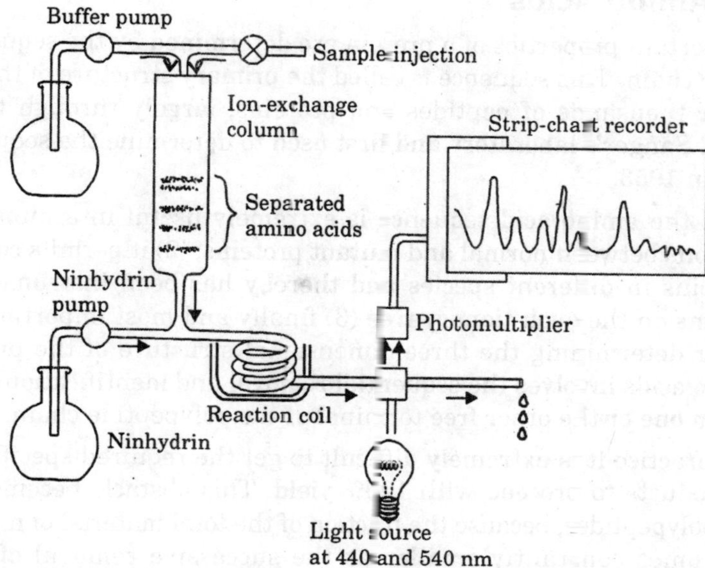

Fig. 5.15. Schematic diagram of an amino acid analyzer. The amino acids are passed through an ion-exchange column and thereby separated. Eluted fractions are mixed and reacted with ninhydrin. The intensity of the resulting colored product is measured in a spectrophotometer, and the results are displayed on a recording chart.

Currently, measurements of amino acid composition are usually carried out on an amino acid analyzer, a device that automates the previously described operations. As illustrated, in the amino acid analyzer consists of an ion-exchange column through which the appropriate eluting buffer is pumped after the amino acids are introduced at the top of the column.

As the separated amino acids emerge, they are ninhydrin solution and passed through a coil of tubing to allow the formation of the colored "ninhydrin reaction product. The separated ninhydrin reaction products then pass through a cell that measures their optical absorbance at 540 and 440 nm and plots the results on a strip-chart recorder. The absorbance is measured at two wavelengths because proline, which is substituted at its amino group, forms a different ninhydrin reaction product, with an absorption-maximum that is correspondingly different from that of the remaining amino acids. Usually the ammo acid analyzer is first standardized by running through it a sample containing known quantities of amino acids to account for any differences in their ninhydrin reaction properties.

In this way it is possible to relate the amount of amino acid present to the amount of colored product formed, as measured by the area under the "peak" produced on the strip-chart recorder. Similarly, the amino acid hydrolysate of a protein of unknown composition can be run through the analyzer, and the relative peak areas can be used to estimate the ratios of the different amino acids present. Conversion ol the relative ratios of amino acids into an estimate of actual composition requires some additional information concerning the protein's molecular weight; for example, an analysis giving relative ratios of Ala (1.0), Gly (0.5), and Lys (2.0.) could correspond to composition Ala_2-Gly Lys_4 or any muLiple thereof.

The required information is usually available, and in any case, an estimation of composition based on a minimum molecular weight of the protein is always possible.

Sequence of Amino Acids

The most important properties of a protein are determined by the sequence of amino acids in the polypeptide chain. This sequence is called the primary structure of the protein. We know the sequences for thousands of peptides and proteins, largely through the use of methods developed in Fred Sanger's laboratory and first used to determine the sequence of the peptide hormone insulin in 1953.

Knowledge of the amino acid sequence is extremely useful in a number of ways : (1) it permits comparisons between normal and mutant proteins; (2) it permits comparisons between comparable proteins in different species and thereby has been instrumental in positioning different organisms on the evolutionary tree (3) finally and most important, it is a vital piece of information for determining the three-dimensional structure of the protein. Determining the order of amino acids involves the sequential removal and identification of successive amino acid residues from one or the other free terminal of the polypeptide chain.

However, in practice it is extremely difficult to get the required specific cleavage reaction of the desired products to proceed with 100% yield. This obstacle becomes significant when sequencing long polypeptides, because the fraction of the total material of minimum polypeptide chain length becomes constantly smaller as the successive removal of terminal residues continues. Conversely, the amino acid released from the polypeptide chain becomes increasingly contaminated with amino acids released from previously unreacted chain. Because of this fundamental chemical limitation, the polypeptide chain must be broken down into sequences short enough for the chemistry to produce reliable results.

The short sequences are then reassembled to obtain the overall sequence. The steps actually involved in protein sequenching are

1. purification of the protein
2. cleavage of all disulfide bonds
3. determination of the terminal amino acid residues
4. specific cleavage of the polypeptide chain into small fragments in at least two different ways
5. independent separation and sequence determination of peptides produced by the different cleavage methods
6. reassembly of the individual peptides with appropriate overlaps to determine the overall sequence.

The first step, protein purification, is discussed. Once the protein is pure, sequence analysis can begin, with cleavage of the disulfide bonds. Cleavage is achieved by oxidizing the disulfide linkages with performic acid. Sometimes this step results in the production of two or more polypeptide chains, in which case the individual chains must be separated. The third step is to determine the polypeptide chain end groups. If the polypeptide chains are pure, then only one N-terminal and one C-terminal group should be detected.

The amino-terminal amino acid can be identified by reaction with fluorodinitrobenzene (FDNB). Subsequent acid hydrolysis releases a colored dinitrophenol (DNP)-labeled amino-terminal amino acid, which can be identified by its characteristic migration rate on thin-layer chromatography or paper electrophoresis. A more sensitive method of end-group determination involves the use of dansyl chloride. Chemical methods for carboxyl end-group determination are considerably less satisfactory. Treatment of the peptide with anhydrous hydrazine at 100°C results in conversion of all the amino acid residues to amino acid hydrazides except for the carboxyl-terminal residue, which remains as the free amino acid and can be isolated and identified chromatographically.

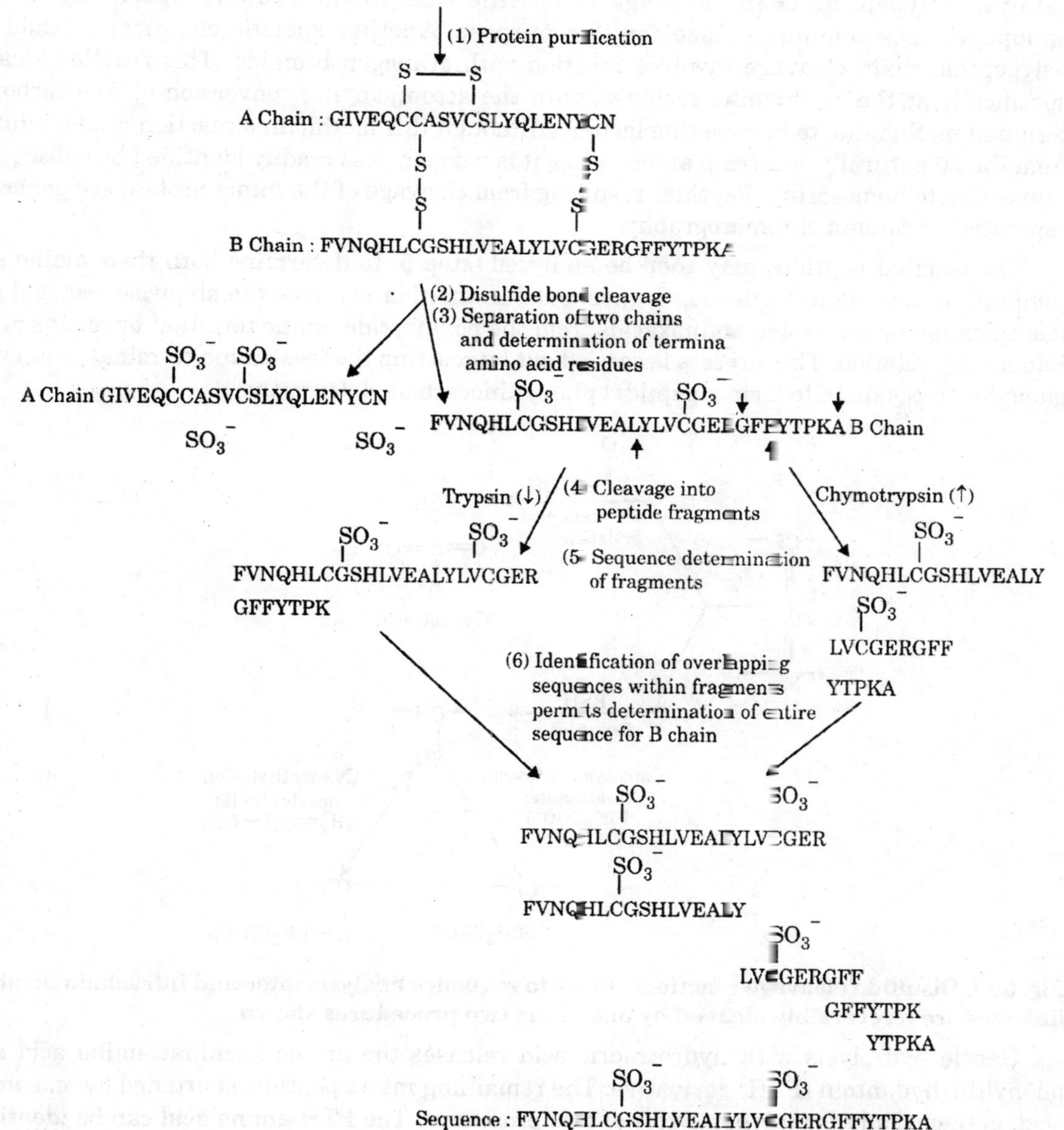

Fig. 5.16. Steps involved in the sequence determination of the B chain of insulin. Amino acids are represented here by their single-letter codes.

Alternatively, the polypeptide can be subjected to limited breakdown (proteolysis) with the enzyme carboxypeptidase. This results in release of the carboxyl-terminal amino acid as the major free amino acid reaction product. The amino acid type can then be identified chromatographically. Step 4 involves breaking down the polypeptide chain into shorter, well-defined fragments for subsequent sequence analysis.

Fragmentation can be achieved by the use of endopeptidases, which are enzymes that catalyze polypeptide chain cleavage at specific sites in the protein. Specificity of four endopeptidases commonly used for this purpose. Another specific chemical method for polypeptide chain cleavage involves reaction with cyanogen bromide. This reaction cleaves specifically at the methionine residues, with the accompanying conversion of free carboxyl-terminal methionine to homoserine lactone. Although this methionine reaction product differs from the 20 naturally occurring amino acids, it is nevertheless readily identified by subsequent conversion to homoserine. Peptides resulting from cleavage of the intact protein are generally separated by column chromatography.

The isolated peptides may then be analyzed (step 5) to determine both their amino acid composition and their sequence. Sequence determination involves the stepwise removal and identification of successive aminoxacids from the polypeptide amino terminal by means of the Edman degradation. This process is carried out by reacting the free amino-terminal group with phenylisothiocyanate to form a peptidyl phenylthiocarbamyl derivative.

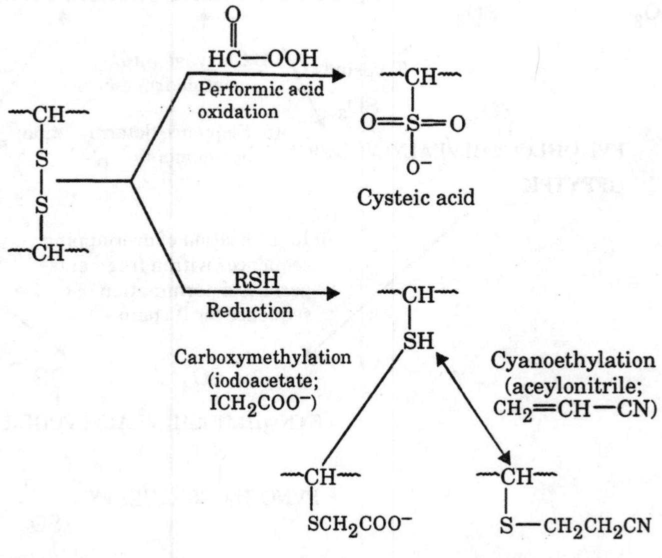

HC(=O)—OOH
Performic acid oxidation
—CH—S—S—CH—
—CH—SO_3^-
Cysteic acid
RSH
Reduction
—CH—SH
Carboxymethylation (iodoacetate; ICH_2COO^-)
Cyanoethylation (aceylonitrile; $CH_2{=}CH{-}CN$)
—CH—SCH_2COO^-
—CH—S—CH_2CH_2CN

Fig. 5.17. Disulfide cleavage reactions. Prior to sequence analysis inter-and intrachain disulfide linkages are irreversibly cleaved by one of the two procedures shown.

Gentle hydrolysis with hydrochloric acid releases the amino-terminal amino acid as a phenylthiohydantoin (PTH) derivative. The remaining intact peptide, shortened by one amino acid, is then ready for further cycles of this procedure. The PTH-amino acid can be identified by its properties on thin-layer chromatography. High pressure liquid chromatography (HPLC) is the method of choice where quantitative results on small amounts of material are required. Devices called sequenators are available that automate the Edman degradation procedure.

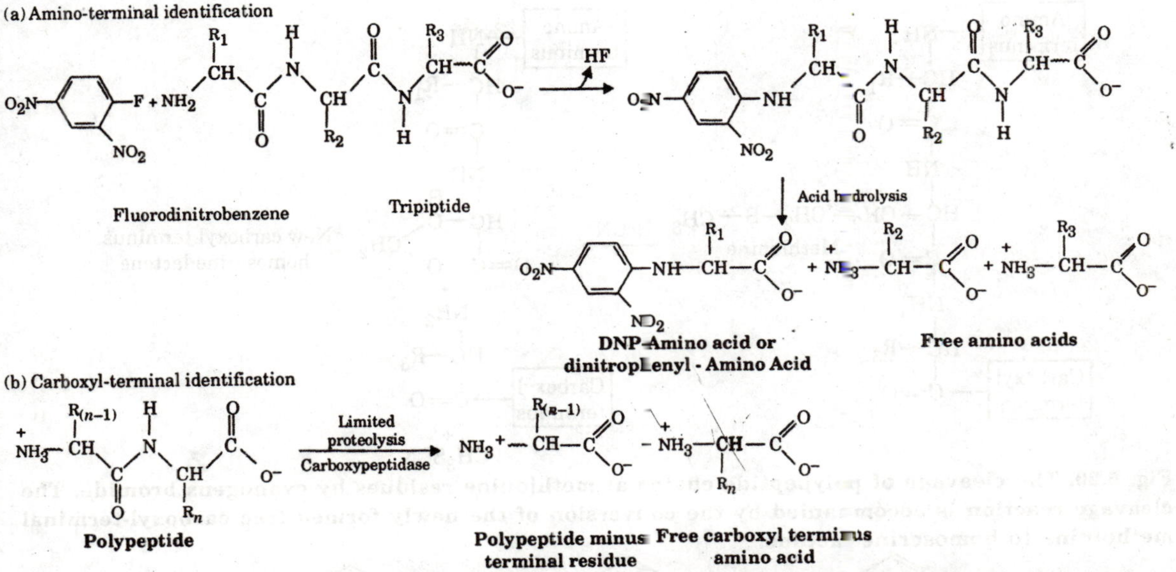

(a) Amino-terminal identification

Fluorodinitrobenzene + Tripiptide —HF→ ... —Acid hydrolysis→ DNP-Amino acid or dinitrophenyl - Amino Acid + Free amino acids

(b) Carboxyl-terminal identification

Polypeptide —Limited proteolysis, Carboxypeptidase→ Polypeptide minus terminal residue + Free carboxyl terminus amino acid

Fig. 5.18. Polypeptide chain end-group analysis. (*a*) Amino-terminal group identification. A more sensitive method, the dansyl chloride method, is described in Methods of Biochemical Analysis 3B. (*b*) Carboxyl-terminal group idenitification. Identification of this amino acid is considerably more difficult.

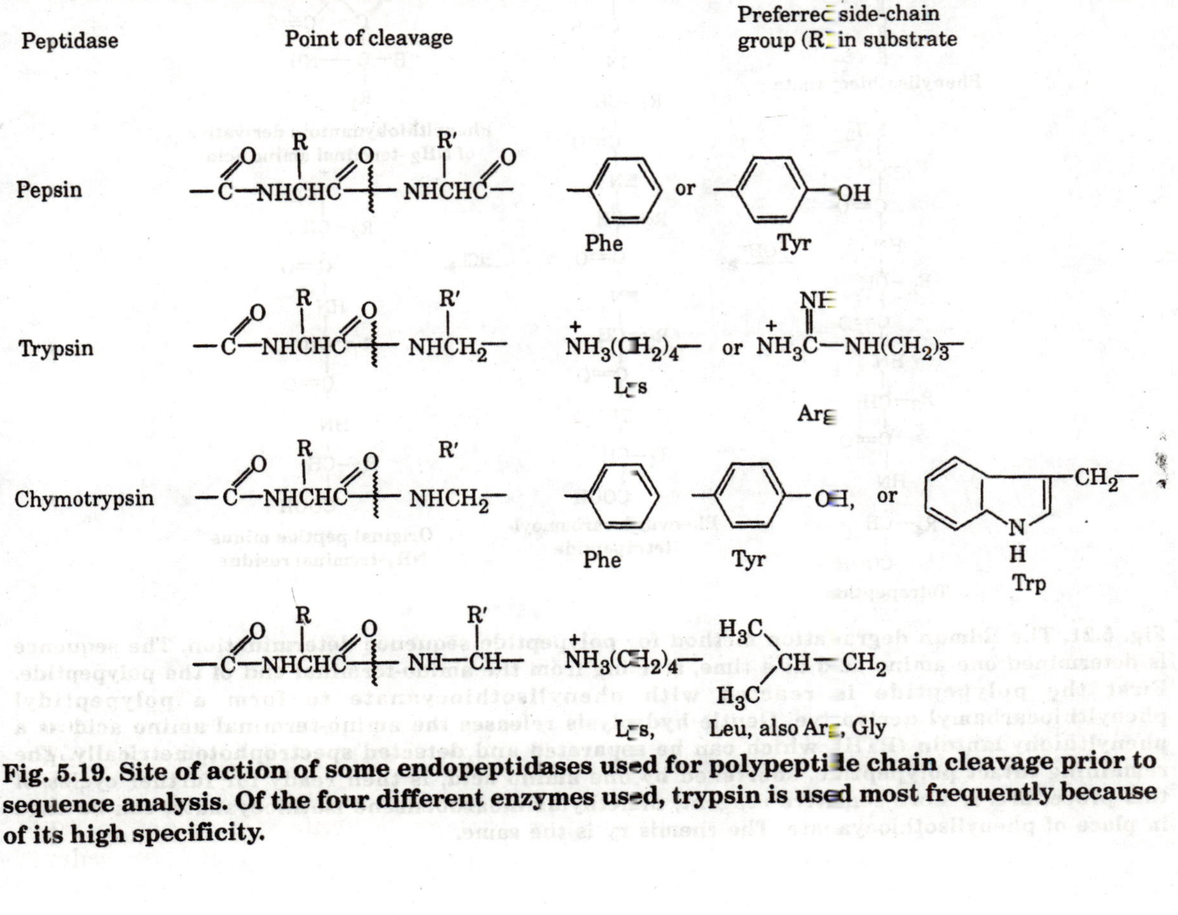

Peptidase	Point of cleavage	Preferred side-chain group (R) in substrate
Pepsin	—C(=O)—NHCHRC(=O)—┼—NHCHR′C(=O)—	Phe or Tyr
Trypsin	—C(=O)—NHCHRC(=O)—┼—NHCH₂—	$\overset{+}{N}H_3(CH_2)_4$— Lys or $\overset{+}{N}H_3C(=NH)—NH(CH_2)_3$— Arg
Chymotrypsin	—C(=O)—NHCHRC(=O)—┼—NHCH₂—	Phe, Tyr, or Trp
	—C(=O)—NHCHRC(=O)—┼—NH—CHR′—	$\overset{+}{N}H_3(CH_2)_4$— Lys, $(H_3C)_2CH—CH_2$— Leu, also Arg, Gly

Fig. 5.19. Site of action of some endopeptidases used for polypeptide chain cleavage prior to sequence analysis. Of the four different enzymes used, trypsin is used most frequently because of its high specificity.

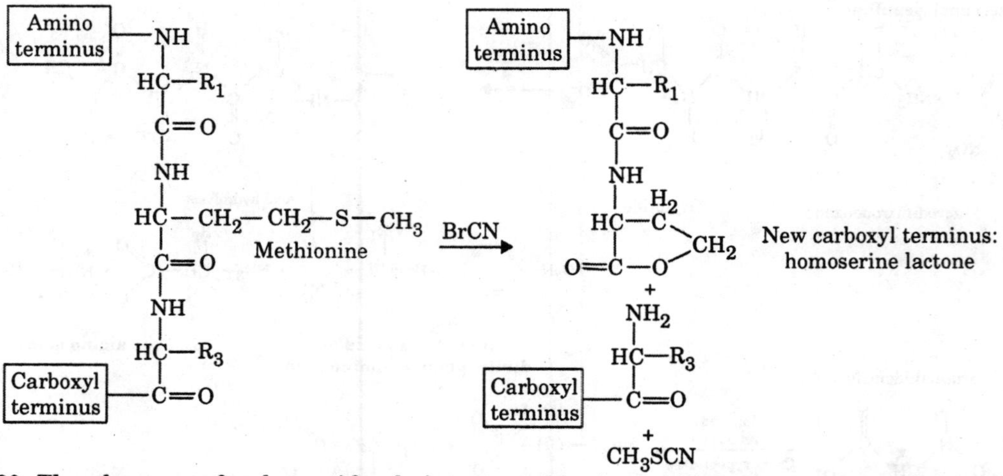

Fig. 5.20. The cleavage of polypeptide chains at methionine residues by cyanogens bromide. The cleavage reaction is accompanied by the conversion of the newly formed free carboxyl-terminal methoinine to homoserine lactone.

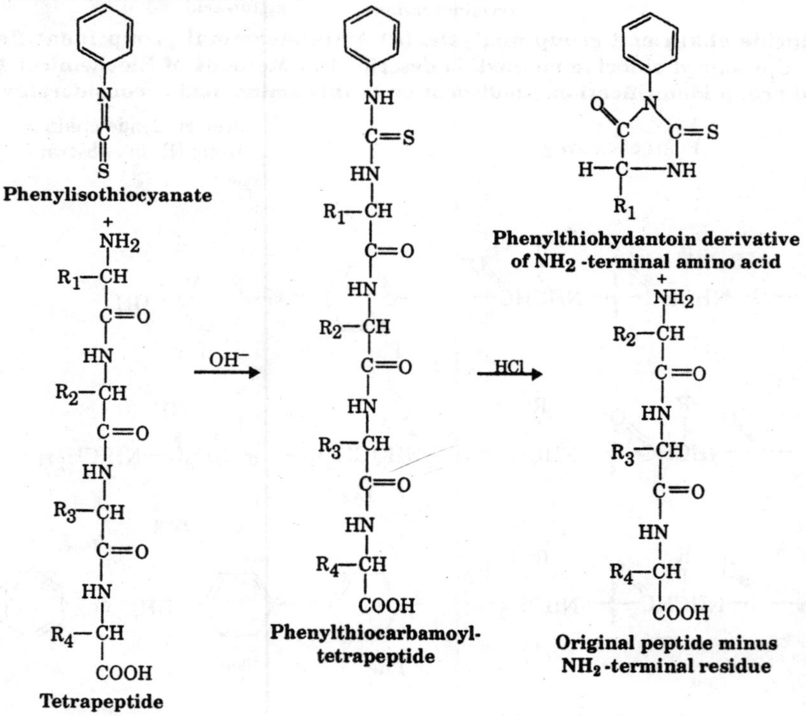

Fig. 5.21. The Edman degradation method for polypeptide sequence determination. The sequence is determined one amino acid at a time, starting from the amino-terminal end of the polypeptide. First the polypeptide is reacted with phenylisothiocyanate to form a polypeptidyl phenylthiocarbamyl derivative. Gentle hydrolysis releases the amino-terminal amino acid as a phenylthiohydantoin (PTH), which can be separated and detected spectrophotometrically. The remaining intact polypeptide, shortened by one amino acid, is then ready for further cycles of this procedure. A more sensitive reagent, dimethylaminoazobenzene isothiocyanate, can be used in place of phenylisothiocyanate. The chemistry is the same.

The success of these devices depends in large part on the covalently linking the peptide to be sequenced to glass beads. Attachment of the peptide through its carboxyl-terminal group to this immobile phase facilitates the complete removal of potentially contaminating reaction products during successive stages of the degradation. Finally, having established the sequences of the individual peptides, it is necessary only to establish how they are connected together in the intact protein (step 6).

It is at this stage that we see why the preceding sequence analysis was performed on peptides obtained by two different specific cleavage methods. This approach makes it possible to piece together the overall sequence, because the two sets of results produce overlapping sequences. That is, the free amino and carboxyl residues of peptides originally interconnected in the intact protein and liberated by one specific cleavage method recur in the internal sequences of the peptides liberated by a second specific method. Once the protein's primary sequence has been determined, the location of disulfide bonds in the intact protein can be established by repeating a specific enzymatic cleavage on another sample of the same protein in which the disulfide bonds have not previously been cleaved.

Separation of the resulting peptides shows the appearance of one new peptide and the disappearance of two other peptides, when compared with the enzymatic digestion product of the material whose disulfide bonds have first been chemically cleaved. In fact, these difference techniques are generally useful in the detection of sites of mutations in protein molecules of previously known sequence, because a single substitution generally affects the chromatographic properties of only a single peptide released during proteolytic digestion. Great progress has been made in recent years in denising procedures for sequencing the DNA that encodes for proteins.

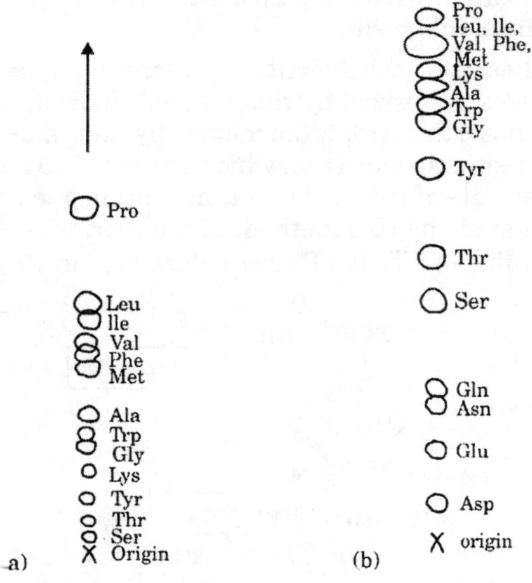

Fig 5.22. Thin-layer chromatography of amino acid-phenylthiohydantoin derivatives on silica gel plates, (*a*) Separation is done in a 98:2 mixture of chloroform and ethanol. (*b*) This is followed by further separation using an 88:2:10 mixture of chloroform, ethanol, and methanol. More sophisticated procedures, using column chromatography, give superior resolution and improved sensitivity. Automated sequencers always use such procedures.

Knowing the sequence of coding triplets in DNA allows us to read off the amino acid sequence of the corresponding protein. Nevertheless, such studies have produced the remarkable observation that some eudaryotic DNA sequences coding for proteins are not continuous but instead contain untranslated intervening DNA sequences. Although these results have profound implications for protein evolution, they obviously confound the general applicability of DNA-sequencing methods for the purposes of protein primary structure determination.

In cases like this, the usual solution has been to isolate the mRNA for the protein and use this to make a DNA carrying the same sequence. This procedure circumvents the intervening sequence problem because the mRNA carries only the coding sequences.

Chemical Synthesis of Peptides

Knowledge about the structure-function interrelationships in proteins and peptides has encouraged bichemists to develop techniques for synthesizing peptides and proteins with predetermined sequences. To synthesize a peptide in the laboratory, we must overcome several problems related to preventing undesired groups from reacting. The amino and carboxyl groups that are to remain unlinked must be blocked; so must all reactive side chains. After blocking those groups to be protected, the charboxyl group is activated. It is of interest that carboxyl-group activation is also employed in natural bisynthesis in the cell.

After peptide synthesis, the protecting groups must be removed by a mild method. The overall process—comprising protection, activation, coupling, and unblocking—is shown in figure 3.23. An important variation of the usual methods of peptide synthesis involves attaching a protected (*t*-butoxycarbonyl group) amino acid to a solid polystyrene resin, removal of the amino protecting group, condensation with a second protected amino acid, and so on. In the last step, the finished peptide is cleaved from the resin. This method has the advantage that cumbersome purification between steps, often resulting in serious losses, is replaced by mere washing of the insoluble resin.

Because each reaction is essentially quantitative, very long peptides, and even proteins, can be synthesized by this method. Indeed, Li synthesized a 39-amino-acid protein hormone, adrenocortiocotropic hormone, by this method, and Robert Merrified synthesized bovine pancreatic ribonuclease, which contains 129 amino acids in a single polypeptide chain. A number of variants of ribonuclease that contain one or more changes in amino acid sequence also have been made by this method. The importance of the Merrifield process was underscored by the awarding of a Nobel Prize to Merrifield in 1984.

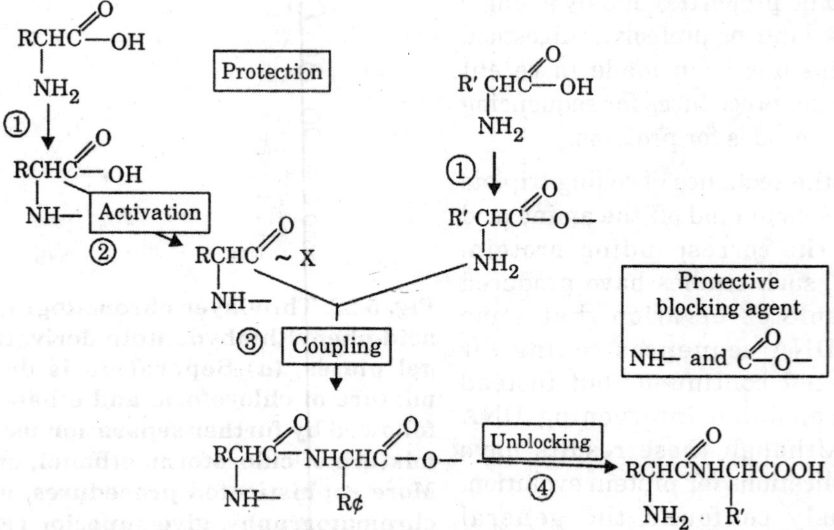

Fig. 5.23. Schematic diagram illustrating the chemical method for peptide synthesis. First the amino acids to be linked are selected. The carboxyl group and the amino group that are to be excluded from peptide synthesis are protected (steps 1 and 1'). Next the amino acid containing the unprotected carboxyl group is carboxyl-activated (step 2). This amino acid is mixed and reacted with the other amino acid (step 3). Protecting groups are then removed from the product (step 4).

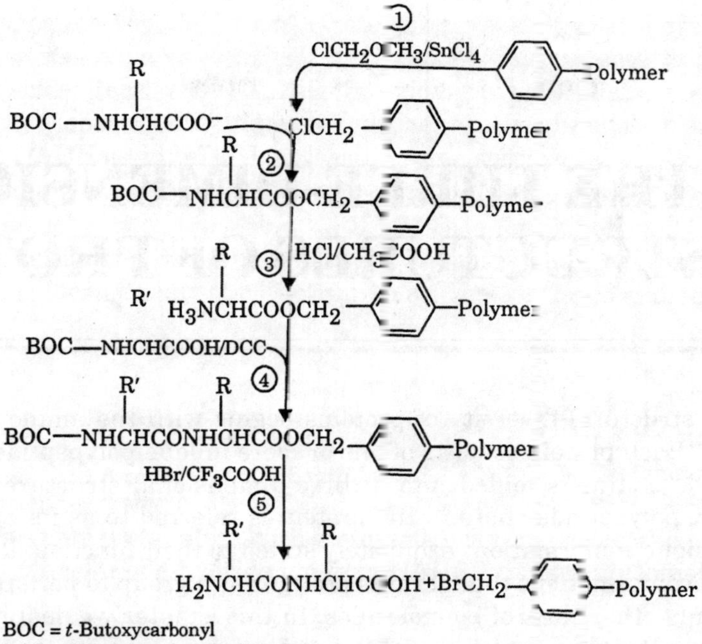

(1) $ClCH_2OCH_3/SnCl_4$ + Polymer-linked benzene ring

$BOC—NH\overset{R}{C}HCOO^-$ + $ClCH_2$—Polymer

(2) $BOC—NH\overset{R}{C}HCOOCH_2$—Polymer

(3) HCl/CH_3COOH

$H_3N\overset{R}{C}HCOOCH_2$—Polymer

$BOC—NH\overset{R'}{C}HCOOH/DCC$

(4) $BOC—NH\overset{R'}{C}HCONH\overset{R}{C}HCOOCH_2$—Polymer

(5) HBr/CF_3COOH

$H_2N\overset{R'}{C}HCONH\overset{R}{C}HCOOH + BrCH_2$—Polymer

BOC = *t*-Butoxycarbonyl
DCC = Dicyclohexylacarbodiimide

Fig. 5.24 Merrifield procedure for solid-state dipeptide synthesis. (1) Polymer is activated. (2) Amino acid containing a *t*-butoxycarbonyl (BOC)-protecting group is carboxyl-linked to the polymer. This amino acid will be the carboxyl-terminal amino acid in the final peptide. (3) The BOC protecting group is removed from the polymer-linked amino acid. (4) A second amino acid, containing a BOC on its α-amino group and a dicyclohexylcarbodiimide (DCC)-activated group, is reacted with the column-bound amino acid to form a dipeptide. (5) The dipeptide is released from the polymer and the BOC-protecting group by adding hydrogen bromide (HBr) in trifluoroacetic acid.

THE THREE-DIMENSIONAL STRUCTURES OF PROTEINS

The enormous structural diversity of proteins begins with the amino acid sequences of polypeptide chains. Each protein consists of one or more unique polypeptide chains, and each of these polypeptide chains is folded into a three-dimensional structure. The final folded arrangement of the polypeptide chain in the protein is referred to as its conformation. Most proteins exist in unique conformations exquisitely suited to their function. It is the availability of a wide variety of conformations that permits proteins as a group to perform a broader range of functions than any other class of biomolecules. In this chapter we deal primarily with the structural properties of proteins, and in the following chapter we consider the functional diversity of proteins.

Traditionally proteins have been divided into two groups : fibrous and globular. Fibrous proteins aggregate to form highly elongated structures having the shape of fibers or sheets. Each protein unit that makes up these aggregated structures is built from a repeating structural motif, giving the molecules a simple structure that is relatively easy to analyze. By contrast, globular proteins are more complex, containing one or more polypeptide chains folded back on themselves many times to give an approximately spherical shape. We consider the relatively simple fibrous proteins first.

Fibrous Protein Structures

Linus Pauling and Robert Corey examined the structures of crystals formed by amino acids and short peptides before they ventured into the world of proteins. From their crystallographic investigations of amino acids and peptides, they formulated two rules that describe the ways in which amino acids and peptides interact with one another to form noncovalently bonded crystalline structures.

These rules laid the foundations for our understanding of how amino acids in protein polypeptide chains interact with one another. Rule number one was that the pelptidyl C—N linkage and the four atoms to which the C and the N atoms are directly linked always form a planar structure as though the C—N linkage is a double bond rather than a single bond as normally written. Pauling reasoned that the C— N linkage is a resonating structure with partial double-bond character, so that it locks the peptide grouping into a planar conformation. This property is extremely important because it greatly reduces the flexibility in the polypeptide chain. With this grouping in a rigid conformation, the only flexibility remaining in the polypeptide backbone results from rotation about the carbon that joins adjacent peptide planar groups. The second rule that Pauling and Corey formulated was that peptide carbonyl and amino groups always form the maximum number of hydrogen bonds.

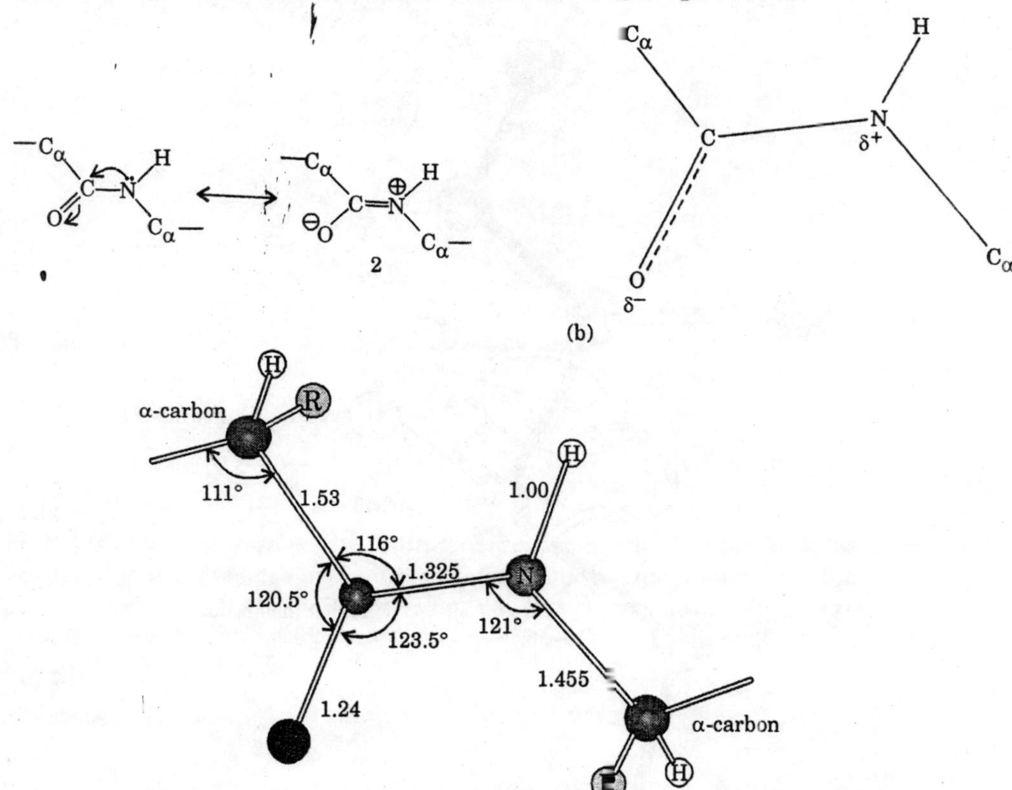

Fig. 6.1. Resonance and the planar structure of the peptide bond. *(a)* Two major hybrids contribute to the structure of the peptide bond. In structure 1 the C—N bond is a single bond with no overlap between the nitrogen lone electron pair and the carbonyl carbon. The carboxyl carbon is *sp*²-hybridized and is therefore planar, whereas the nitrogen is *sp*³-hybridized and pyramidal. By contrast, in structure 2 there is a double bond between the amide nitrogen and the carbonyl carbon; also, the nitrogen atom bears a charge of + 1 and the carbonyl oxygen bears a charge of –1. Both the carboxyl carbon and the amide nitrogen are *sp*²-hybridized, both are planar, and all six atoms lie in the same plane. *(b)* The structure of the peptide bond is a compromise between the two resonating hybrids, structures 1 and 2. *(c)* Dimensions of the peptide bond and surrounding linkages. The C—N bond length of 1.325 Å is significantly less than the length of a single C—N bond, 1.47 Å.

Recall that in water hydrogen bonds are formed between a partially unshielded proton from one molecule and an oxygen atom that originated from another molecule. Nitrogen atoms can also serve as H bond acceptors. The attraction between the H bond donor and the H bond acceptor is strongest along the lone pair orbital axis of the acceptor atom. As a rule the angle between an N or O acceptor and an N—H or O—H donor is close to 180°. Thus a hydrogen bond brings two interacting groups close together and orients them in a certain way. This second rule further limits the number of conformations available to polypeptide chains. Following their investigations on amino acid and peptide crystals Pauling and Corey turned their attention to the x-ray diffraction patterns of a number of fibrous proteins.

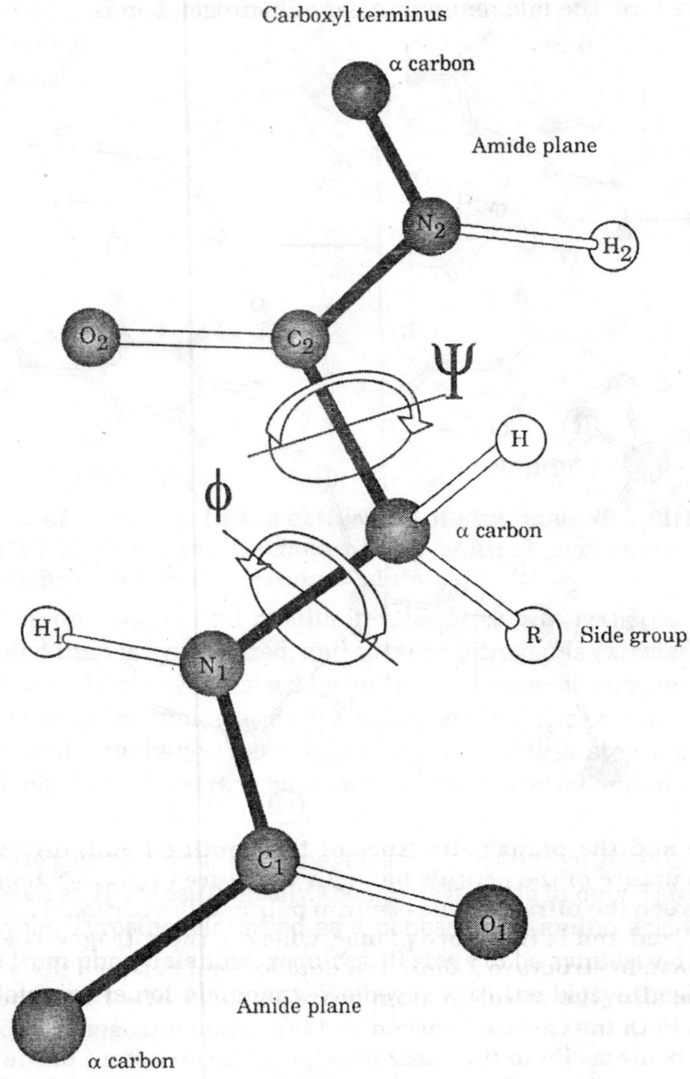

Fig. 6.2. Basic dimensions of a dipeptide. The conformational degrees of freedom of a polypeptide chain are restricted to rotations about the single-bond connections between the adjacent planar transpeptide groups to C_α, that is, the C_α—C_2 and C_α—N_1 single bonds. The corresponding rotations are represented by γ and ϕ, respectively, which have values of 180° for the fully extended configuration shown.

A vast number of fibrous proteins exist in nature, but the majority of them give diffraction patterns that fall into one of three types : the α pattern, the β pattern, and the collagen pattern. Fiber diffraction data only give information about the repeating units of a protein structure because of the lack of three-dimensional order in the fibers; but because fibrous protein polypeptide chains are arranged in simple repetitious units the overall structure could be deduced. The first type of diffraction pattern, which was observed for a subgroup of the keratins, was consistent with a helical arrangement of the polypeptide chains; this pattern indicated that there are two regularly repeating units along the helix axis : one with a repeat of 5.4 Å and one with a repeat of 1.5 Å.

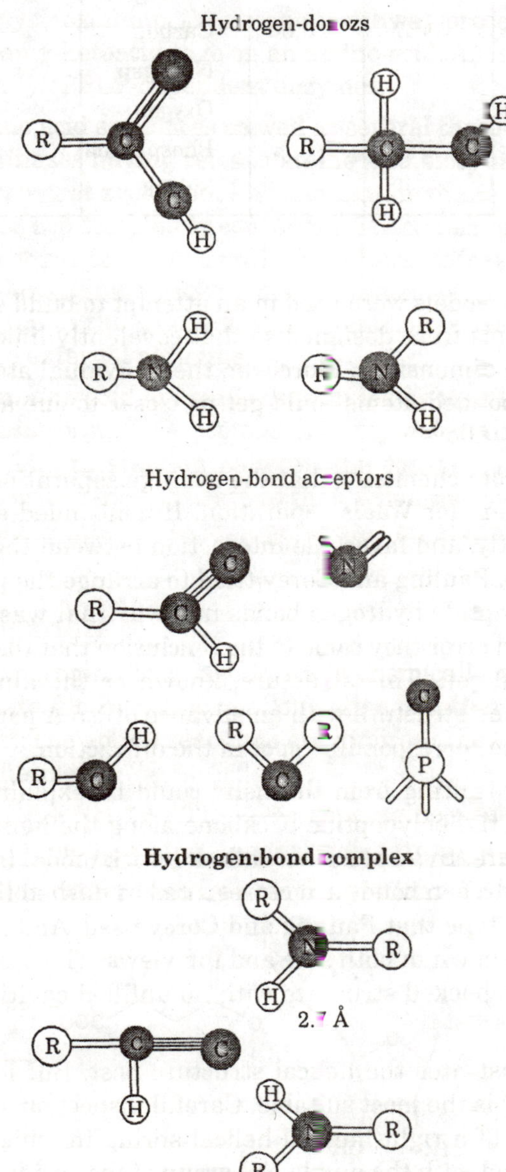

Fig. 6.3. Major hydrogen-bond donor and acceptor groups found in proteins. Note that the angle between the O acceptor and the N—H donor is 180° in the hydrogen-bond complex.

Table 6.1. Radii for Covalently Bonded and Nonbonded Atoms

Element	*Covalent Bond Radii (in Å)*			Elemen	*Van der Waals Radii (in Å)*
	Single Bond	*Double Bond*	*Triple Bond*		
Hydrogen	0.30			Hydrogen	1.2
Carbon	0.77	0.67	0.60	Carbon	2.0
Nitrogen	0.70			Nitrogen	1.5
Oxygen	0.66			Oxygen	1.4
Phosphorus	1.10			Phosphorus	1.9
Sulfur	1.04			Sulfur	1.8

Molecular Models

Space-filling molecular models were used in an attempt to build structures compatible with the x-ray data. These models were designed so that covalently linked atoms were accurately spaced according to known dimensions. Moreover, the individual atoms made as hard spheres were of a size so that nonbonded atoms could get no closer to one another than their van der Waals radii would normally allow.

It should be recalled from chemistry that the average separation between two nonbonded atoms is known as their van der Waals separation. If nonbonded atoms get closer than this they repel each other greatly, and favorable interaction between them falls less rapidly if the distance is larger than this. Pauling and Corey tried to arrange the polypeptide chains so as to maximize the number of peptide hydrogen bonds in a way that was consistent with the x-ray diffraction data. By trial and error they came to the conclusion that the most acceptable structure was a right-handed helical coil. This structure, known as the alpha (α) helix, has a rigid, regularly repeating backbone structure with an advance of 1.5 Å per amino acid residue along the helix axis explaining the corresponding spot in the diffraction.

The diffraction spots resulting from the helix could be explained by a spacing of 5.4 Å between adjacent terms of the polypeptide backbone along the helix axis. We see three ways of representing this structure. In *(a)* we see a ball-and-stick model in which the planar groups are highlighted and the hydrogen bonds are represented as dashed lines. In *(b)* we see a space-filling model similar to the type that Pauling and Corey used. And in (*c*) we see a wire model. The latter two models are shown in both side and top views. The space-filling model *(b)* shows that the α helix is a tightly packed structure with no unfilled cavities whether one looks at a profile or down the helix.

The wire model (*c*) illustrates the helical structure best. But for purposes of discussion, the ball-and-stick model (*a*) is the most suitable. Careful inspection shows that the polypeptide backbone follows the path of a right-handed helical spring in which each residue's carbonyl group forms a hydrogen bond with the amide NH group of the residue four amino acids further along the polypeptide chain. All residues in the α helix have nearly identical conformations so they lead to a regular structure in which each 360° of helical turn incorporates approximately 3.6 amino acid residues and rises 5.4 Å along the helix axis direction.

This rise accounts for the observed advance per amino acid residue along the helix axis of 5.4/3.6 = 1.5 Å. Although alternative helical arrangements having different hydrogen-bonding patterns and different geometries are possible the α helix is by far the most commonly observed.

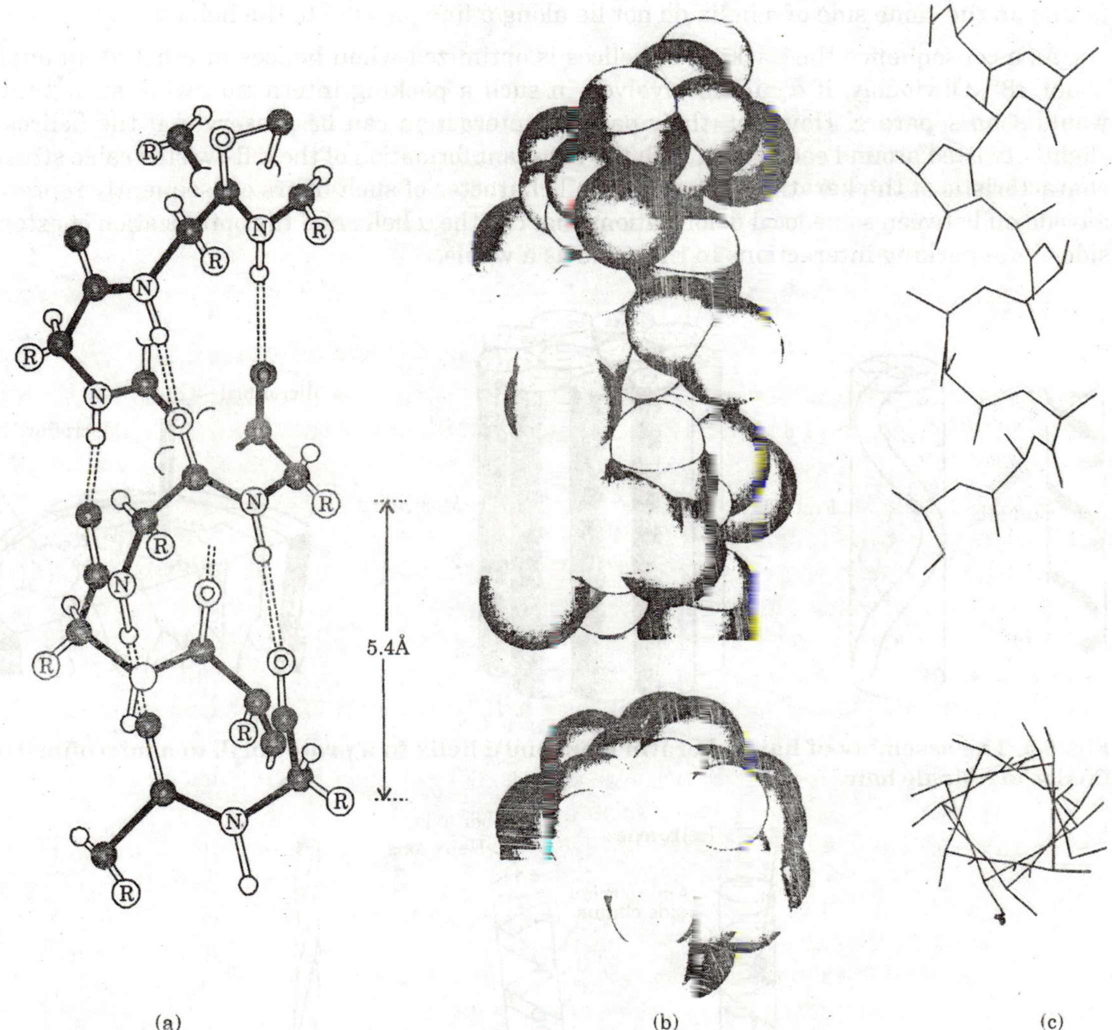

Fig. 6.4. Three ways of projecting the α-helix. (*a*) This simple ball-and-stick model highlights the planar peptides. The interpeptide hydrogen bonds are shown by dashed lines, and the amino acid side chains are indicated by R groups. Approximately two turnings of the helix are shown. There are about 3.6 residues per turn. In (*b*) and (*c*) we see identical projections of a side view using space-filling and wire models, respectively. The space-filling models use van der Waals radii for the atoms.

The unique stability of the α helix is related to the formation of regularly arranged hydrogen bonds between all the carbonyl and amino groups and to the tight packing achieved in folding the chain to form the structure. Fibrous proteins in which the α helix is a major structural component are found in hair, scales, horns, hooves, wool, beaks, nails, and claws—proteins referred to as α-keratins. When individual α helices aggregate in this side-by-side fashion, they usually form long cables in which the individual helices are spirally twisted so that the resulting cable has an overall left-handed twist. The formation of a cable with twisted α helices results from optimization of packing among the amino acid side-chain residues between helices. In how the side-chain residues of an α helix are arranged in a spiral fashion so that residues falling on the same side of a helix do not lie along a line parallel to the helix axis.

As a consequence the packing of helices is optimized when helices interact at an angle of about 18°. Obviously, if α hilices involved in such a packing interaction were straight, they would soon separate. However, their packing interaction can be preserved if the helices are slightly twisted around each other, with the resultant formation of the left-twisted cable structure characteristic of the keratins. The coiled-coil character of such fibers consequently represents a trade-off between some local deformations that coil the α helix and the optimization of extended side-chain packing interactions in the cable as a whole.

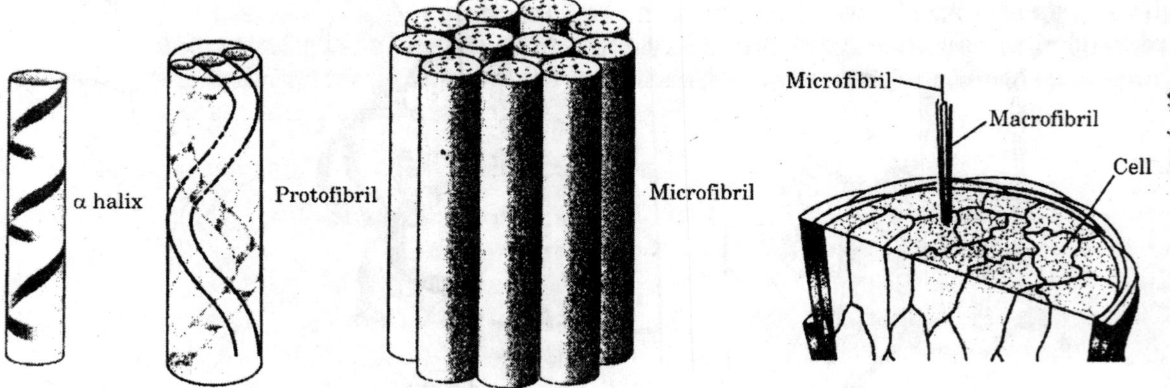

Fig. 6.5. The assembly of hair α-keratin from one α helix to a protofibril, to a microfibril, and finally, to a single hair.

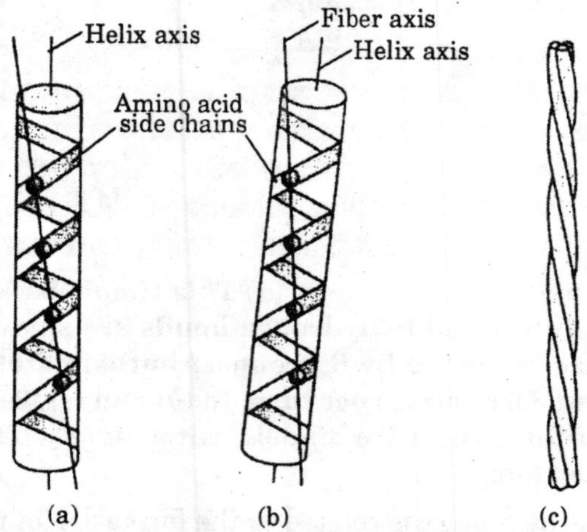

Fig. 6.6. Coiling of α helices in α-keratins. Residues on the same side of an α helix form rows that are tilted relative to the helix axis. Packing helices together in fibers is optimized when the individual helices wrap around each other so that rows of residues pack together along the fiber axis. Helices in coiled coil (c) are oriented in parallel.

The springiness of hair and wool fibers results from the tendency of the α-helical cables to untwist when stretched and spring back when the external force is removed. In many forms of keratin, the individual α helices, or fibers, are covalently linked by disulfide bonds formed between cysteine residues of adjacent polypeptide chains. In addition to giving added strength to the fibers, the pattern of these covalent interactions serves to influence and fix the extent of curliness in the hair fiber as a whole. Chemical reactions and mechanical processes involving reductive cleavage, reorganization, and reoxidation of these interhelix disulfide bonds form the basis of the "permanent wave."

The β-Keratins Form Sheetlike Structures

Pauling and Corey noticed that certain fibrous proteins give radically different diffraction patterns and behave macroscopically as sheets rather than fibers. Diffraction patterns of fibrous proteins from different sources reveal a repeat along the direction of the extended polypeptide chain of either 13.0 or 14.0 Å, suggesting two closely related structures. Pauling and Corey interpreted these patterns as resulting from extended polypeptide chains lying side by side either in a parallel or an antiparallel fashion.

Although both of these structures exist in nature, the antiparallel sheet is more common in fibrous proteins. This structure is known as the antiparallel β-pleated sheet. Regular hydrogen bonds form between the peptide backbone amide NH and carbonyl oxygen groups of adjacent chains. The β sheet can be extended into a multistranded structure simply by adding successive chains in the appropriate directions to the sheet.

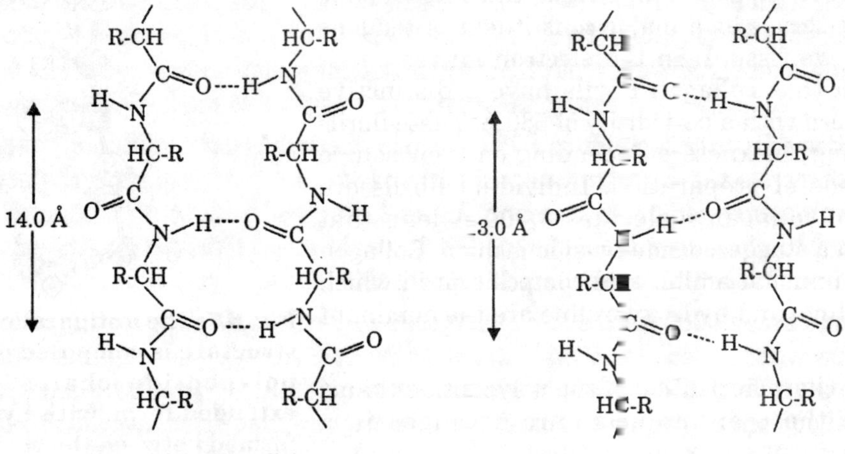

(a) Antiparallel — 14.0 Å

(b) Parallel — 13.0 Å

Fig. 6.7. Two forms of the β-sheet structure : (*a*) the antiparallel and (*b*) the parallel β sheet. The advance per two amino acid residues is indicated for each structure.

Parallel and antiparallel β-pleated sheets are both composed of polypeptide chains that have conformations pointing alternate R groups to opposite sides of the sheet but that have their peptide planes nearly in the sheet plane to allow for good interchain hydrogen bonding. Nevertheless, the chain conformation that produces the best interchain hydrogen bonding in parallel sheets is slightly less extended than that for the antiparallel arrangement. As a result, the parallel sheet has both a shorter repeat period 6.5 Å (versus 7.0 for the antiparallel structure) and a more pronounced pleat. The best known β-keratin structure in nature is that found in certain silks. Silks are composed of stacked anti-parallel β sheets.

Sequence analysis of silk proteins shows them to be composed largely of glycine, serine, and alanine, in which every alternate residue is glycine. Since the side-chain groups of a flat antiparallel sheet point alternately upward and downward from the plane of the sheet, all the glycine residues are arranged on one surface of each sheet and all the substituted amino acids are on the other. Two or more such sheets can consequently be packed intimately together to form an arrangement of stacked sheets in which two adjacent glycine-substituted or alanine-substituted sheet surfaces interlock with each other.

Owing to me extended conformations of the polypeptida chains in the β sheets and the interlocking of the side chains between sheets, silk is a mechanically rigid material that resists stretching.

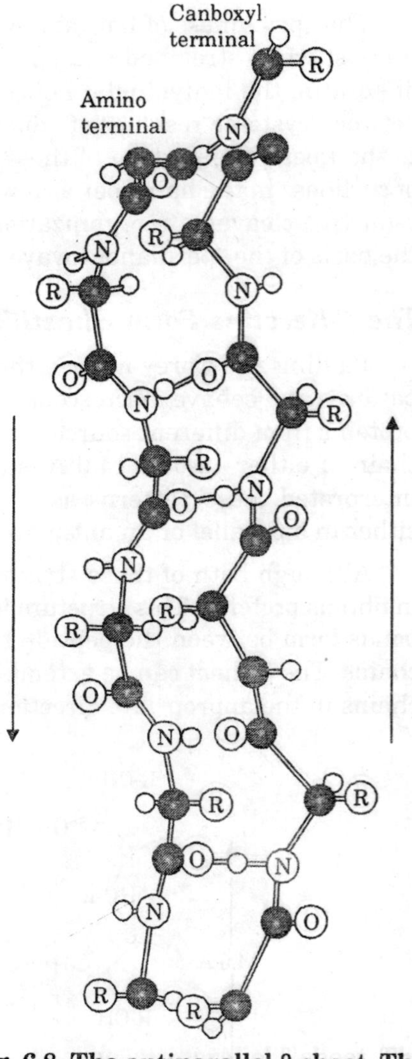

Fig. 6.8. The antiparallel β sheet. This structure is composed of two or more polypeptide chains in the fully extended form, with hydrogen bonds formed between the chains. Hydrogen bonds are shown as dashed lines.

Triple-Stranded

Collagen is a particularly rigid and inextensible protein that serves as a major constituent of tendons and connective tissues. In the electron microscope it can be seen that collagen fibrils have a distinctive banded pattern with a periodicity of 680 Å. These fibrils are of varying thicknesses depending on their source and the mode of preparation. Individual fibrils are composed of collagen molecules 3,000 Å long that aggregate in a staggered side-by-side fashion. Collagen has a most unusual amino acid composition in which glycine, proline, and hydroxyproline are the dominant amino acids.

Further characterization of the polypeptide chains shows that these amino acids are arranged in a repetitious tripeptiae sequence, Gly-X-Y, in which X is frequently a proline and Y is frequently a hydroxyproline. This unusual amino acid sequence and the unique diffraction pattern of collagen were strong indications for a totally different type of fibrous protein. Pauling attempted to determine the structure of collagen by his molecular model approach but failed. At a meeting in Cambridge, England, in 1958 where the correct structure was presented, Pauling jested that the structure he had proposed might be more stable than the one found in nature. The natural structure was found by Ramachandran. The repeating proline residue excluded the possiblility that the polypeptide chains in collagen could adopt either an α-helical or a β-sheet conformation.

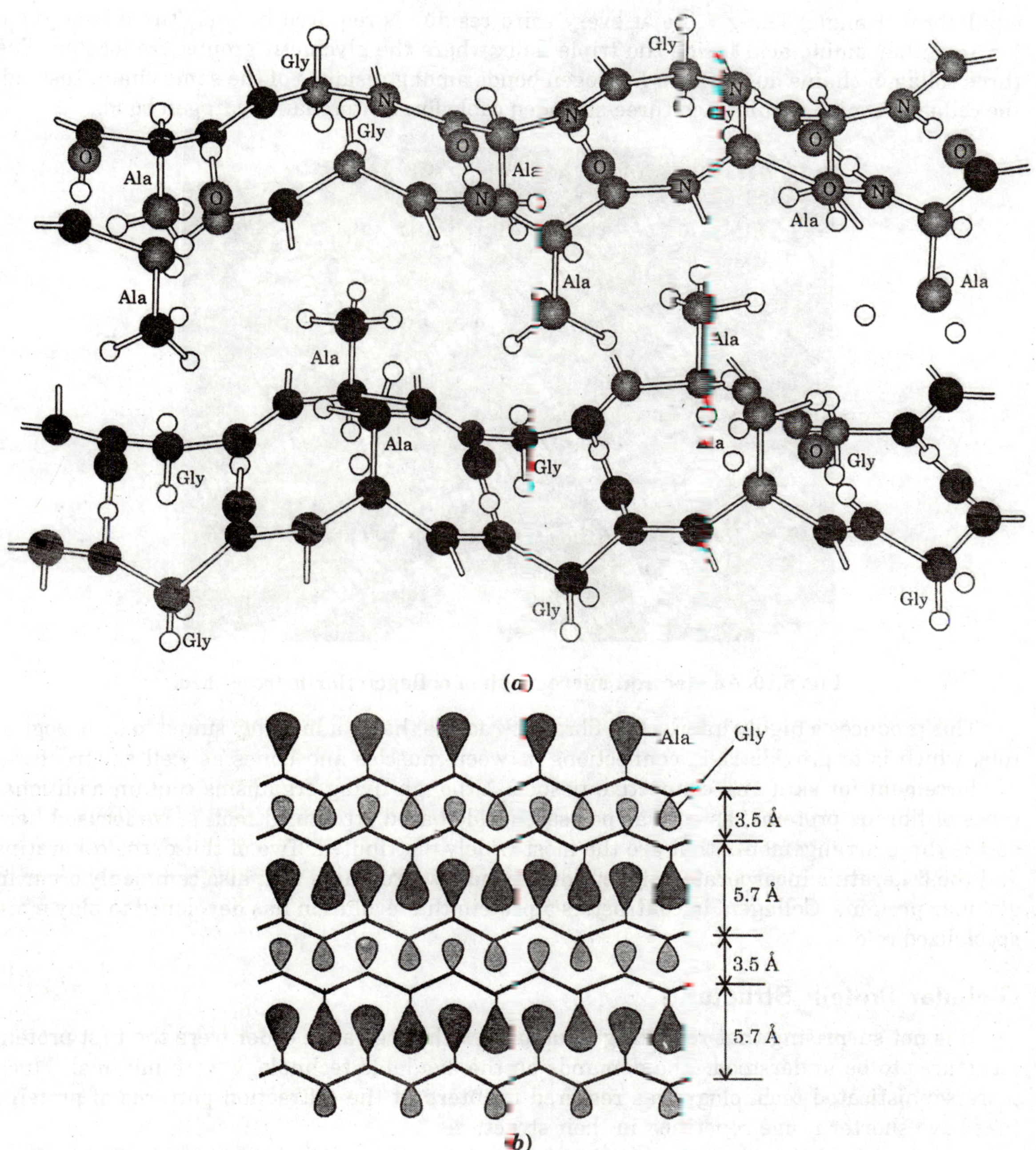

Fig. 6.9. The three-dimensional architecture of silk. ***(a)*** **The side chains of one sheet nestle quite efficiently between those of neighbouring sheets.**

Instead, individual collagen polypeptide chains assume a left-handed helical conformation and aggregate into three-stranded cables with a right-handed twist. When viewed down the polypeptide chain axis, the successive side-chain groups point toward the corners of an equilateral triangle. The glycine at every third residue is required because there is no room for any other amino acid inside the triple helix where the glycine R groups are located. The three collagen chains do not form hydrogen bonds among residues of the same chain. Instead, the collagen chains within each three-stranded cable form interchain hydrogen bonds.

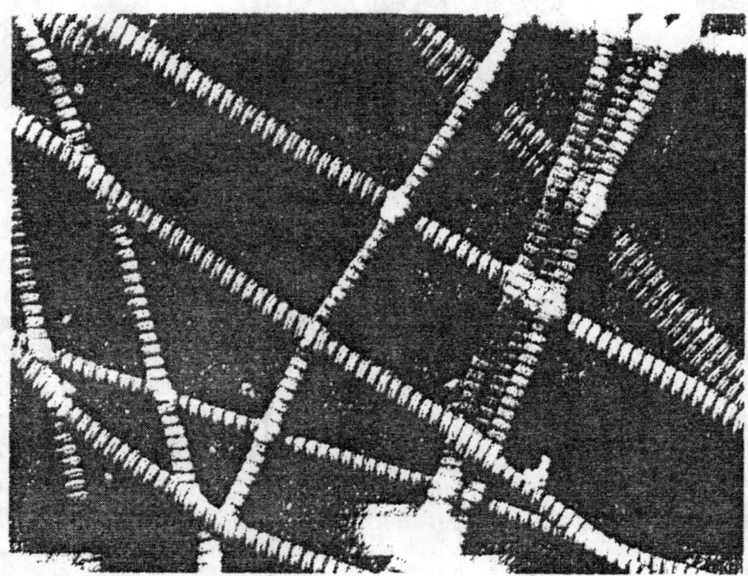

Fig. 6.10. An electron micrograph of collagen fibrils from skin.

This produces a highly interlocked fibrous structure that is admirably suited to its biological role, which is to provide rigid connections between muscles and bones as well as structural reinforcement for skin and connective tissues. Although living organisms contain additional types of fibrous proteins, as well as polysaccharide-based structural motifs, we focused here on the three arrangements that are the most widely distributed. Two of these, the α-keratins and the β-keratins incorporate polypeptide secondary structures that also commonly occur in globular proteins. Collagen, in contrast, is a protein that evolution has developed to play more specialized role.

Globular Protein Structures

It is not surprising that repeating structures with long-range order were the first protein structures to be understood. The demands on the available technology were minimal. Much more sophisticated technology was required to interpret the diffraction patterns of proteins that have shorter range repetition in their structures.

Among the crystallographers who struggled over these structures were two who emerged as the major pioneers : John Kendrew and Max Perutz. In the 1950s they led their research teams to determine the first of such structures—myoglobin and hemoglobin. Since that time enormous advances in protein chemistry and computer technology have systematized the necessary research and greatly reduced the amount of work and time required to determine a protein structure.

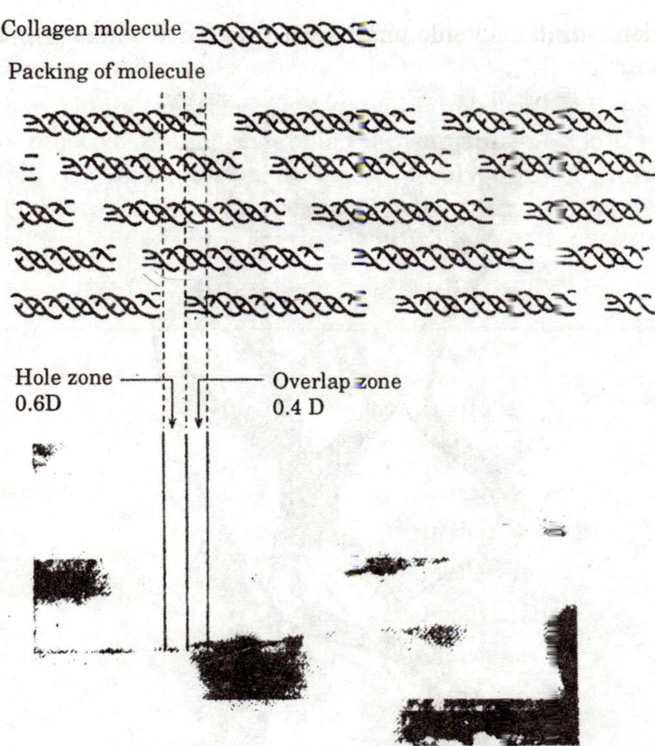

Fig. 6.11. The banded appearance of collagen fibrils in the electron microscope arises from the schematically represented staggered arrangement of collagen molecules *(above)* that results in a periodically indented surface. D, the distance between cross striations is ≈ 680 Å so that the length of a 3,000-Å-long collagen molecule is 4.4D.

Accurate structural determinations have now been made for more than 500 different proteins. From this wealth of data, patterns of structure are becoming apparent that suggest, among other things, that the overall folding arrangements of proteins may some day be predictable from the amino acid sequences of the polypeptide chains.

Folding of Globular Proteins

In the analysis of fibrous protein structures little mention was made of the importance of their interaction with water in determining the final folded structures of the proteins. This is because most of the side chains in fibrous proteins are exposed to water except when they interact with each other to form multimolecular aggregates. Then the relative between other similar side chains and water becomes a major issue.

In the case of globular proteins the interaction of amino acid side chains with water is a major issue from the start because globular proteins have many of their amino acid side chains buried in the interior of their folded structures. Hence, in our analysis of the structure of globular proteins we must be aware of the structural considerations that are important in the determination of fibrous proteins but also of additional considerations, first raised, that relate to the interaction of the amino acid side chains with water. We may think of the structure of globular proteins at four levels.

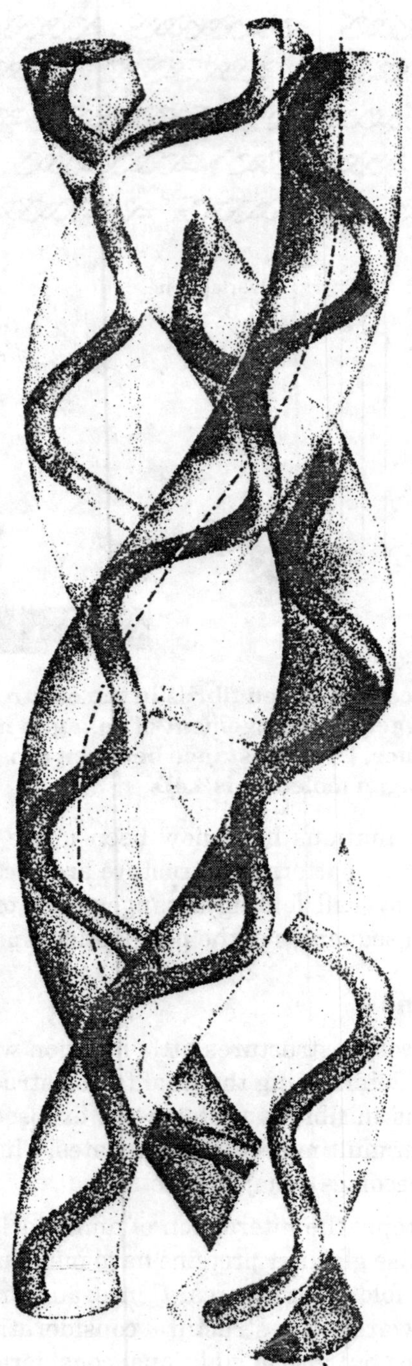

Fig. 6.12. The triple helix of collagen.

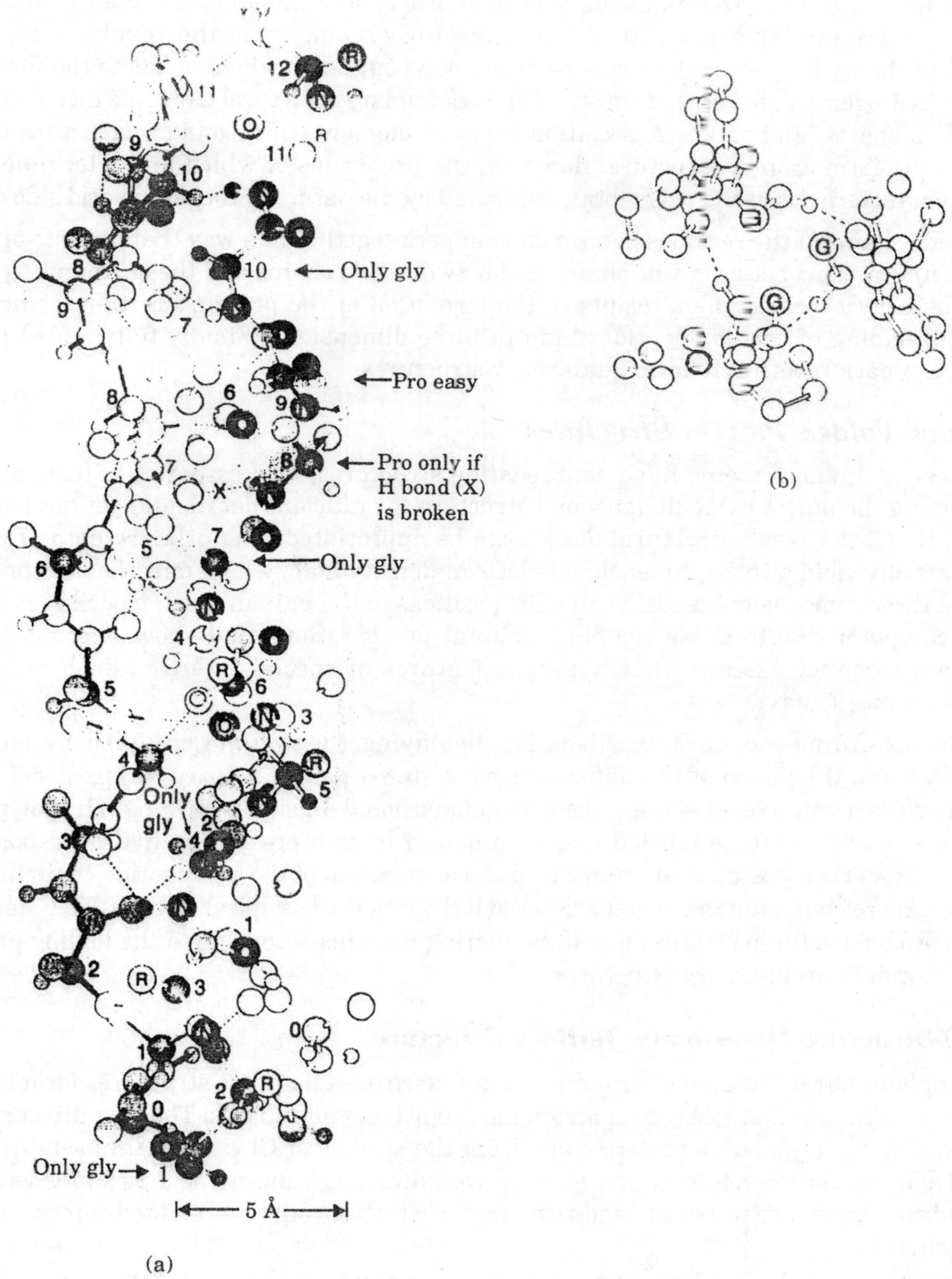

Fig. 6.13. The basic coiled-coil structure of collagen. Three left-handed single-chain helices wrap around one another with right-handed twist. (*a*) Ball-and-stick single-collagen chain. (*b*) View from the top of the helix axis. Note that glycines are all on the inside. In this structure the C=O and N—H groups of glycine protrude approximately perpendicularly to the helix axis so as to form interchain hydrogen bonds.

Ultimately, the entire three-dimensional structure is determined by a protein's amino acid sequence, or primary structure. At a higher level of organization, the regularly repeating geometry of the hydrogen-bonding groups of the polypeptide backbone leads to the formation of regular hydrogen-bonded secondary structures. Secondary structural elements include regions of α helix, β sheets, and bends. Association between elements of secondary structure in turn results in the formation of structural domains, the properties of which are determined both by chiral characteristics of the polypeptide chain and by the nature of the amino acid side chains.

The side chains of the residues in each domain pack together in a way that tends to optimize favorable interactions between side chains and between side chains and the surrounding water. Further association of domains results in the formation of the protein's tertiary structure—the overall folding of the polypeptide chain in three dimensions. Finally fully folded protein subunits can pack together to form quaternary structures.

Visualizing Folded Protein Structures

Because globular proteins have nonrepeating structures, it is essential to have a means for displaying the entire three-dimensional structure in sufficient detail, and yet not too much detail, so that the overall structural design can be appreciated. The primary data of protein crystallography yield a three-dimensional electron-density map, which must be interpreted in terms of a three-dimensional model of all atom positions in the protein. Such modeling is usually done by computer graphics. We see four different presentations that show selected parts of the protein ribonuclease and that highlight features of special interest. Each method of presentation has its advantages.

The space-filling model is excellent for displaying the volume occupied by molecule constituents and the shape of the outer surface. A stereo pair of a space-filling model. When the pair is viewed with stereo glassgs, the three-dimensional illusion is striking. The polypeptide chain, with N and O atoms labeled and with dotted lines representing hydrogen bonds. In addition, all α-carbon positions are numbered. An abstraction of the polypeptide chain in which the β strands are characterized as flat arrows and the α helices as spiral ribbons. This simplified style has proved useful in classifying and comparing proteins according to the folding patterns of their secondary and tertiary structures.

Primary Structure Determines Tertiary Structure

Throughout our discussion of protein structures we assumed that structures form because they represent the most stable way of arranging the polypeptide chains. The first direct support for this notion for a globular protein came from the studies of Christian Anfinsen. Anfinsen unfolded pancreatic ribonuclease, an enzyme containing 124 amino acid residues with four disulfide bridges, and then found conditions under which it could be refolded into its original native structure.

First, the enzyme was denatured in a solution containing a hydrogen-bond-breaking reagent (urea) and 2-mercaptoethanol (a thiol reagent that reduces disulfides to sulfhydryls), thus cleaving the covalent cross-links. These conditions have been used since as a general means of denaturing proteins in a manner that completely disrupts the conformation without causing precipitation. The reduced, denatured ribunuclease was enzymatically inactive.

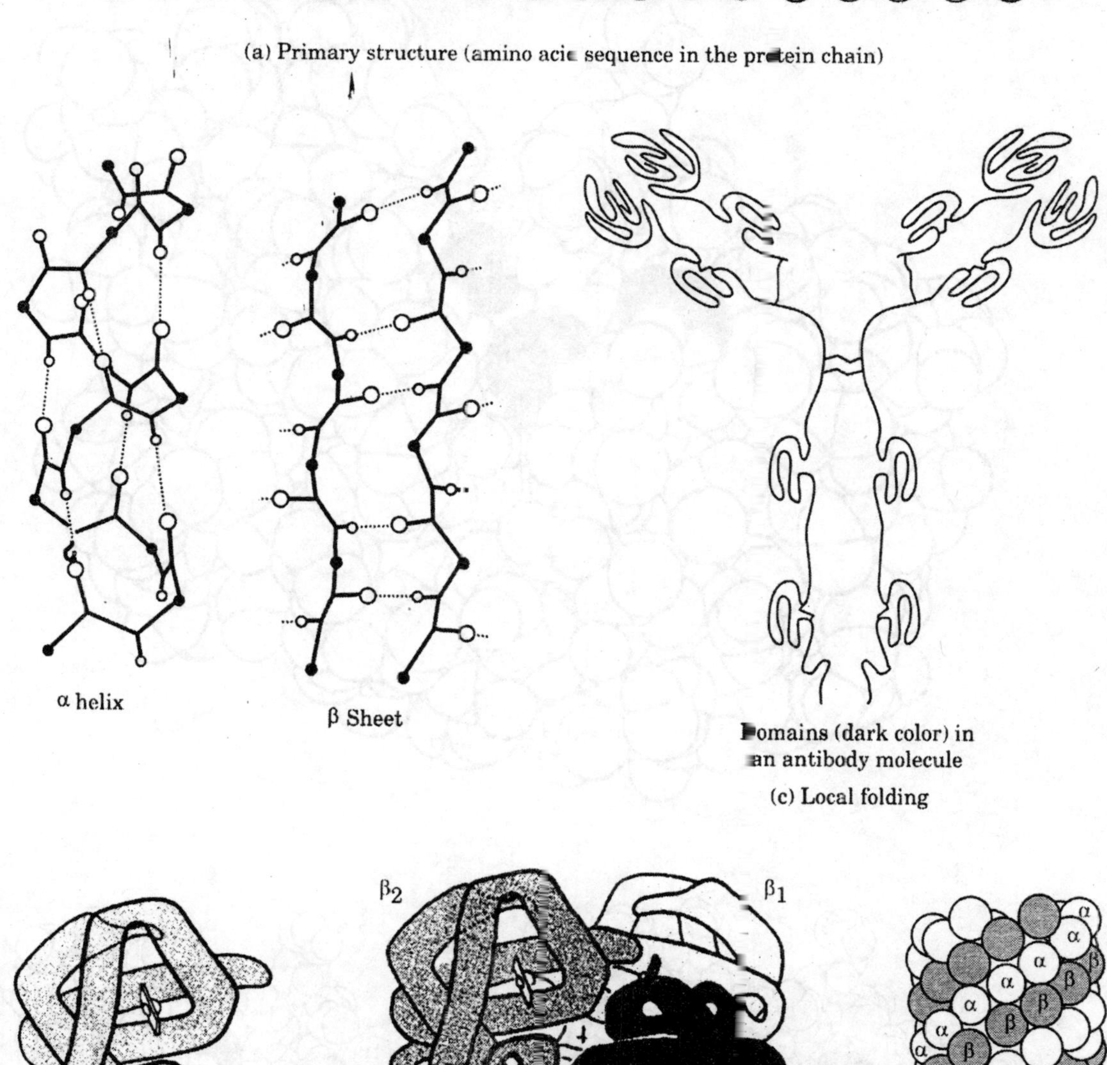

(a) Primary structure (amino acid sequence in the protein chain)

Domains (dark color) in an antibody molecule

(c) Local folding

One complete protein chain (β chain of hemoglobin)

(d) Tertiary structure

The four separate chains of hemoglobin assembled into an oligomeric protein

(e) Quaternary structure

β (white) and β (color) tubulin molecules in a microtubule

(f) Quaternary structure

Fig. 6.14. Hierarchies of protein structures.

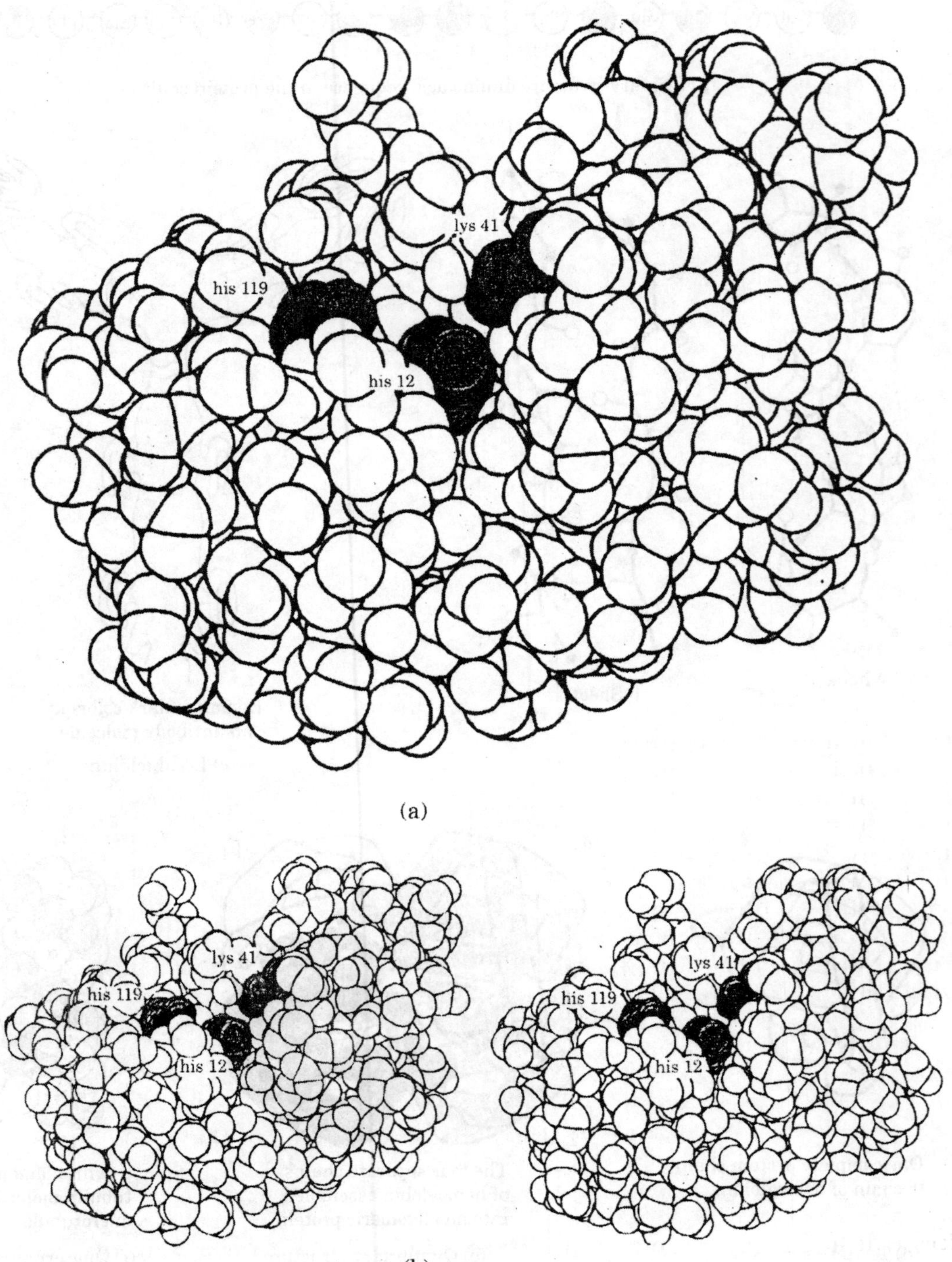

(a)

(b)

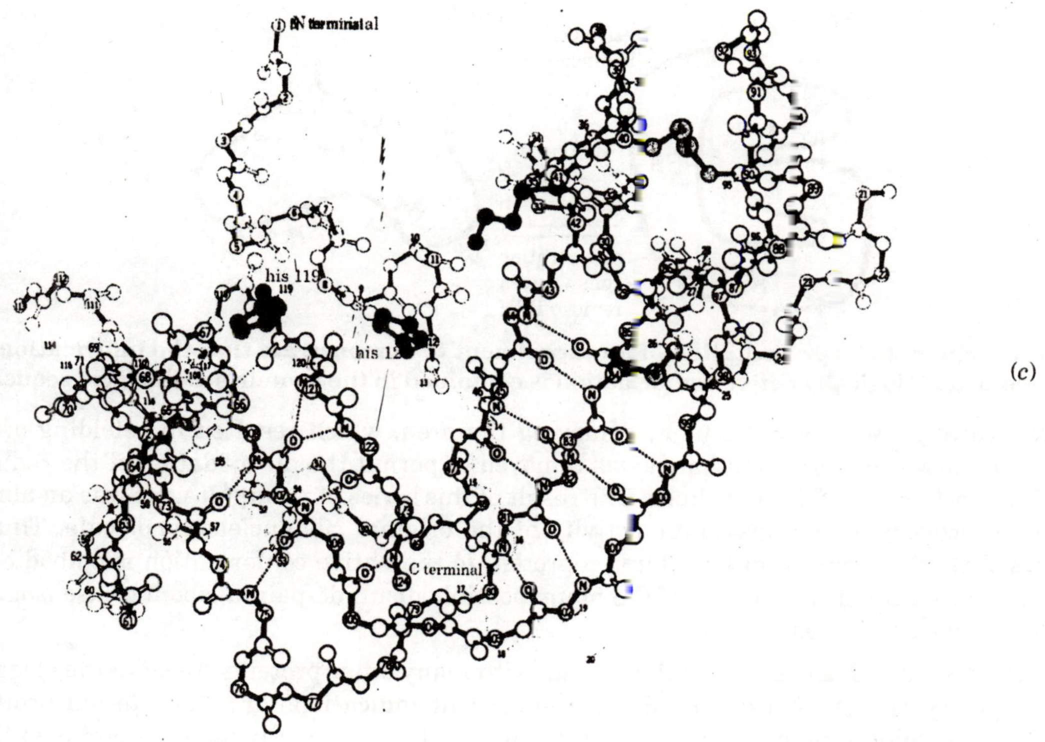

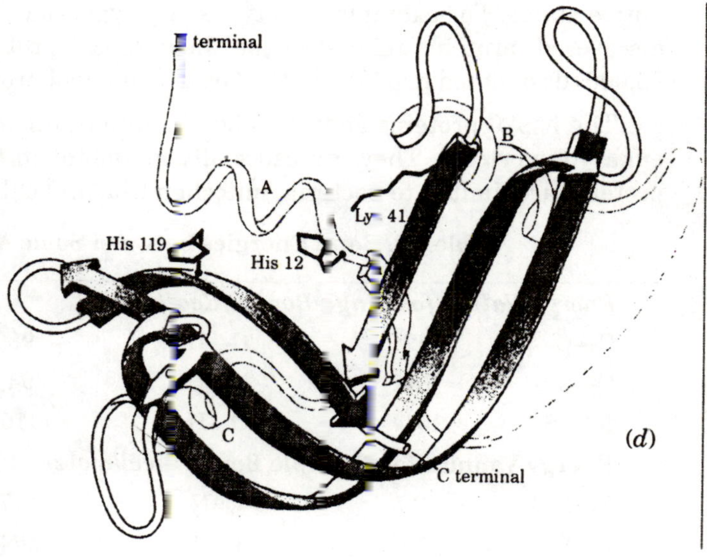

Fig. 6.15. Four ways of representing the three-dimensional structure of the protein pancreatic ribonuclease. *(a)* Space-filling model, *(b)* The stereo pair of space-filling models illusion can be seen without stereo glasses by the following method. With your eyes about 10 inches from the page, stare at the drawings below as if you were looking straight ahead at a far-away object. A double image forms and the central pair drifts together and fuses. Then the illusion becomes apparent. Adjust the page, if necessary, so that a horizontal line is perfectly parallel with the eyes. As an aid to seeing one image with each eye, use two cardboard tubes from toilet paper rolls. Close the right eye and focus the left eye on the left image. Do the same for the other eye. Now, with both eyes together, the three-dimensional illusion should appear, *(c)* Ball and stick model. Dotted lines represent hydrogen bonds, and numbers indicate α-carbon positions. *(d)* Ribbon model. The β-strands are represented by the flat red arrows. The α helices are shown as spiral ribbons.

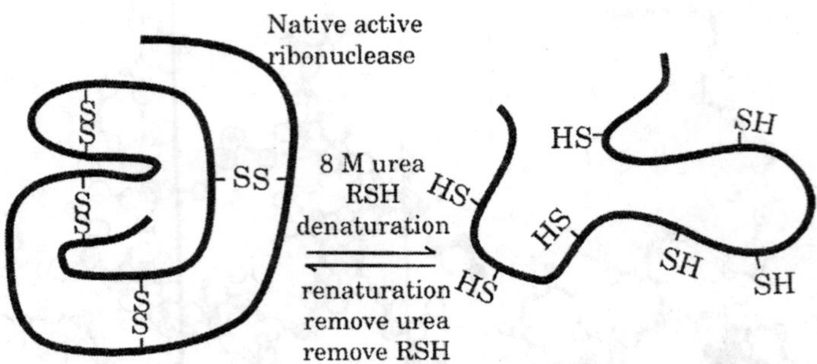

Fig. 6.16. Schematic representation of an experiment to demonstrate that the information for folding into a biologically active conformation is contained in the protein's amino acid sequence.

Renaturation was carried out by removing the area, which resulted in refolding of the protein. Finally the mercaptoethanol was removed to permit the air oxidation of the reduced disulfides back to disulflde cross-links. The result of this series of manipulations was an almost complete recovery of the enzymatic activity of the original ribonuclease molecule. Thus it appears that the information for folding a protein to the native conformation is embodied in the amino acid sequence, because of the many possible disulfide-paired ribonuclease isomers, only one was formed in major yield.

Further studies have given similar results with many other proteins. Despite the elegance and simplicity of Anfinsen's experiment, the current indications are that special proteins accelerate the folding process and may return unfolded or incorrectly folded proteins to their native states. For example, two types of polypeptide-chain-binding (PCB) proteins have been discovered: proteins related to the "heat shock" protein with a subunit molecular weight of 70,000 (designated hsp70) and the GroEL family of proteins.

The hsp70 proteins appear to be ubiquitous, occurring in bacteria, mitochondria, and the eukaryotic cytosol. They ate especially abundant in the endoplasmic reticulum. The GroEL proteins are limited to bacteria, mitochondria, and chloroplasts.

Table 6.2. Bond Energies between Some Atoms of Biological Interest

Energy Values for SingleBonds (Kcal/mole)					
C—C	82	C—H	99	S—H	81
O—O	34	N—H	94	C—N	70
SS—S	51	O—H	110	C—O	84
Energy Vanlues for Multiple Bonds (kcal/mole)					
C=C	147	C=N	147	C=S	108
O=O	96	C=O	164	N=N	226

Secondary Valence Forces are the Glue that Holds Polypeptide Chains Together

The studies of Anfinsen not only showed that folding can be a spontaneous process predetermined by the primary amino acid sequence, but they suggested that folded structures are thermodynamically more stable. Thus the folding of globular proteins should be understandable in terms of the types of forces that exist between polypeptide chains and water. When forces involved in determining protein conformation are being considered, we can usually ignore covalent bond energies, despite their large magnitude. This is because covalent bonds (except for disulfide bonds) are not made or broken when polypeptide chains fold into their native three-dimensional conformations. The bonds that are affected (made or broken) on folding are, by and large, of the noncovalent type involving secondary valence forces.

Typically the intermolecular bond energies between noncovalently linked atoms range in value from 0.1 to 6 kcal/mol. The intermolecular forces between noncovalently linked atoms may be grouped into four categories : hydrophobic effects, electrostatic forces, van der Waals forces, afad hydrogen bonds (H bonds).

Hydrogen Bonds

Many characteristics of H bonds have already been discussed because of the major role they play in stabilizing fibrous proteins. Additional aspects of H bonds to be considered relate more closely to the structures of globular proteins. Evidence for the existence of an H bond comes from the observation of a decreased distance between dqnor and acceptor groups forming the H bond.

Thus from the van der Waals radii given in table 4.1, we can calculate the distances between nonbonded H and O atoms (2.6 Å) and between nonbonded H and N atoms (2.7 Å). When an H bond is present, this distance is usually reduced by about 0.8 A in both cases. Polypeptides carry a number of H-bond donor and acceptor groups, both in their backbone structure and in their side chains.

Water also contains a hydroxyl H-donor group and an oxygen H-acceptor group for making H bonds. Formation of the maximum number of H bonds between a polypeptide chain and water requires the complete unfolding of the polypeptide chain. However, it is not obvious that such an unfolding would result in a net energy gain. The reason is that water is a highly H-bonded structure, and for every H bond formed between water and protein, an H bond within the water structure itself must be broken. The compromise followed by most proteins is to maximize the number of intramolecular H bonds between the backbone peptide groups but to keep most of the potential H-bond-forming side chains of the protein near the protein-water interface where they can interact directly with water

Van der Waals Forces

Van der Waals interactions are of two types : one attractive and one repulsive. Attractive van der Waals forces involve interactions among induced dipoles that arise from fluctuations in the electron charge densities of neighboring nonbonded atoms. Such interactions amount to 0.1–0.2 kcal/mol; despite their small size, the large number of such interactions that occur when molecules come close together makes such interactions quite significant.

Van der Waals forces favor close packing in folded protein structures. Repulsive van der Waals interactions occur when non-covalently bonded atoms or molecules come very close together. An electron-electron repulsion arises when the charge clouds between two molecules begin to overlap. If two molecules are held together exclusively by van der Waals forces, their average separation is governed by a balance between the van der Waals attractive and repulsive forces. This distance is known as the van der Waals separation. Some van der Waals radii for biologically important atoms are given in table 6.1. The van der Waals separation between two nonbonded-atoms is given by the sum of their respective van der Waals radii.

Hydrophobic Effects

Hydrogen bonds and van der Waals forces are of major importance in determining the secondary structures formed by fibrous proteins. To understand the complex folded structures found in globular proteins additional types of interactions between amino acid side chains and water must also be considered. The so-called hydrophobic effects, that lead to the interaction of hydrophobic groups in proteins, are the hardest type of rpncovalent interactions to appreciate. Whereas H bonds and van der Waals forces relate primarily to enthalpic factors, hydrophobic effects relate primarily to entropic factors.

Furthermore, the entropic factors mainly concern the solvent not the solute. As a first step to understanding the nature of these effects it is useful to consider the interaction between a small hydrophobic molecule, like hexane and water. It is tempting to ascribe the low water solubility of hexane to van der Waals attractive forces between these small hydrophobic molecules.

However, thermodynamic measurements indicate that this is not the case. The hydrophobic molecule hexane has a small favorable enthalpy for solution in water. In spite of this, hexane is poorly soluble in water due to an unfavorable entropic factor. What is this mysterious factor? Owing to the weak enthalpic interactions between hexane and water, the water withdraws slightly in the region of the apolar hydrophobic molecule and forms a relatively rigid hydrogen-bonded network with itself. The network effectively restricts the number of possible orientations of water molecules directly surrounding the dissolved hexane molecuIes.

This ordering of water constitutes an energetically unfavorable entropic effect. The eliame type of entropic effect plays a major role in directing the folding of globular proteins. About half of the amino acid side chains in proteins are hydrophobic (*e.g.*, alanine, valine, isoleucine, leucine, and phenylalanine). Entropic effects strongly favor internal locations for these side chains where they are free from contacts with water (*e.g.*, which shows the location of polar and apolar side chains in cytochrome *c*). The native folded state of a globular protein reflects a delicate balance between opposing energetic contributions of large magnitude. Whereas entropic factors favor the packing of hydrophobic side chains into the interior regions of globular proteins, enthalpic factors favor placing hydrophilic side chains on the surface of the protein where they can interact with water. In the limited number of cases where data are available, the overall entropy of folding appears to be slightly negative (unfavorable) and the overall enthalpy is also slightly negative (favorable). On balance, folding is opposed by the entropy change but favored by the enthalpy change and occurs in most cases because the latter factor outweighs the former.

Electrostatic Forces

Electrostatic forces are of three main types : charge-charge interactions, charge-dipole interactions, and dipole-dipole interactions. The energy of interaction between two charges Q_1 and Q_2 is proportional to the product of the charges and inversely proportional to the distance R between them table 6.3):

Table 6.3. Energy Dependence of Interaction on the Distance of Separation of the Interacting Species

Range of Interaction	*Type of Interaction*
$1/R$	Charge–charge
$1/R^2$	Charge–diple
$1/R^3$	Dipole–dipole
$1/R^6$	Van der Waals (dipole-induced dipole) attractive forces
$1/R^{12}$	Van der Waals repulsive forces

$$\text{energy of interaction} \propto \frac{Q_1Q_2}{R}$$

In solution this interaction is reduced by the dielectric constant of the surrounding medium:

$$\text{energy of interaction} \propto \frac{Q_1Q_2}{\in R}$$

If the two charges in question are buried within a protein, their interaction energy can be substantially increased because the dielectric constant in the regions inaccessible to water is much lower than the dielectric constant of water.

The long-range nature of charge–charge interactions has led to the speculation that such forces can be important in accelerating interaction between proteins, between proteins and nucleic acids, and between proteins and small molecules, such as coenzymes and substrates. Favorable charge-charge interactions between oppositely charged amino acids are less significant in determining protein folding than are the ion-dipole interactions between the charged groups of amino acid side chains and water. With very few exceptions side chains containing charged groups as well as polar side chains are located on the protein surface at the protein-water interface.

Functional Units of Tertiary Structure

Within a single folded chain or subunit, contiguous portions of the polypeptide chain often fold into compact local units called domains, each of which might consist, for example of a four-helix cluster or a "barrel" or an antiparallel β sheet. Sometimes the domains within a protein are very different from each other, as within the protease papain, but often they resemble each other very closely, as in rhodanese. The separateness of two domains within a subunit varies all the way form independent globular domains joined only by a flexible length of polypeptide chain, to domains with tight and extensive contact and a smooth globular surface for the outside of the entire subunit, as in the proteolytic enzyme elastase.

An intermediate level of domain separateness, characterized by a definite neck or cleft between the domains, is found in phosphoglycerate kinase.

Domains as well as subunits can serve as modular bricks to aid in efficient assembly of the native conformation. Undoubtedly, the existence of separate domains is important in simplifying the protein-folding process into separable, smaller steps, especially for very large proteins. There is no strict upper limit on folding size. Indeed, known domains vary in size all the way from about 40 residues to more than 400.

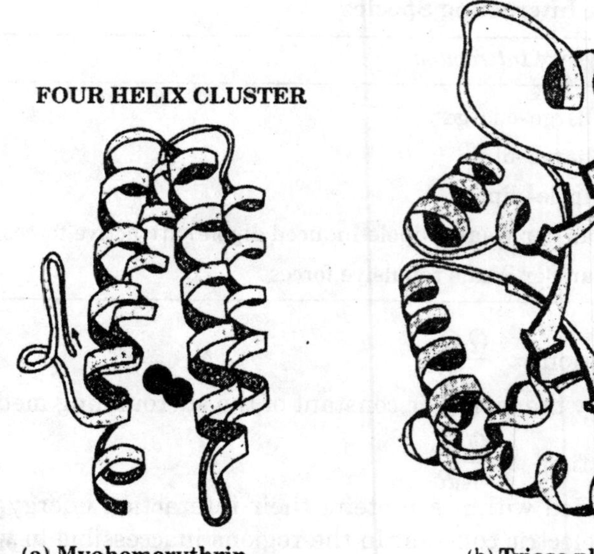

(a) Myohemerythrin (b) Triose phosphate isomerase

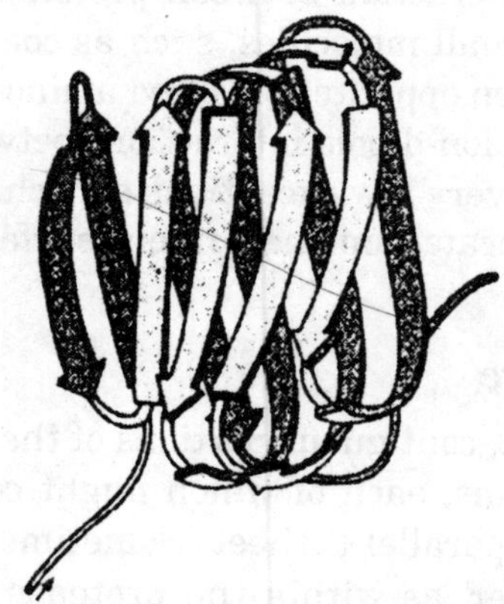

(c) Tomato bushy stundt virus domain 3

Fig. 6.17. Examples of three types.of structural domains found in many proteins: *(a)* the four-helix cluster; *(b)* the β barrel; and *(c)* the antiparallel β sheet.

Furthermore, it has been estimated that there may be more than a thousand basically different types of domains. Another important function of domains is to allow for movement. Completely flexible hinges would be impossible between subunits because they would simply fall apart. However, flexible hinges can exist between covalently linked domains.

Limited flexibility between domains is often crucial to substrate binding, allosteric control (discussed in chapter 9), or assembly of large structures. In hexokinase, the two domains within the individual subunits hinge toward each other on binding of the substrate glucose, enclosing it almost completely. In this manner glucose can be pound in an environment that excludes water as a competing substrate.

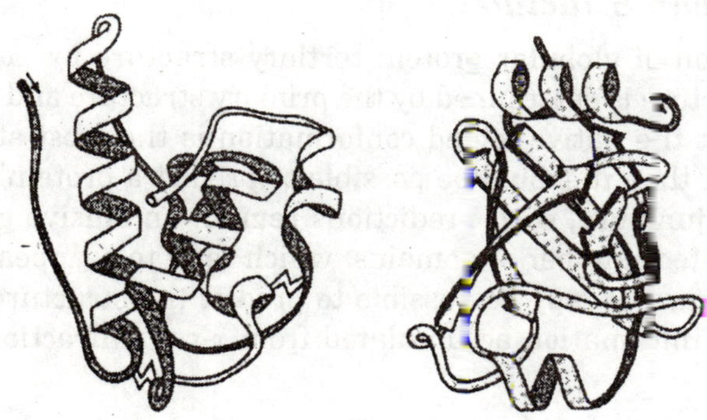

Fig. 6.18. Papain, a protein in which the domains are very different from each other.

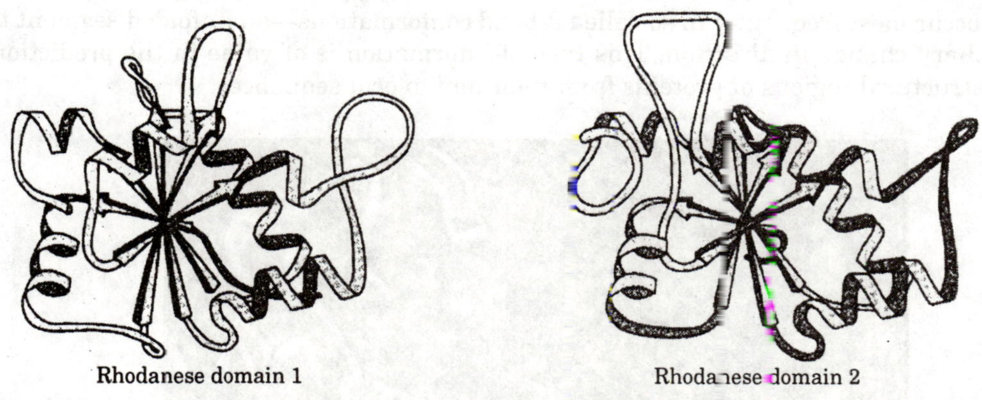

Rhodanese domain 1

Rhodanese domain 2

Fig. 6.19. Rhodanese domains 1 and 2 as an example of a protein with two domains that resemble each other extremely closely. Rhodanese is a liver enzyme that detoxifies cyanide by catalyzing the formation of thiocyanate from thiosulfate and cyanide.

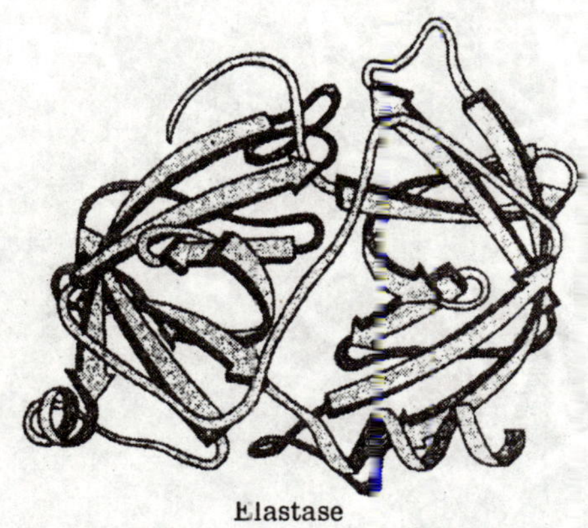

Elastase

Fig. 6.20. Schematic backbone drawing of the elastase molecule, showing the similar β-barrel structures of the two domains.

Predicting Protein Tertiary Structure

We began the discussion of globular protein tertiary structure by pointing out that the secondary and tertiary structure is determined by the primary structure and that this is probably a reflection of the fact that the native folded conformation is the most stable structure that can be formed. If this is so, then it should be possible to predict a protein's structure from its primary sequence. At this juncture, such predictions remain an elusive goal. However, most proteins are made of a limited number of domains, which tend to reappear in many different proteins. Since this is the case, it may be possible to predict the structures of many proteins in the future by using the information accumulated from *x*-ray diffraction studies of related proteins.

It is clear that certain amino acids tend to form particular secondary structures. Acid, methionine, and alanine appear to be the strongest α-helix formers, whereas valine, isoleucine, and tyrosine are the most probable β-sheet formers. Proline, glycine, asparagine, aspartic acid, and serine occur most frequently in so-called β-bend conformations—an unfolded segment that permits a sharp change in direction. This type of information is of value in the prediction of secondary structural regions of proteins from their amino acid sequences.

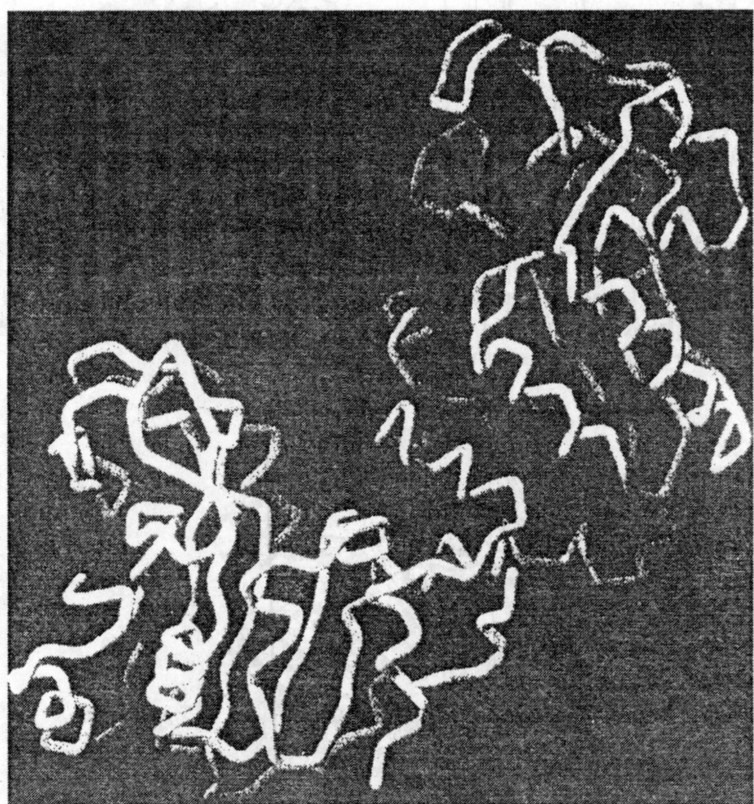

Fig. 6.21. The dumbbell domain organization of phosphoglycerate kinase, with a relatively narrow neck between two well-separated domains.

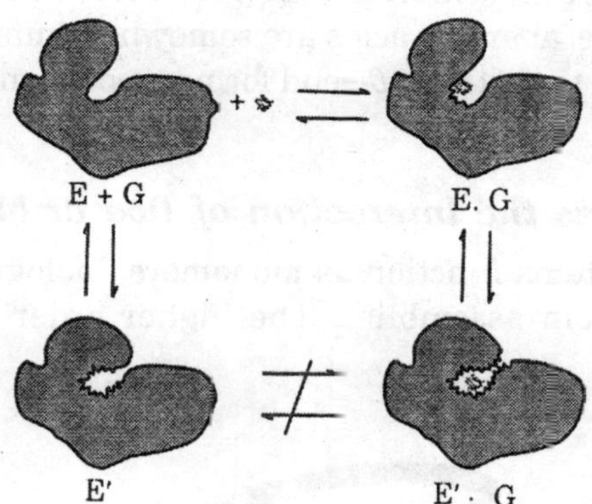

Fig. 6.22. Schematic representation of the change in conformation of the hexokinase enzyme on binding substrate. E and E' are the inactive and active conformations of the enzyme, respectively. G is the sugar substrate. Regions of protein or substrate surface excluded from contact with solvent are indicated by a crinkled line.

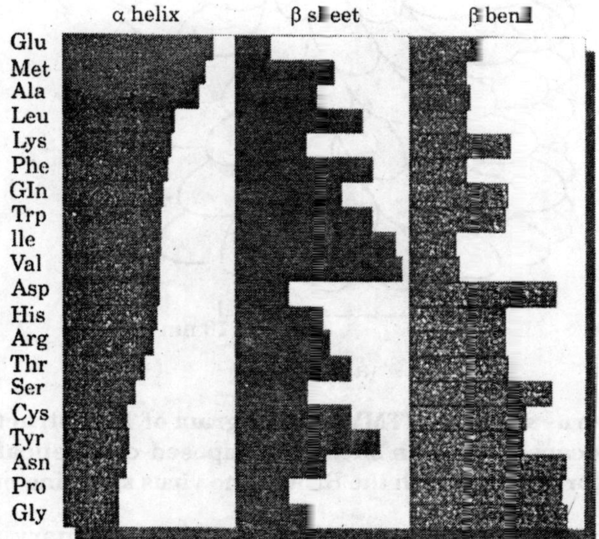

Fig. 6.23. Relative probabilities that any given amino acid occurs in the α-helical, β-sheet, or β-hairpin-bend secondary structural conformations.

The observed frequencies of occurrence of an amino acid in a given conformation provide estimates of the probabilities that the same amino acid behaves similarly in a sequence the actual secondary structure of which is unknown. To predict the secondary structure from the sequence, it is consequently necessary only to plot the probabilities for the individual amino acids sequentially, or better, to plot a local average over a few adjacent residues. Plotting such an average accounts for the cooperative nature of secondary-structure formation.

The sequences Gly-Pro-Ser and Ala-His-Ala-Glu-Ala, for example, give high joint probabil-ities for being, respectively, in β-bend and α-helical conformations. However, comparisons of predicted versus directly observed polypeptide conformations give mixed results. This situation is a consequence of two facts: that several amino acids are somewhat ambiguous in their secondary-structure-fonning tendencies, and that strong β-bend formers occasionally turn up in the middle of α helices.

Quaternary Structure Involves the Interaction of Two or More Proteins

Although many globular proteins function as monomers, biological systems abound with examples of more complex protein assemblies. The higher order organization of globular

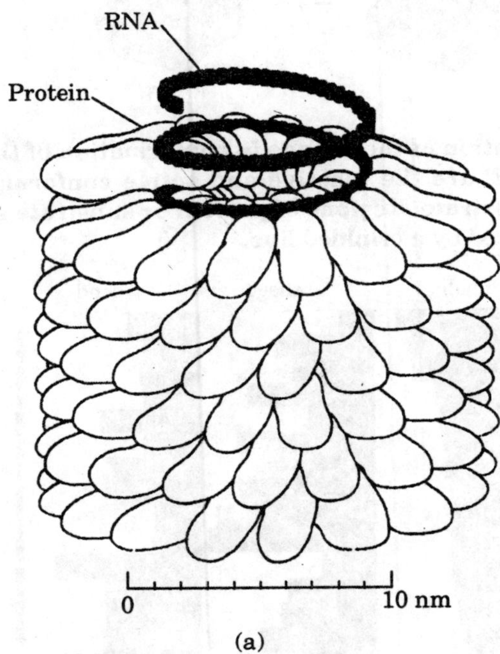

Fig. 6.24. Tobacco mosaic virus structure (TMV). (*a*) Diagram of TMV structure, an example of a helical virus. The nucleocapsid (protein shell) is composed of a helical assembly of 2,130 identical protein subunits (protomers) with the RNA of the virus spiraling on the inside.

subunits to form a functional aggregate is referred to as quaternary structure. Protein quaternary structures can be classified into two fundamentally different types. The first involves the assembly of proteins (sometimes referred to as subunits because they constitute a part of the final structure) that have very different structures. Examples range from the hormone insulin, which has two different subunits, to complex assemblies such as ribosomes, which contain 20 or more nonidentical protein sub-units in addition to one or more RNA components. The organization of quaternary structures depends on the specific nature of the interactions between each molecular sub-unit and its neighbors. Each intermolecular interaction generally occurs only once within a given aggregate arrangement, so that the overall complex structure has a highly irregular geometry. A second, commonly observed pattern of quaternary structure is for a molecular aggregate to have multiple copies of the same kind of subunits. Owing to the recurrence of specific structural interactions between the subunits, such aggregates typically

Table 6.4. Molecular Weight and Subunit Composition of Selected Proteins

Protein	*Molecular Weight*	*Number of Subunits*	*Function*
Glucagons	3,300	1	Hormone
Insulin	11,466	2	Hormone
Cytochrome *c*	13,000	1	Electron transport
Ribonuclease A (pancreas)	13,700	1	Enzyme
Lysozyme (egg white)	13,900	1	Enzyme
Myoglobin	16,900	1	Oxygen storage
Chymotrypsin	21,600	1	Enzyme
Carbonic anhydrase	30,000	1	Enzyme
Rhodanese	33,000	1	Enzyme
Peroxidase (horseradish)	40,000	1	Enzyme
Hemoglobin	64,500	4	Oxygen transport
Concanavalin A	102,000	4	Unknown
Hexokinase (yeast)	102,000	2	Enzyme
Lactate dehydrogenase	140,000	4	Enzyme
Bacteriochlorophyll protein	150,000	3	Enzyme
Ceruloplasmin	151,000	8	Copper transport
Glycogen phosphorylase	194,000	[illegible]	Enzyme
Pyruvate dehydrogenase (*E. coli*)	260,000	[illegible]	Enzyme
Aspirate carbamoyltransferase	310,000	12	Enzyme
Phosphofructokinase (muscle)	340,000	4	Enzyme
Ferritin	440,000	24	Iron storage
Glutamine synthase (*E. coli*)	6000,000	12	Enzyme
Satellite tobacco necrosis virus	1,300,000	60	Virus coat
Tobacco mosaic virus	40,000,000	2,1[illegible]0	Virus coat

form regular geometric arrangements. Some outstanding examples of such aggregates are the rod-like viruses and the polyhedral viruses.

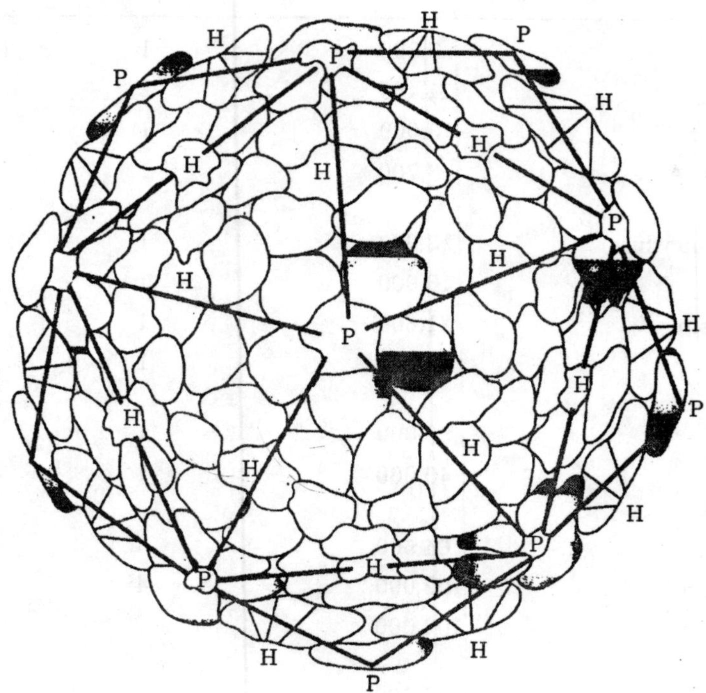

Fig. 6.25. The structure of the capsid (protein shell) for an icosahedral virus such as tomato bushy stunt virus. Pentons (P) are located at the 12 vertices of the icosahedron. Hexons (H), of which there are 20, form the edges and faces of the icosahedron. Each penton is composed of five protein subunits and each hexon is composed of six protein subunits. In all, the structure contains 180 protein subunits.

In a typical rodlike virus, the helical aggregation of the protein-coat subunits form a cylindrical container for the virus's nucleic acid. The assembly of symmetrical aggregates in a polyhedral virus reflects the structural stabilization that occurs when all the subunits interact in geometrically similar ways, that is, essentially like the atoms ma salt crystal.

However, one of the surprising results from x-ray crystallography is that such assemblies often contain distinctly different types of quaternary interactions, even between chemically identical subunits. In the 180-subunit polyhedral protein coat of tomato bushy stunt virus, for example, groups of 5 subunits are in contact around a fivefold symmetry axis, whereas other groups of 6 subunits have a distinct but similar contact around a three-fold axis. The versatility that permits such non-equivalent associations allows assembly of larger and more complex structures and may be even more common in biological structures too large to have been examined crystallographically.

7 ENZYMATIC PROTEINS

Restriction endonucleases are among the most accurate enzymes known. Consequently, changing their very high sequence specificity is one of the scientific goals in studying these enzymes. This review focuses on protein engineering regarding Type II restriction enzymes. They comprise the vast majority of the about 3600 entities listed in REBASE today, although the number of those studied in detail is far smaller. First, we will discuss different approaches to changing the recognition of a given sequence starting with rational design using site-directed mutagenesis and progressing to random mutagenesis combined with in vivo selection. Then, we will describe attempts to lengthen the recognition sequence by designing additional enzyme contacts to DNA. As the dimeric nature of all restriction enzymes doubles the effect of each amino acid exchange, we will also discuss engineering of subunit composition. Finally, we will speculate on possible future directions of protein engineering of restriction enzymes.

Table 7.1 gives an overview on enzymes, their mutants and properties that are mentioned in this review.

Modifying Contacts of Restriction Enzymes to their Recognition Sequence

The first cocrystal structure of a DNA-binding protein in complex with its binding sequence to be solved was the one of the EcoRI restriction endonuclease. It featured only six amino acid residues contacting one base pair each which led to proposals of how the specificity of this enzyme could be changed. The revised structure shows a much more complicated network of interacting residues as outlined in Fig. 7.1. We will use the EcoRI endonuclease as the prime example in this review.

Table 7.1. Overview on enzymes, their mutants and properties

Enzyme	*Mutant*	*Methodology*[a]	*Properties*
Enhanced cleavage activity (Sect. 2)			
PvuII	D34G	rand.mut.	Loss of DNA binding for central GC base pair, still only cleaves its recognition site CAGCTG
EcoRI	I197A	site dir.mut.	Specific activity five times higher than wt-enzyme, although not involved in DNA recognition

Enzyme	Mutant	Methodology	Properties
BamHI	C54A	rat.des.	Higher specific activity but no direct DNA contact
HindIII	E86K	rat.des.	Higher specific activity but no direct DNA contact
Recognition of modified sequences (Sect. 2)			
EcoRV	T94V	rat.des.	Cleaves Sp-methylphosphonate four orders of magnitude faster than unmodified substrate
EcoRI	Q115A or A142G	site dir.mut.	Cleavage of GAAUTC as fast as GAATTC
EcoRV	N188Q	rat.des.	Prefers GATAUC over GATATC
EcoRI	G140A	site dir.mut.	Prefers GAAUTC or GAATUC over GAATTC about ten times at low activity level
EcoRI	M137Q	site dir.mut.	GAATT5mC is preferred ten times over GAATTC but not in 5 mCGAATT5 mCG
BamHI	N116H/SII8G	in vivo select.	GGmATCC is strongly preferred over GGATCC
Recognition of altered sequences (Sect. 2)			
EcoRI	G140S/ NN141S/ R145K	site dir.mut.	Attempted change of recognition from GAATTC to GGATCC, almost no cleavage activity or binding activity, unstable proteins
EcoRI	M137Q/ Arg200/ Arg203	site dir.mut.	Attempted change of recognition from GAATTC to CAATTG, cleavage activity but no change in specificity
Enhancement of relaxed specificity (Sect. 2)			
BsoBi	D246A	sat.mut.	Prefers CCCGGG (70-fold in binding and 100-fold reduced cleavage) over CTCGAG of the former CYCGRG recognition
BscBI	A32T/ T40P/ Q140L/ D246A	rand.mut.	Tenfold higher cleavage activity compared to D246A but still with enhanced speciicity
Eco57I	T862N	in vivo select.	Generation of a relaxed specificity CTGRAG from CTGAAG (selection for methylase)
BstYI	K133N/ S172N	in vivo select. rand.mut.	Some mutants with preference for AGATCT instead of RGATCY; combination produces mutants with only sixfold less cleavage activity for AGATCT and no cleavage for GGATCC
Lengthening of recognition sequence (Sect. 3)			
EcoRV	A181E	sat.mut.	Preference for TGATATCA tenfold in Oligodeoxynucleotide and plasmid

Enzyme	*Mutant*	*Methodology*[a]	*Properties*
EcoRV	K104R	rat. Des.	Preference for TGATATCA tenfold in oligodeoxynucleotide and plasmid
EcoRV	A181E/ K104R	rat.des.	Preference for TGATATCA of the single mutants did not result in more specificity for the double mutant
EcorRV	N92T/ S183A/ T222S	dir.ev.	25-fold preference for AT flanked GATATC sited
EcoRV	R226AorV	Ala scan.,sat.mut.	Preference for nonflexible surrounding Sequence of GATATC
EcoRI	I197Q	site dir.mut	Possible additional contact outside of the recognition sequence GAATTC led to no enhancement of specificity
EcoRI	A138N/ M137Q/ I197Q	site dir.mut.	More active than I197Q but still no enhancement in specificity
EcoRI	A138N/ GlyAB/ I197Q	site dir.mut.	Cleaves ony TGAATTCA on λ DNA but no real contact made to the outer base pair; instable protein, stabilized with Ca^{2+} it cleaves the canonical site again
EcoRI	K130E	site dir.mut.	Higher selectivity due to flexible or nonflexible neighboring sequences
EcoRI	Myb133	rat.des.	Needs Ca^{2+} for stabilization and no longer cleaves GAATTC on λ DNA, cleavage of AAGGAATTCCTA identified with a random substrate pool, protein only partially folded
Changed subunit composition (Sect. 4)			
EcoRI	L158D-I230K	site dir.mut.	Heterodimeric protein L158D-I230K shows increase in specific activity compared to either homodimer, still very unstable enzyme
AlwI	N.AlwI	rat.des.	Monomeric enzyme, recognizes its specific site but nicks a single strand with high specificity
Single-chain nuclease (Sect. 4)			
PvuII	Single-chain PvuII	rat.des.	Nearly as active as the homodimeric wild-type enzyme
EcoRI	EcoFus	rat.des.	Deleted 65 C-terminal amino acids and 72 N-terminal amino acids, detectable cleavage activity but not to be purified
EcoRI	EcoRI	NCNC rat.des.	Bridging linker between C-and N-terminus, not to be purified due to linker cleavage

[a]Ala scan.: alanine scanning; dir. ev.: directed evolution scheme; in vivo select.: in vivo selection; rand. mut.: random mutagenesis; rat. des.: rational design; sat. mut: saturation mutagenesis; site dir. mut.: site-directed mutagenesis.

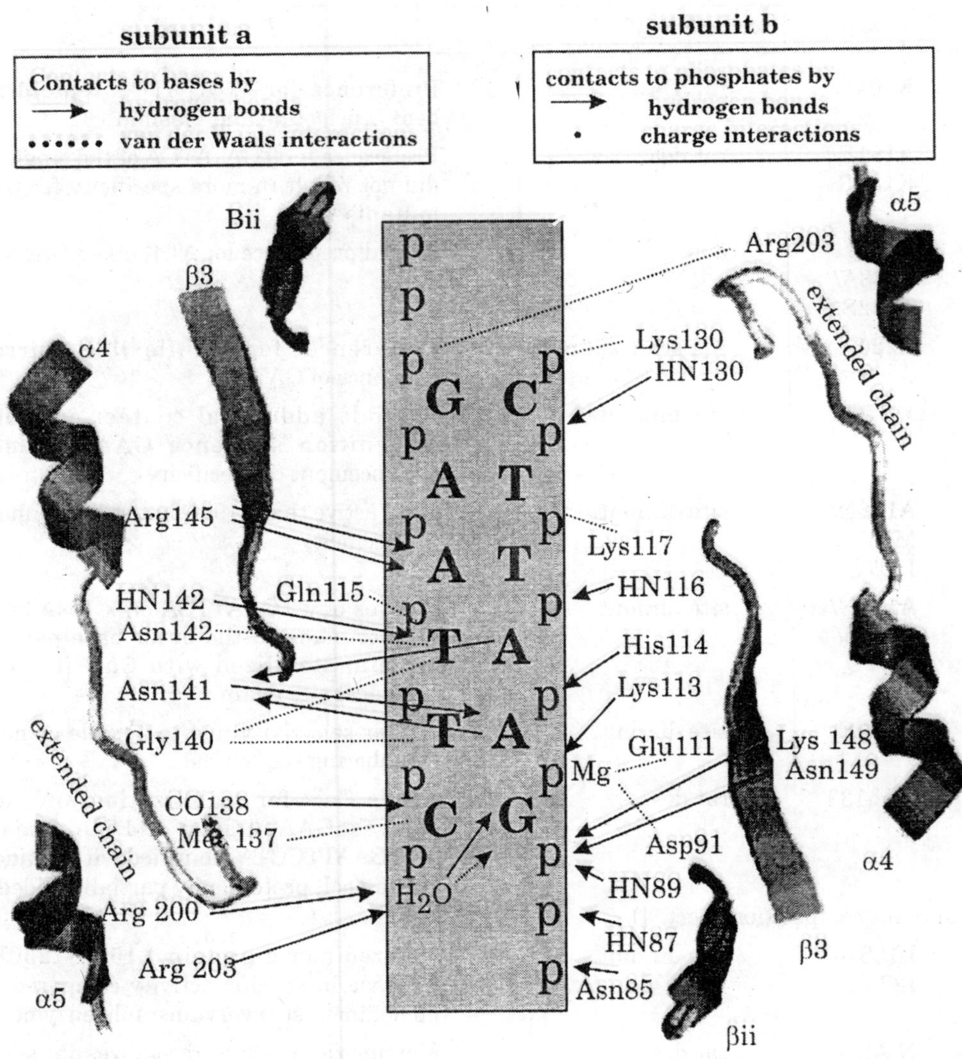

Fig. 7.1. Interaction of restriction endonuclease EcoRI with its recognition sequence. Each subunit establishes the same contacts but to get a clearer picture in subunit *a* the phosphate and in subunit *b* the base contacts are omitted. Furthermore, all contacts between amino acid residues which are of course important for coupling of restriction to catalysis and for subunit interactions are not shown

Each subunit of the homodimeric enzyme forms nine hydrogen bonds and five hydrophobic interactions to the bases of the recognition sequence (on the left side of Fig. 7.1) and 12 phosphate contacts (on the right side of Fig. 7.1). Because of the intrinsic symmetry of both the enzyme and palindromic recognition sequence, 52 specific interactions are made overall. To establish so many contacts in the small space of six base pairs the major groove of the DNA has to be widened by kinking the phosphodiester backbone. Nevertheless, all these contacting residues are close to each other and interwoven by direct and water-mediated interactions which are omitted, clarity. Similar complexity is found in other cocrystal structures of restriction endonucleases. Therefore, this redundant and over-determined recognition that seems to be a prerequisite for the very high specificity of these enzymes, has to be overcome to perform successful protein engineering. At least, every effort has to be embedded in this network of contacts.

Site-Directed Mutagenesis

Site-directed mutagenesis is used primarily to verify the contacts seen in the cocrystal structure. However, changing specificity was also always intended, because not only conservative changes or substitutions by alanine were made, but side chains were introduced which should have the potential of establish-ing different contacts. We and others have mutated all residues depicted. The results of all these mutagenesis experiments suggest a general rule: The change of a contacting residue results in a sometimes very large drop in catalytic activity but not in a change of specificity. This has to be expected because mutant enzymes are immediately toxic to their host by cleaving sequences which are not protected by the accompanying methylase.

In the Type II systems, protection is achieved usually by a separate enzyme specific for the same sequence. Therefore, during evolution a strong selection pressure reduced all potentialities to cleave unprotected sequences even when misincorporation during translation creates mutated enzymes. This explains both parts of the rule: The large drop in catalytic activity is due to a strict coupling of recognition and catalysis and the redundancy results in a tolerance to the loss of some specific contacts. A good example is the PvuII endonuclease in which Asp34 interacts with the guanine of the central base pair in the recognition sequence CAGCTG.

The D34G mutant was isolated as a nearly inactive enzyme with strong DNA binding. Closer inspection of its DNA-binding properties showed that it has completely lost its specificity for the central two base pairs but still cleaves only the canonical sequence with four orders of magnitude reduced activity. However, there are also deviations from the general rule. Firstly, not all mutations lead to a reduction of cleavage activity.

The EcoRI single mutant I197A has a five times higher specific activity than the wild-type enzyme. It is not involved in DNA recognition and must have some structural role although this is not obvious in the crystal structure. The same is true for the C54A mutant of BamHI and possibly for the E86K mutant of HindIII, an enzyme whose structure is not yet known. Therefore, the, low cleavage rate of restriction enzymes can be slightly improved.

A larger enhancement should have consequences on accuracy as is found in another class of mutants mostly at non-contacting residues. They cleave faster at sequences deviating by one base pair from the recognition sequence. This behavior of so-called promiscu-ous mutants is better described as interference with the strict coupling of recognition and catalysis as the inherent low inaccuracy of the enzymes is strongly enhanced while the canonical cleavage is not necessarily faster. Secondly, the few cases of changed specificity have to be seen as deviations from the rule as well.

The finding that the EcoRV endonuclease cleaves oligodeoxynucleotide substrates with a Sp-methylphosphonate at GATATpC as fast as the unmodified sequence suggests that the contacting residue Thr94 might contact both substrates by simply rotating its methyl group to the Sp-methylphosphonate or its hydroxyl group to the normal phosphate oxygen. This led to the design of the T94V mutant which cleaves the Sp-methylphos-phonate four orders of magnitude faster than the unmodified substrate. This is the best change in specificity found for restriction enzymes, although for a completely artificial recognition sequence. In several instances, the recognition of thymine bases could be modified such that a discrimination against uracil is no longer possible or even uracil is the preferred substrate.

In EcoRI the mutants Q115A and A142G cleave at GAAUTC as fast as at GAATTC while the wild type enzyme shows a tenfold difference in favor of the canonical sequence. The EcoRV mutant N188Q with a similar behavior at GATAUC was generated.

The EcoRI mutant G140A even prefers the GAAUTC or GAATUC sequence about ten times although at a low activity level. The added methyl group on the protein not only needs space but also can hydrophobically interact with the C5 position of a pyrimidine. We use this mutant as a starting point to establish recognition of a different base pair by mutating also AsnHI and Arg145 which contact the complementary adenine; Gerschon 2000; Vennekohl, unpubl.

The main difficulty is that each additional mutation reduces the catalytic activity further. At such a low level, the incorporation of mutations with a promiscuous effect only slightly enhances the cleavage rate. Additionally, these triple or quadruple mutants tend to be more unstable which not only hampers purification but may also be responsible for a low level of unspecific nuclease activity interfering with specific DNA cleavage. One mutant produces a transient band pattern, but to date it could not be identified which sites are cleaved. Another hydrophobic contact in EcoRI recognition led also to a specificity change.

Among the mutants for Met 137 contacting the cytosine residue, only M137Q shows no reduced cleavage activity. The impaired interaction is possibly compensated by a repositioning of the neighboring Alal38 for which mutations with a promiscuous effect are already known. Cleavage experiments with oligodeoxynucleotide substrates containing modified bases established a preference of this mutant for a methyl group at C5 of the pyrim-idine. The best substrate was GAATTSmC which is preferred ten times over GAATTC by the mutant. This is exactly opposite for the wild type enzyme which is inhibited by the cytosine methylation. As 5-methyl cytosine is a natural base in eukaryotic DNA in the sequence context CpG, we also tested the palindromic extension of the EcoRI recognition sequence 5 mCGAATTS mCG.

However, the effect of the additional methyl group next to the 5′-site of the recognition sequence counteracts the preference of the mutant. However, we also use the M137Q mutant as a starting point to change the outer base pair recognition of the EcoRI sequence. This is especially difficult because the guanine contacts are mediated by a water molecule positioned by the long amino acid residues Arg200 and Arg203. Lysine residues at these positions have the potential to reorient the water in context with the M137Q mutation. Unfortunately, the double and triple mutants show no change in specificity. We are left with the fact that in the case of restriction enzymes with a well-defined recognition sequence we only were able to change specificity to bases not normally found in DNA. Obviously, it would be advantageous to test DNA recognition without a large and rigid enzyme framework. For EcoRI we showed that an oligopeptide with the sequence of the extended chain which harbors most of the base-contacting residues, able at high concentrations to protect the EcoRI sites in plasmid DNA against cleavage.

One of the rea-sons for low affinity binding of the peptide was argued to be the necessity of peptide dimerization on the DNA to mimic an EcoRI dimer bound to the recognition sequence. Therefore, we synthesized a bidentate peptide with two covalently joined extended chain sequences and variants of it (Vennekohl 2000), which still have very low binding constants to GAATTC containing oligodeoxynucleotide. Peptide libraries comprising variants of the extended chain motif show an increase in unspecific binding especially if the positive charge of the peptides was raised. Furthermore, we showed in cleavage protection assays, which allow for much higher peptide concentrations, that the sequence specificity of the bidentate peptide is relaxed to GAN$_{(1or2)}$TC and, therefore, cannot mimic an EcoRI recognition exactly.

For enzymes with degenerate recognition sequences it may be easier to enhance specificity to a more defined sequence.

The BsoBI restriction endonuclease recognizes CYCGRG where Y stands for both pyrimidines and R for both purines. This actually leads to the recognition of the palindromic sequences CCCGGG and CTCGAG as well as the asymmetric sequence CTCGGG (which is identical to CCCGAG because of the antiparallel DNA double strand). In the cocrystal structure, Asp246 contacts the purine via a water molecule. Saturation mutagenesis at this residue led to the mutant D246A which prefers CCCGGG (Zhu et al., 2003) This is due to a 70-fold stronger binding of this sequence in the context of a 100-fold reduced cleavage activity overall. Thus CTCGAG is nearly not cut at all. For improvement of the catalytic activity of the mutant the authors used random mutagenesis and the SOS induction assay in which a reporter gene shows the induction of the SOS response by multiple DNA cleavage events.

Most of the secondary mutations were outside the DNA-protein interface as we described before for other restriction enzymes.' Some of the more active secondary mutants lost the new sequence preference but one quadruple mutant retained it in the context of a tenfold higher cleavage activity. This work is promising and also demonstrates the power of random mutagenesis followed by in vivo selection.

IN VIVO SELECTION SYSTEMS

For selection of BamHI specificity mutants, cells with a spectinomycin resistance gene controlled by an antisense RNA were used. Represser binding to the antisense promoter allows the cells to become resistant against the antibiotic. This system worked fine for the identification of catalytic site residues by omitting the methylase gene in the cells as well as for the identification of mutations of contacting residues which still allow binding to the canonical sequence.

Reactivation of the catalytic center was possible for 14 out of 17 of these mutants. Some peculiarities in the cleavage reaction led to a closer inspection of the N116H/S118G double mutant which in contrast to the wild type enzyme strongly prefers GGmATCC over GGATCC in binding and cleavage. For selection of new sequence specificities the binding sequence has to be changed. Using this approach, some BamHI mutants have been selected to bind to the sequences AGATCT, GCATGC, as well as CAGCTG instead of GGATCC. But the reactivation of the catalytic center proved to be unsuccessful or nothing interesting could be selected. This raises the question whether a selected mutant which binds a different sequence can be activated again to cleave at this sequence. The D34G mutant of PvuII described above argues against this, too. However, there is also an example in favor of the feasibility of this approach.

A subclass of the Type II restriction enzymes combines the catalytic centers of endonuclease and methylase activity in one polypeptide chain. Sequence specificity is controlled by the same DNA-binding domain. Rimseliene et al. (2003) have used Eco57I to establish a general scheme for their specificity change. After inactivation of the catalytic center of the endonuclease the DNA-binding domain is mutagenized and the resulting plasmid ensemble amplified in vivo where they are target for methylation by the mutant they code for. The selection is performed in vitro by cleaving the purified plasmid pool with a restriction endonuclease of a different specificity. GsuI was used in this case. It cleaves at CTGGAG which deviates in one base pair from the recognition sequence CTGAAG of Eco57I. Only those plas-mids stay uncut which are methylated at the new sequence. These are used for reactivation of the endonuclease.

The resulting single mutant T862N exhibites a relaxed specificity CTGRAG recognizing both the sequences of Eco57I as well as of GsuI. Actually this scheme is a selection for a methylase with new sequence specificity but it shows that reactivation of endonucleolytic activity after a specificity change is possible. The selection scheme becomes more elaborate if one tries to directly select for an active restriction enzyme. A prerequisite for this is to protect the cells against the newly generated cleavage activity. Therefore, selection is only possible for those specificities for which a methylase gene is already at hand. The counterselection against the wild-type specificity as well as all undesired ones is obvious, as these activities kill the cell.

The positive selec-tion for the desired specificity needs sorting out of all those cells harboring the vast majority of inactive enzymes produced by random mutagenesis. We and others used challenging phages to kill the cells which cannot restrict. In principle, this should work fine, but it needs a mutant enzyme with high cleavage activity as well as the ability for efficient facilitated diffusion to find its cleavage sites quickly. If the mutant is about 100-fold less active than the wild-type enzyme restriction drops significantly. Taking into account the results from the site-directed mutagenesis experiments described above, it is very unlikely to find specificity mutants with high enough cleavage rates. We do not know of any positive results from such selection experiments.

Again, the situation seems to be better for specificity enhancement of restriction enzymes with degenerate sequences. BstYI recognizes RGATCY. Its gene was randomly mutagenized and introduced into cells harboring solely, the BglI methylase which is specific for one of the three BstYI sequences: AGATCT (Samuelson and Xu 2002). Mutant genes were isolated from the surviving clones and reintroduced in cells with a SOS reporter gene to select for active endonucleases. Some of the mutants showed a preference for AGATCT. By combining their mutations, an enzyme was created which is only sixfold less active than the wild type but no longer cleaves GGATCC.

LENGTHENING OF RECOGNITION SEQUENCES

Considering the results described so far, it seems to be easier to enhance sequence specificity than to design a new one. For a restriction enzyme with nondegenerate recognition this means to leave the evolved contacts intact and to add some specificity outside. Actually restriction enzymes with longer recognition sequences are very useful, as the DNAs manipulated in vitro become longer. At the moment, only 13 restriction enzymes with eight base-pair recognition sequences are listed in REBASE. From cocrystal structures of EcoRV it was predicted that the

A181E mutant should prefer CGATATCN and the K104R mutant NGATATC(G/T). Again, site-directed mutagenesis was used to verify such predictions.

Saturation mutagenesis of AlalSl gave several mutants with slight preferences for outside base pairs. A181E in contrast to the predictions, preferred TGATATCA tenfold in oligodeoxynucleotide as well as in plasmid substrates. Lanio et al. (2000) extended these studies by mutations of Lysl04. Again, they did not attain the expected speci-ficities. Combining the A181E mutation with K104R, which exhibits a slight preference for the same sequence, did not result in a more specific double mutant. It is not strictly proven that these mutants establish a base contact. It also may be that they are influenced by the sequence context of the cleavage site and by that become more selective. Selectivity mutants can not be seldom for EcoRV.

A directed evolution scheme of three successive rounds with about 500 manually inspected mutants yielded several mutants with preferences for AT-flanked cleavage sites including one triple mutant with a 25-fold preference. While from this study it is not clear which property of the recognition sequence leads to the preferential cleavage, Wenz et al. (1998) have directly tested oligodeoxynucleotides with nonflexible oligo-dA oligo-dT or flexible oligo-dAT flanking sequences. They started with the R226A mutant obtained in an alanine scanning for possible phosphate contacts. This mutant shows some preference for a nonflexible surrounding which is enhanced by a factor of 100 in the R226 V mutant. For the creation of a lengthened recognition sequence based on the EcoRI endonuclease, we chose Ile 97. This residue is ideally positioned above the major groove outside the recognition sequence.

However, not any mutation led to an enhancement of specificity. We considered three possible reasons for this. First, the enzyme might not need an additional contact as long as all interactions to the normal recognition sequence are made. Therefore, we interfered with canonical recognition by secondary mutations, but all mutants behaved as if Ile 197 was not changed.

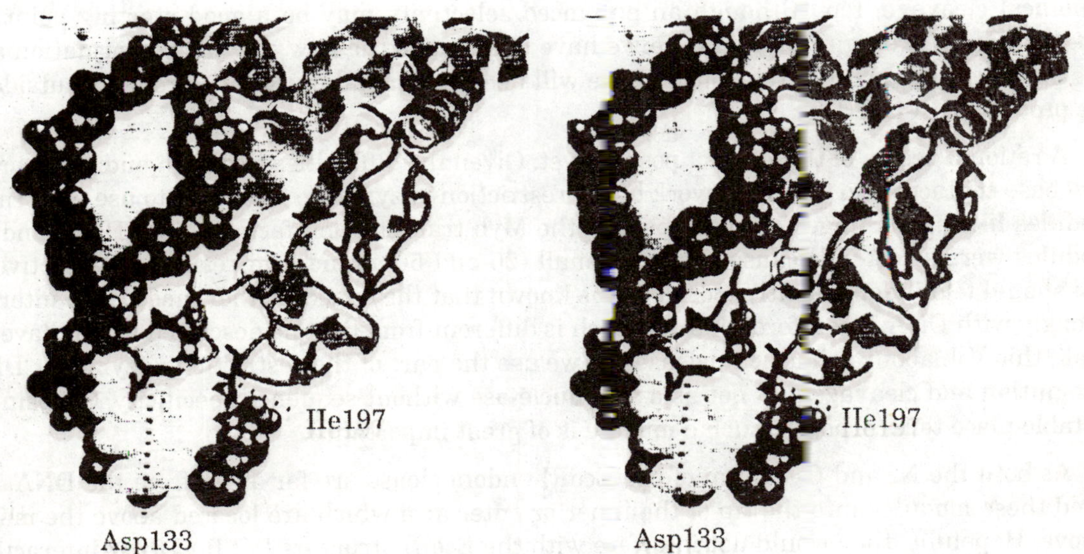

Fig. 7.2. Stereo-view of Aspl33 and Ile197 on top of the major groove which were used to lengthen the recognition sequence of restriction endonuclease EcoRI.

Second, a longer side chain such as glutamine or arginine may be so flexible that equivalent interactions may be made with varying base pairs. Therefore, we tried to integrate He 197 mutations into the existing network of interacting residues by introducing M137Q and A138N mutation which points toward Ilel97. Some of the triple mutants are even more active than the I197A mutant described above but showed no indication of enhanced specificity. In mutants combining these two approaches, the properties of the mutations interfering with normal recognition dominated. The third reason may be that Ile197 is too far away from the bases to establish contacts even with long side chains.

As it is located at the tip of a short loop, we introduced glycine residues before (position A) and/or behind (position B) Ile197 to move it deeper into the major groove. Cleavage activity was good for the construct with the two glycine residues (GlyAB/I197Q), but again only at the canonical sites. As the former reasons also apply for this mutant, we added mutations to interfere with normal interaction or to connect the glutamine to the other interacting residues. This last mutant A138N/GlyAB/I197Q cleaves only two of the five EcoRI sites in bacteriophage X DNA. These two are the only EcoRI sites which are palindromic up to eight base pairs TGAATTCA. This looks promising but does not prove a real contact to the outer adenine.

We already have found several mutants which cleave slowly at one or more sites of X DNA. We studied this phenomenon by using a plasmid pool harboring all possible sequence surroundings at an EcoRI site varying three base pairs before and behind it. The mutants tested exhibit a higher selectivity due to flexible or nonflexible neighboring sequences whose influence is ignored by the wild-type enzyme. Several rounds of selection were needed to find the selectivity preferences. In contrast to this behavior, a mutant establishing a real base contact should give a clear result in a single selection round. This is not the case with the mutant described above. Successive rounds of selection failed because the mutant is only active for a few days after purification. Stabilization of enzyme structure with Ca^{2+} ions is possible, but this stabilizes canonical cleavage, too. Although an enhanced selectivity may be a good starting point for lengthening the recognition sequence we have to stabilize the new enzyme conformation and not the canonical one. For that purpose we will need additional mutations probably outside of the protein-DNA interface.

A rational design of these is not possible yet. Given the difficulties to design and incorporate new base contacts into the framework of the restriction enzyme we also tried to use preformed modules like a zinc finger or one repeat of the Myb transcription factor. These DNA-binding modules were selected because they are small (26 and 50 amino acid residues, respectively) and should fold independently. Besides, it is known that they need further modules to interact strongly with DNA. Therefore, this approach is different from the one described in Kandavelou et al., this Vol. about chimeric nucleases as we use the part of the restriction enzyme for DNA recognition and cleavage and not as a sole nucle-ase without sequence specificity. Choosing a suitable place to incorporate such a module is of great importance.

As both the N- and C-termini of the EcoRI endonuclease are far away from the DNA, we fused these modules into the tip of the inner or outer arm which are located above the major groove. Hopefully, they would not interfere with the EcoRI structure but find their interaction partner in the DNA. The Myb 133 fusion needs Ca^{2+} ions for stabilization but did not cleave the EcoRI sites on X DNA. Using the plasmid pool described above, we identified the partially palindromic sequence AAGGAATTCCTA as recognition sequence in a single selection round. With an oligodeoxynucleotide substrate.this preference could be verified under conditions where we used vast excess of unspecific DNA.

Therefore, the fusion mutant is still able to cleave GAATTC if the sequence is isolated in an oligodeoxynucleotide substrate but not in the presence of many other sequences. Circular dichroism spectroscopy revealed that not all mutant molecules are properly folded as the amount of secondary structure is less than anticipated for complete folding of both EcoRI and Myb part. Again, additional changes are necessary to get a stable mutant with a new specificity.

CHANGING THE SUBUNIT COMPOSITION

All orthodox Type II restriction enzymes work as dimers of identical subunits. Therefore, two single strands can be cleaved in a single binding event. Recognition of a palindromic sequence makes evolution of specific contacts more economical as only half of them have to be evolved. Thus, study of these enzymes becomes more difficult because each mutation has double effect as it automatically occurs twice in the homodimer. Furthermore, understanding of communication between the two subunits is a complicated task. Several attempts to generate heterodimeric enzymes were made. The simplest way is to use two genes with fusions to different affinity tags and to coexpress them in one cell. Purification using both affinity materials consecutively sorts out homodimeric proteins and yields heterodimeric enzymes.

In principle, this opens the way to asymmetric recognition which may be easier to engineer than a symmetrical one. However, due to the use of largely identical sequences of the two genes, the danger of recombinational modification arises. This led to high instability of two EcoRI genes cloned into a polycistronic message. We tried to destabilize the homodimeric interface and simultaneously stabilize a heterodimeric protein. As in most enzymes, the monomer-monomer interface of EcoRI is mainly hydrophobic. We mutated two of these residues (Leu158 and/or Ile230) which contact each other from both subunits to aspartate or lysine.

The L158D mutation destabilizes the homodimer to a large extent, while the other single mutants turned out to be dimeric. In vitro formation of heterodimers is complicated by the fact that the I230K mutant is a rather stable dimer by itself. Thus, longer incubation times are necessary and the amount of heterodimeric enzyme is hard to determine. One indication of heterodimer generation is an increase in specific activity as the L158D monomer is a nickase and sequence recognition is completely lost. Some preferences for cleavage of distinct phosphodiester bonds seem to reflect the flexibility of the sugar-phosphate backbone and the sequence dependent fine structure of the double helix, which leads to preferential cleavage of many so-called unspecific nucleases. The Type IIS enzyme AlwI gave a completely different result. It is homologous to the monomeric but sequence-specific nickase N.BstNBI.

Swapping of the dimerization domain resulted in a monomeric N.AlwI which still recognizes its sequence but nicks a single strand with high activity. Obviously the composition of Type US enzymes out of a DNA-binding domain specific for the whole recognition sequence and a separated nuclease domain is helpful in these experiments. The second nuclease domain is needed only for cleavage of the other single strand. Another way to attain monomeric restriction enzymes is to fuse the subunits into a single chain. This was done successfully for the PvuII endonuclease where the C-terminus of one subunit is close to the N-terminus of the other. The single chain PvuII is nearly as active as the wild type enzyme. In case of EcoRI, fusion is more complicated as C- and N-terminus of both subunits are far away from each other.

Therefore, we fused the two subunits where they come close, deleting 65 residues from the C-terminal end of the first part and 72 residues from the N-terminal end of the other. Although cleavage activity is detectable in crude extracts, the deletions impeded purification in large scale. We also tried to bridge the distance between C- and N-terminus with a linker comprising 21 amino acid residues. This linker is sensitive to cellular proteases which again impedes purification of the fusion protein. This single-chain EcoRI is very active as we experienced unexpected difficulties during its construction.

Normally, for all cloning of EcoRI mutants, we use a gene with a good but strongly repressible promoter (p_L) and a bad ribo-somal binding sequence which allows for cloning in cells without the methylase gene. Even though this works for wild-type EcoRI, it was not possible for the fusion construct. Obviously, the low transcription rate which always occurs is tolerated by the cells as long as the few mRNA molecules are translated slowly to inactive monomers, as those have to find each other to form an active enzyme. The single-chain EcoRI does not need to dimerize and immediately cleaves DNA of its host cell. Even one active restriction enzyme may kill the cell. This could be the reason that restriction enzymes evolved to be dimeric. During the phase of establishing a new restriction-modification system, the endonuclease is poorly expressed while the methylase prevails. In this critical phase the need to bring together two monomers to assemble an active endonuclease may be helpful.

FUTURE DIRECTIONS (IN VITRO EVOLUTION)

We have described that despite very large efforts, engineering of restriction endonucleases was rewarded by little success. It appears that these enzymes are fine-tuned conformational machines which tolerate only few structural deviations. On the other hand, the structural homology of some of these enzymes even in the small sample of structures solved today shows that they are evolutionarily related. This can also be deduced from their primary sequences although with much less confidence, which indicates that larger changes are necessary to evolve an enzyme with a new specificity. Nowadays, we have tools at hand to perform evolution in vitro.

In principle, one has to produce a library of mutant genes to obtain a large variety of proteins and to select afterwards those enzymes with the desired specificity and activity. Because the amount of all possible mutants for a typical protein is enormous, these steps have to be repeated several times. However, in vitro two main difficulties arise. Firstly, phenotype and genotype have to be coupled. As we want to select the properties of proteins and reverse translation is not possible, we have to link each enzyme physically to its mRNA which can be reverse transcribed to the gene. This is one of the reasons for creating single-chain restriction endonucleases.

The most popular methods only work with monomeric proteins. These are phage display, ribosome display and mRNA-protein fusions. The first has the disadvantage that the" phages are produced by cells. It is better to perform all steps in vitro because transformation into cells reduces the diversity from 10^{13} to about 10^{8}. The second difficulty is the selection step itself. The properties of many restriction endonucleases suggest the following procedure: In the absence of Mg^{2+}, many of these enzymes bind strongly to their recognition sequence.

Variants with other specificity or with low binding strength can be washed away. Then catalytically efficient variants can be eluted by adding Mg^{2+}. Therefore, one can easily select for an efficient restriction enzyme. However, one will start with very inefficient enzymes which have to be optimized in later steps. Not to lose promising candidates in the first selection rounds may be difficult. Furthermore, this selection scheme is not compatible with ribo-some display, as high Mg^{2+} concentrations are used to fix protein and mRNA on the ribosome. Thus, the enzyme will dissociate from the ribosome during the first selection phase. Despite all difficulties, we are convinced that in vitro evolution experiments will tell us a lot more about restriction enzymes as was the case with other systems.

Especially the contribution of residues outside the protein-DNA interface will show us their importance for activity and specificity.

8 ENGINEERING AND APPLICATIONS OF CHIMERIC NUCLEASES

Each human cell contains about 3×10^9 base pairs (bp) within its genome. With the first draft sequence of the human genome now available, biologists estimate that there are about 30,000–40,000 different genes within the genome. This is fewer than originally anticipated, but still a huge number. These genes code for all of the human body's proteins. Simple mutations within the coding region of critical genes can lead to the formation of abnormal proteins, resulting in disease phenotypes, premature death, or failure of an embryo to develop. Furthermore, mutations that affect I the regulatory region of genes can result in aberrant gene expression within cells, and give rise to cancer phenotypes. The 'Holy Grail' of the Human Genome Project is 'Gene Therapy', that is, how genes might someday be used, i modified, or even changed to correct human disease. Gene therapy is based on a simple concept : Mutations within the coding or regulatory regions of certain critical genes give rise to disease phenotypes.

Correction of these mutations to normal alleles will result in the reversion of disease phenotypes to normal phenotypes. This is particularly true of monogenie diseases. Thus, gene therapy provides a new paradigm for treating human disease by correcting the causative genetic defect. However, the implementation of this novel concept into clinical practice and therapeutic reality has proven to be extremely difficult; and, the progress has been very slow. This is because targeting a particular defective gene for correction among 30,000 or more human genes within the genome requires exquisite sequence-specificity of the gene therapy vectors, which are the vehicles used to deliver the therapeutic genes into cells. This problem is further compounded by the fact that these genes and their regulatory sequences, which account for about 5% of the human genome, are buried within the context of a 3×10^9-bp DNA sequence.

Current gene therapy vectors lack the required sequence specificity that is necessary for the targeted correction of a defective gene within the genome. As a result, the therapeutic genes are delivered and inserted randomly within the genome that is at many sites other than the location of the genetic defect. Such random insertions of the therapeutic gene at unwanted locations are mutagenic, especially when the insertions occur in critical genes. Over time, these insertions can and do give rise to cancer or other disease phenotypes. Many of the difficulties associated with gene therapy are likely to be overcome if one could insert the corrected version of the mutation at the precise location of the genetic defect within the genome. How then does one achieve targeted correction of a gene defect within cells? Biologists have known for a while that when a defined chromosomal break is introduced at a unique site within a genome, homologous recombination is induced at that site in a large fraction of cells in a population.

The challenge then, so far unfulfilled, is to develop a general means of introducing a double strand break (DSB) uniquely at a given locus in the genome to induce homologous recombination. In this chapter, we discuss the progress towards targeted correction of a genetic defect. This involves two steps : (1) engineering of chimeric nucleases, the molecular tools necessary to make a chromosomal DSB (double-strand break) at a chosen site within the human genome; (2) application of these molecular tools for targeted chromosomal cleavage and correction of a genetic defect by inducing homologous recombination at that locus in cells.

Engineering of Chimeric Nucleases

In order to make a unique chromosomal DSB within a mammalian genome, we need restriction enzymes that recognize DNA sequences of 16 bp or more in length. Such restriction enzyme sites will occur once every 4^{16} (= 4.3 × 10^9) bp on average, which is about once per human genome. Most bacterial enzymes typically recognize short palindromes that are 4 to 8 bp in length. This means that they will cut DNA on average once every 4^4 to 4^8(=256 to 65,536) bases depending on the restriction enzyme used. For instance, one can expect an 8-bp site to occur about 45,776 times within the human genome. Obviously, bacterial restriction enzymes are not useful for the task at hand, which is to introduce a targeted chromosomal break at a site of our own choosing within the human genome. Over a decade ago, our laboratory set out to engineer chimeric nucleases, the molecular tools necessary for introducing a targeted chromosomal DSB, within a mammalian genome.

The research started with the study of FokI restriction endonuclease, a bacterial Type US restriction enzyme. FokI recognizes the nonpalindromic pentadeoxyribonucleotide, 5′-GGATG-3′:5′-CATCC-3′, in duplex DNA and cleaves 9/13 nucleotides downstream of the recognition site. It does not recognize any specific sequence at the site of cleavage. This property implies the presence of two separate protein domains within FokI: one for sequence-specific recognition of DNA and the other for the endonuclease activity. Once the DNA binding domain is anchored at the recognition site, a signal is transmitted to the endonuclease domain, probably through allosteric interactions, and the cleavage occurs. We reasoned that one may be able to swap the FokI recognition domain with other naturally occurring DNA-binding proteins that recognize longer DNA sequences or other designed DNA-binding motifs to create chimeric nucleases. In this way, Type US enzymes like FokI are probably ideal candidates for engineering novel sequence specificities.

Functional Domains in FokI Restriction Endonuclease

Several laboratories including ours have independently cloned the FokI restriction-modification system from *Flavobacterium okeanokoites*. These labs have also reported overproduction and purification of FokI endonuclease. As a first step, we probed the domain structure of FokI by limited proteolysis using trypsin. Our studies on proteolytic fragments of FokI endonuclease revealed an N-terminal DNA-binding domain and a C-terminal domain with nonspecific DNA cleavage activity. Waugh and Sauer (1993, 1994) showed that single amino acid substitutions decouple the DNA-binding and strand scission activities of FokI endonuclease.

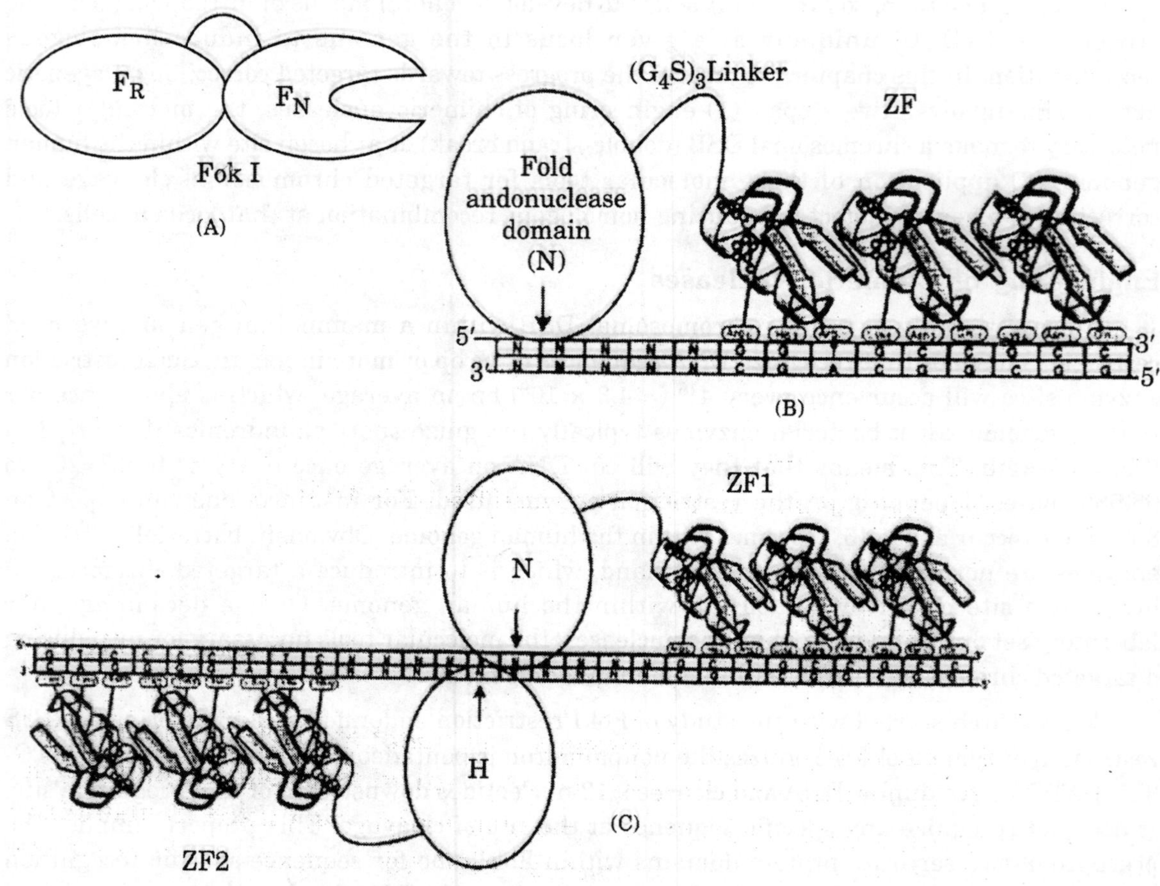

Fig. 8.1. A Schematic representation of FokI endonuclease and ZFN. A Domain structure of FokI; B ZFN bound to its 9-bp cognate site; and C A pair of ZFN with different specificities bound to their target sequence. Since ZFN requires two copies of the 9-bp recognition sites in an inverted orientation in order to dimerize and produce a DSB, it effectively has an 18-bp recognition site. F_R FokI recognition domain; F_N or *N* FokI cleavage domain; *ZF1* and *ZF2* zinc finger proteins.

The mode of DNA-binding by FokI was analyzed by DNA foot-printing methods. These studies showed a lack of protection at the cleavage site of DNA by FokI. Wah et al. (1997, 1998) reported the crystal structures of FokI and FokI-bound to its cognate site, which confirmed the modular nature of FokI endonuclease. The endonuclease domain appears to be sequestered in a "pig-gyback" fashion by the recognition domain. This is consistent with the DNA foot-printing analysis. Thus, the crystal structure of FokI endonuclease is in complete agreement with the model derived from rigorous biochemical studies.

Chimeric Nucleases

The modular nature of FokI endonuclease suggested that it might be feasible to engineer chimeric nucleases by fusing other DNA-binding proteins to the cleavage domain of FokI. The obvious next step was to create novel chimeric nucleases by simply swapping the DNA recognition domain of FokI with other DNA-binding proteins that recognize longer DNA sequences. This indeed proved to be the case. Several novel fusion nucleases were generated in our lab by linking the nonspecific FokI cleavage domain to other DNA-binding proteins.

The latter include the three common eukaryotic DNA-binding motifs namely the helix-turn-helix motif, the zinc finger motif and the basic helix-loop-helix protein containing a leucine zipper motif (bzip). Our lab reported the creation of the first chimeric nuclease by fusing the *Drosophila Ubx* homeodomain to the FokI cleavage domain. This was followed by the fusion of zinc finger motifs and the N-terminal 147 amino acids of the yeast Gal4 protein respectively, to the cleavage domain of FokI. Such engineered nucleases were shown to make specific cuts very close to their expected recognition sequences in vitro. Since then other labs have reported making FokI nuclease domain fusions with various other DNA-binding proteins. FokI fusions have been used to identify high-and low-affinity binding sites of transcription factors in vitro, to study recruitment of various factors to the promoter sites in vivo using the method called protein position identification with a nuclease tail (PIN-POINT) assay and to analyze Z-DNA conformation-specific proteins.

Chimeric nucleases now form a novel class of engineered nucleases in which the non-specific cleavage domain of FokI is fused to other DNA-binding motifs. The most important chimeric nucleases are those based on zinc-finger DNA-binding motifs.

Zinc Finger Binding and Specificity

The modular structure of zinc finger domains (ZF) and modular recognition by zinc finger proteins make them the most versatile of DNA recognition motifs for designing artificial DNA-binding proteins. Each zinc finger consists of about 30 amino acids and folds into a ββα-structure, which is stabilized by the chelation of a zinc ion by the conserved Cys_2His_2 residues. Each finger typically recognizes a 3-bp DNA sequence by inserting the α-helix into the major groove of DNA. Binding of longer DNA sequences is achieved by linking several of these zinc finger motifs in tandem. Each finger, because of variations of certain key amino acids in the α-helix of one-zinc finger to the next, makes its own unique contribution to DNA-binding affinity and specificity.

Because they appear to bind as independent modules, zinc fingers can be linked together in a peptide designed to bind a predetermined DNA site. Although more recent studies suggest that there might be a synergistic interaction between adjacent zinc fingers and that the zinc finger-DNA recognition is more complex than originally perceived, it still appears that zinc finger motifs will provide an excellent framework for designing DNA-binding proteins with a variety of new sequence-specificities. In theory, one can design a zinc finger for each of the 64 possible triplet codons, and using a combination of these fingers, one could design a protein for sequence-specific recognition of any segment of DNA. At present, the rules relating to zinc finger sequences/DNA-binding preferences and redesigning of DNA-binding specificities of zinc finger proteins are well understood only for the 5′-GNN-3′ triplets.

Although numerous structural studies of zinc finger protein complexes with DNA are available, the information is not yet sufficient for rational design of zinc finger domains that bind to any given triplet. The creation of zinc finger chimeric nucleases (ZFN) that recognize

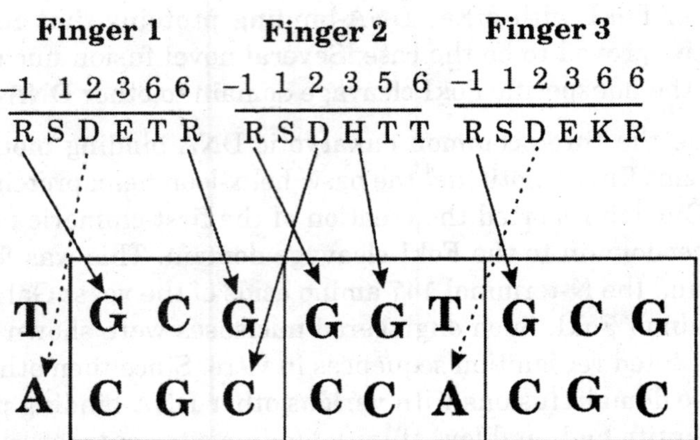

Fig. 8.2. Key base contacts deduced from the crystal structure of Zif 268-DNA complex. Each finger makes contact with its target 3-bp site. In addition, Asp^2 at position 2 in each finger makes contact with base outside the 3-bp site.

and cleave any target sequence depends on the reliable creation of zinc finger proteins (ZF) that can specifically recognize a target sequence. A phage display selection method could be used to identify zinc finger proteins that bind to a particular DNA site. This is the method of choice for designer zinc finger proteins. Briefly, a library of zinc fingers with randomized amino acids at specific positions (–1 to 6) within the α-helix is fused to a viral coat protein of a phage that is used to infect bacteria.

The virus is then allowed to replicate and reproduce in culture to make more copies. Since the fusion proteins are displayed on the surface of the virions, they can be enriched for highly specific and functional zinc finger proteins that bind a chosen target substrate with high affinity. Three different selection methods based on the phage display-parallel selection, sequential selection, and bipartite selection-have been reported in literature using Zif268-derived phage libraries for selection of designer zinc fingers. These libraries are constrained by the target site overlap problem because of the presence of Asp^2 at position 2 of the α-helix in each of the three-zinc finger domains of the Zif268. The Asp^2 of all three fingers make a cross-strand contact to a base outside the canonical triplet site. This precludes the presence of a simple general recognition code that will facilitate the rational design of zinc finger-based DNA-binding domains. This also makes selection of zinc finger proteins by using phage display libraries more difficult.

Three strategies have since been developed to overcome this problem : (1) zinc fingers 1 to 3 can be sequentially selected in the context of neighboring zinc fingers; (2) targets can be selected libraries with simultaneous randomization of amino acid residues in two adjacent fingers; and (3) bipartite selection which com-bines the sequential selection with the parallel selection technique. However, these approaches do not provide for the selection of zinc finger domains that function as independent, modular recognition motifs. They select for targets that are dependent on their inter-domain interaction for their sequence specificity. An alternate approach based on a bacterial two-hybrid system has been described for the isolation of a candidate zinc finger protein in an in vivo context from a randomized Zif268-derived mutant library of > 10^8

members in size. This method is analogous to the yeast two-hybrid system previously described by Hochschild and coworkers to study protein-protein interactions. In the bacterial two-hybrid system, zinc finger-DMA interactions are required for cell growth and survival.

However, this approach also suffers from the same limitation as the phage display methods in that it does not allow for selection of zinc finger motifs that act as inde-pendent, modular recognition domains since it also utilizes a Zif268-derived mutant library. In collaboration with Dr. Marc Ostermeier's lab, we are developing a double-reporter, one-hybrid system for rapidly selecting zinc finger proteins and improving their sequence specificities. This system will also allow for identification of zinc finger motifs that act as independent modular units. This will be done by using a mutant zinc finger library that is based on consensus back-bone framework for each and every finger within the protein; and by limiting the amino acid at position +2 of the α-helix of each finger to a glycine residue thus eliminating the cross-strand base contact that occurs outside the 3-bp site in the Zif268 derived libraries due to the presence of Asp^2. The one hybrid, double reporter system that is being developed in our lab is based on the one hybrid system originally described by Hochschild and coworkers that is used to detect protein-DNA interactions.

In their system, the gene for a DNA binding domain is fused to a subunit of *E. coli* RNA polymerase. The fusion is then used to activate transcription from a lac-derived promoter provided the binding site for the DNA-binding domain is present and centered at the –63 position. The reporter gene under the control of the lac-derived promoter is β-galactosidase. Thus, this method provides a way to screen and assay for interaction between the DNA-binding domain and various DNA-binding sites positioned at –63. Our system is designed to create one hybrid, double reporter system for evaluating and evolving DNA-binding specificity of zinc finger proteins.

The gene coding for the zinc finger is fused to a subunit of *E. coli* RNA polymerase. The fusion protein is then used to activate transcription of a reporter gene under the control of a lac-derived promoter provided the zinc finger binding site is placed at an appropriate distance upstream of the promoter. In our system, we incorporate two separate operons each containing one reporter gene under the control of a lac-derived promoter. The only difference between the two operons is the nature of the reporter gene and the target zinc finger binding sites, which are placed upstream of the promoter. In this way, binding of a zinc finger protein to two different sites can be evaluated simultaneously. We plan to employ two different reporter systems as well.

In the first, the antibiotic resistance genes coding for chloramphenicol and tetracycline, respectively, will be placed under the control of lac-derived promoters on separate operons. In the second, we have chosen two reporter genes namely the green fluorescent protein (GFP) and the red fluorescence protein (dsRED) to be placed under the control of lac-derived promoters on separate operons. The fluorescent system offers selection of zinc finger mutants with improved sequence specificity using quantitative flow cytome-try and fluorescence activated cell sorting. Recent advances suggest that a combination of design and selection is best suited to identify custom zinc finger DNA-binding proteins for known target sites. So far, only zinc finger modules that specifically recognize 5′-GNN-3′ and 5′-ANN-3′ triplets have been identified through design and selection strategies.

Studies are underway to determine the zinc finger recognition preferences for the other triplets, 5′-CNN-3′ and 5′-TNN-3′, respectively. In many instances, the current knowledge of zinc finger preferences for 5′-GNN-3′ and 5′-ANN-3′ is more than sufficient for designing and/or selecting a zinc finger protein to target a specific site within a gene of the human genome.

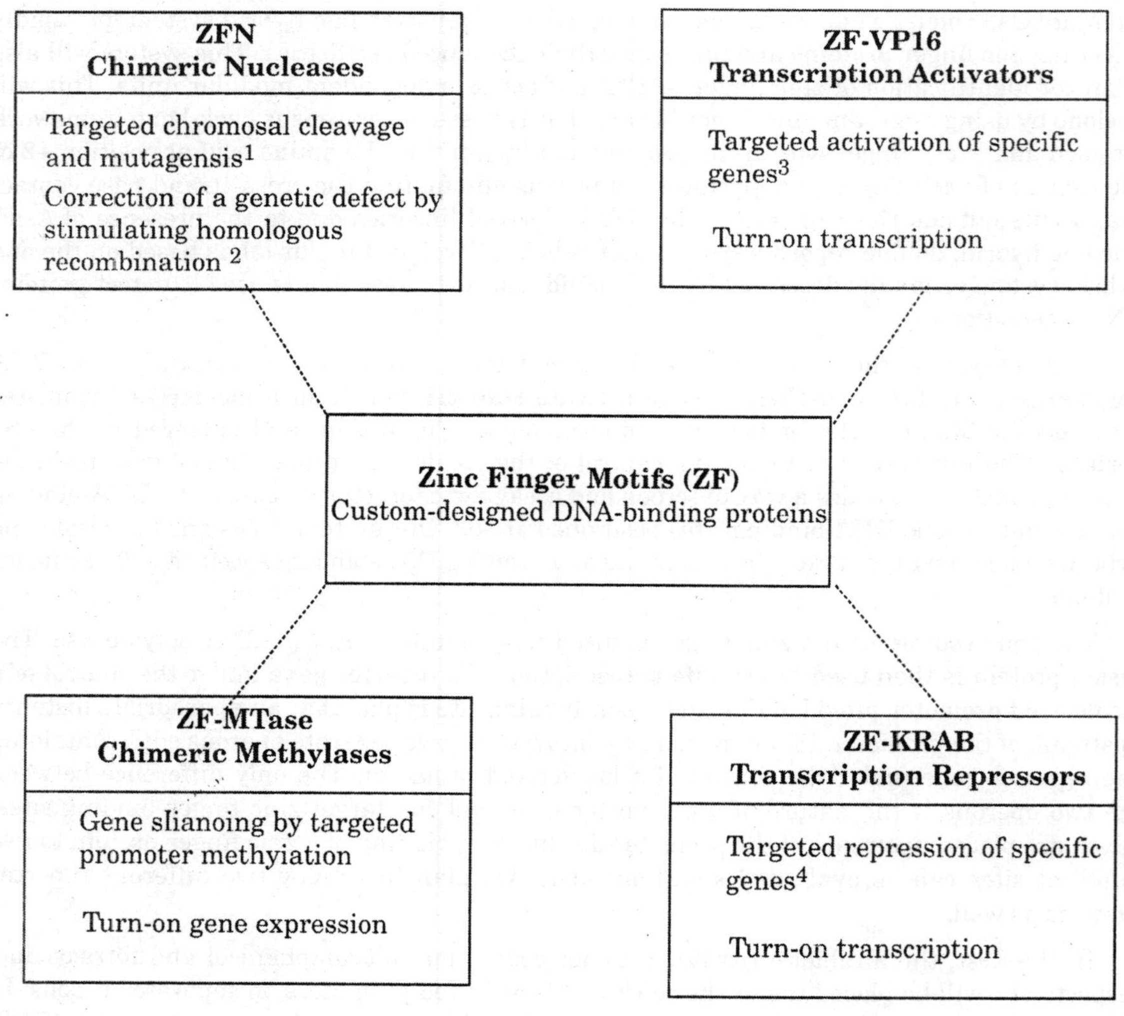

Fig. 8.3. Zinc finger protein platform technology. Additional functional domains like the FokI cleavage domain (N), activator domain (VP16), represser domain (KRAB) and methylases (Mtase) can be fused to the designed zinc finger motifs (ZF) to form, chimeric proteins that act as chimeric nucleases (ZFN), transcriptional activators (ZF-VP16), transcriptional repressers (ZF-KRAB) and targeted methylases (ZF-Mtase), respectively.

The ability to design or select zinc fingers with desired specificity implies that DNA-binding proteins containing zinc fingers will be made to order. Therefore, we reasoned that one could design "artificial" nucleases (ZFN) that will cut DNA at any preferred site by making fusions of zinc finger proteins (ZF) to the cleavage domain (N) of FokI endonuclease. We have been successful in engineering several novel ZFN by fusing three-zinc finger proteins to the cleavage domain of FokI. In these fusions, the three Cys_2His_2 zinc finger DNA-binding domain is at the N-terminus and the FokI nuclease domain is at the C-terminus; they are connected by a flexible 15 amino acid $(Gly_4Ser)_3$ linker to achieve efficient DSB at the predicted 9-base pair recognition site.

We have shown that the fusion of FokI cleavage domain to the zinc finger motif does not alter the sequence specificity of the zinc finger protein and it does not change its binding affinity significantly. It must be made clear that sequence specificity of the engineered ZFN is only as good as the zinc finger motif that was used to make it. If a zinc finger shows affinity for degenerate sites, then, the engineered ZFN will cut at those degenerate binding sites albeit with lower affinity. Although, our lab was the first to fuse another functional moiety (FokI cleavage domain) to zinc finger proteins, our long-term focus and efforts have remained on the development of chimeric nucleases.

Other functional domains, like activator domains and represser domains have been fused to the designed zinc finger motifs to form hybrid proteins that act as transcription activators and transcription repressers within cells. More recently, engineered transcription factors have been shown to stimulate blood vessel growth and wound healing by the induction of the gene coding for VEGF-A (vascular endothelial growth factor A) in mice. Cytosine methylation has been targeted preferentially to pre-determined sequences by attaching CpG-specific DNA methyltransferase to Zif 268 three-zinc finger proteins. It appears that important applications in medicine and biological research arising from zinc finger platform technology are almost certain to follow.

Mechanism of Cleavage by Zinc Finger Chimeric Nucleases (ZFN)

A three-zinc finger protein recognizes nine base pairs. An issue relevant to gene targeting is the ability of the cleavage reagent to distinguish the chosen target from all other sequences in a complex genome. In theory, any sequence shorter than 16-bp would be present in multiple copies in the human genome. In particular, any given 9-bp sequence is predicted to occur 10^4 times within the human genome. Therefore, the three-zinc finger nucleases are expected to make about 10,000 DSB within the human genome. How then does one achieve a unique, targeted chromosomal DSB within a genome? Our study on the mechanism of cleavage by the three-zinc finger nucleases indicates that the cleavage domain must dimerize in order to cut DNA.

Therefore, efficient cleavage requires two zinc finger-binding sites situated in close proximity in an inverted orientation. Such a configuration of binding sites orients the nuclease domains towards each other promoting dimerization. As expected, the spacer length between the inverted repeats is critical for the dimerization of the nuclease domains to occur. This in turn is determined by the $(Gly_4Ser)_3$ linker, which connects the zinc fingers to the FokI cleavage domain. Thus, ZFN prefer inverted repeats as their substrate to make a DSB. However, at high enzyme concentrations, single binding sites are cut; one molecule binds at its cognate site, while the second probably binds non-specifically nearby on the DNA facilitating the dimerization of the FokI catalytic domain.

The first clue for FokI dimerization came from the crystal structure of the FokI endonuclease, which revealed the interacting interface between the catalytic domains. In addition, kinetic studies on the mechanism of cleavage by FokI further supported that dimerization of FokI enzyme was required to produce a DSB. Since ZFN requires two copies of the 9-bp recognition sites in a tail-to-tail orientation in order to dimerize and produce a DSB, it effectively has an 18-bp recognition site. In principle, the binding sites need not be identical provided the ZFN that bind both sites are present. We have shown that two ZFN with different sequence specificities collaborate to produce a DSB when their binding sites are appropriately placed and oriented with respect to each other. Since the recognition specificity of zinc fingers can be manipulated experimentally, ZFN could be engineered so as to target a unique site within a genome.

Application of Chimeric Nucleases

The modular structure of FokI endonuclease and zinc finger proteins has made it possible to create artificial nucleases that will cut DNA at a pre-determined site. This approach opens the way to generate many novel ZFN with tailor-made sequence specificities desirable for various applications in biotechnology and medicine.

Recombinogenic Repair

How might these ZFN be used in the correction of a genetic defect? One approach would be to recruit the pre-existing cellular machinery to achieve the goal. Cells of many different organisms use recombination to repair DNA damage, especially DSB. If left un-repaired such DSB in a cellular chromosome would be lethal. Carroll and coworkers (1996) have shown that recombination in frog oocytes proceeds by exonuclease resection and then annealing of complementary strands. This is called single-strand annealing and this same mechanism is thought to operate in mammalian cells. One could utilize this process for gene-targeting experiments.

The essence of this approach is to target the chromosomal DNA for sequence-specific cleavage at or near the site where one wants the recombination to occur. Exonuclease would then resect the DNA at that chromosomal site, generating single-strand tails (Lin et al. 1984). Recombination would proceed by a single-strand annealing mechanism with the exogenous DNA that is present in the cell. The homologous foreign DNA would thus be incorporated at this chromosomal site. If the chromosomal target site is not cleaved, then it would not be accessible for exonucleolytic resection. In this case, the exogenous DNA will be degraded while the target remains unchanged. Thus, making a targeted DSB would greatly stimulate homologous recombination between the exogenous DNA and a chromosomal sequence.

Such experiments have been performed using Group I intron-encoded homing endonucleases in yeast, cultured mammalian and plant cells. Although the target sites for these enzymes range from 15-40-bp in length, they exhibit a broad range of sequence specificities and cutting frequencies because of their binding to degenerate sites. In addition, their sequence specificities are not amenable to easy manipulation. As a result they have only a limited use in gene-targeting experiments. Recently, we have shown in collaboration with Dana Carroll's lab that the three-zinc finger chimeric nuclease, Zif-QQR-F_N finds and cleaves its target in vivo. This was tested by microinjection of DNA substrates and the enzyme into frog eggs. Double-strand cleavage required an inverted repeat of the 9-bp target.

When the appropriate sites were placed in the recombination substrate, this DNA was cleaved in frog oocytes by the injected enzyme and homologous recombination ensued. These microinjection experiments have shown that greater than 90% of the substrate was cleaved in vivo and almost all cleaved substrates underwent homologous recombination. Although, in this proof-of-principle experiment, an extrachromosomal target was used to test the idea, two important conclusions can be drawn: (1) chimeric nucleases find and cleave their targets within cells; and (2) the targeted DSB stimulates homologous recombination at the site of cleavage, provided an undamaged template of the homolog is available for recombinogenic repair within cells. This is what normally occurs in cells : when one copy of a gene on one chromosome is damaged, it is repaired using the other undamaged allele on the paired chromosome as a template.

Chromosomal Cleavage and Mutagenesis

The obvious next step was to show specific cleavage of a chromosomal target within cells by chimeric nucleases. Recently, Dana Carroll's lab reported specific chromosomal cleavage and mutagenesis in *Drosopila melanogaster* (Bibikova et al. 2002). They designed a pair of ZFN that targeted the yellow (Y) gene on the X-chromosome of *Drosophila*. The nucleases were cloned under the control of a heat shock promoter and then introduced into *Drosophila* genome via P element-mediated transformation. Only when both ZFN were expressed in the developing larvae did somatic mutations occur specifically in the Y gene (yellow gene of *Drosophila* X-chromosome).

The yellow mosaics were observed in 46% of males expressing both ZFN. No yellow mosaics were observed in control larvae that expressed a single ZFN or without heat shock. Germline Y mutations were recovered from 5.7% of males but none from the females tested. This was as expected in female fruit flies because the targeted DSB would be repaired by recombination with the uncut Y homolog. In males, only simple ligation via nonhomologous end joining (NHEJ) is available to repair the chromosomal damage. DNA sequence analysis of the mutants revealed that all mutations were small deletions and/or insertions localized to the targeted site, which was typical of repair by NHEJ.

Many of these mutations led to frame shifts or deletion of essential codons of the Y gene. As a result, the male larvae emerging from the heat shock experiment showed yellow patches in the otherwise dark posterior abdomen. In subsequent experiments, Carroll's lab has recovered *y* mutations in female fruit flies, but always at a much lower frequency than that in male flies. This suggests that if the cleavage efficiency of ZFN was high enough to cut both homologues of the Y gene simultaneously, one can produce female germline mutations. Since accurate repair from the homolog appears not to be fully efficient, it might be possible to induce germline mutations in autosomal recessive genes using ZFN (Bibikova et al. 2002). Furthermore, since DSB stimulates mutagenic repair that essentially operates in all organisms, targeted cleavage by ZFN could facilitate generation of directed mutations in a variety of cells and organisms.

In a corollary experiment, Bibikova et al. (2003) have shown in *Drosophila,* that gene targeting is enhanced using designed zinc finger nucleases. The ZFN-induced gene-targeting frequencies are about ten fold greater than those observed without target cleavage. Several important conclusions can thus be drawn from this study : (1) chimeric nucleases can be used to target and cleave a specific chromosomal site within cells; (2) organisms that express ZFN, like the *Drosophila* embryos in this case, are viable and they go on to mature into adults; (3) chimeric nucleases can be used to generate directed mutations in organisms and cells, including targeted germline mutations in autosomal recessive genes—provided the ZFN sequence-specificity and cleavage is high enough to target both copies of the alleles within the genome; and (4) gene targeting—the process of gene replacement by homologous recombination—is enhanced through targeted DSB using designed ZFN. Similar experiments are underway in our lab using mammalian cells.

We plan to correct a tyrosinase gene mutation in albino mouse melanocytes using ZFN. The long-term goal is to correct a tyrosinase gene mutation in albino mouse embryonic stem cells. Melanocytes are melanin pigment-producing cells of the skin. Tyrosinase is a key enzyme for melanin synthesis and pigmentation and it catalyzes the conversion of L-tyrosine to melanin. Melanocytes from albino mice contain a homozygous point mutation (TGT to TCT) in the tyrosinase gene. As a result, the amino acid cysteine is changed to a serine in the tyrosinase enzyme. Correction of this point mutation should restore the tyrosinase enzyme activity; and hence confer pigmentation to the albino melanocytes, resulting in dark cells. We have engineered ZFN to target a specific site within the tyrosinase gene for cleavage.

In vitro studies indicate that as expected ZFN bind and cleave the target site contained in a plasmid. These constructs are currently being tested in cell culture studies. Continued expression of ZFN is toxic to the melanocytes, which results in cell death. This warrants controlled expression of ZFN in these cells as in the case of experiments done using *Drosophila* larvae. After completion of this study, we plan to perform similar experiments using mouse embryonic stem cells with the eventual goal of generating dark progeny from albino mouse parents.

Gene Targeting in Human Cells Using ZFN

Gene targeting is a powerful technique to incorporate genetic changes in mammalian genomes. However, this process is very inefficient even in murine embryonic stem cells where it has been successfully applied to produce transgenie mouse "knockouts". Only one cell among 10^6 cells treated with donor DNA has the desired mutant phenotype. Powerful selection techniques are then needed to enrich for the cells carrying the directed mutation. Recently, Porteus and Baltimore (2003) have described a system based on the correction of a mutated GFP gene to study gene targeting in somatic cells. They have shown that gene targeting using ZFN is stimulated over 2000-fold by targeted chromosomal cleavage. This corresponds to a targeted correction rate of 3-5% of the cells. This is very encouraging indeed since these targeting rates are in the range of being therapeutically useful.

Future Experiments

The next logical step is to consider gene targeting using chimeric nucleases as a form of gene therapy for the correction of genetic defects that give rise to human diseases, particularly, in the treatment of monogenic diseases, which arise from deleterious mutations in a single gene. It would entail the following steps:

1. Identify a target site within the gene of interest. Inverted sequences of the form $(NNC)_3$...$(GNN)_3$ separated anywhere between 6 to 16 bases make for excellent targets. The target sequence could be within several hundred base pairs from the mutation site for gene conversion.
2. Design and/or select zinc finger proteins (ZF) that recognize the target.
3. Convert the engineered ZF to zinc finger chimeric nucleases (ZFN).
4. Deliver the ZFN and normal gene directly into the nucleus either by microinjection or through viral vectors; ZFN direct targeted chromosomal DSB and stimulate homologous recombination (through recombinogenic repair) with the exogenously provided normal gene. To achieve targeted chromosomal cleavage and mutagenesis by NHEJ (through mutagenic repair) only ZFN need to be delivered into the cells.
5. Monitor for homologous recombination at the target site.

Direct and rigorous experimentation is required to unequivocally establish the induction of homologous recombination at the targeted chromosomal site using human cell lines. Furthermore, it is important to show that unwanted recombination or insertions of the therapeutic gene do not occur at nonhomologous sites at other parts of the genome due to residual cleavage by ZFN. The genomic integrity and stability of treated cells need to be intact at all other loci except at the targeted chromosomal DSB after the ZFN treatment. These studies should be followed by an application to correct a genetic defect in a transgenic animal model.

The focus of our lab is now on two such research projects. In the first, the goal is to achieve targeted chromosomal cleavage and mutagenesis of the CCR5 (P chemokine receptor 5) gene in human cell lines. In the second, our aim is to attempt targeted correction of a gene mutation in the CFTR (cystic fibrosis transmembrane conductance regulator) gene associated with cystic fibrosis in humans. Both of these projects are described briefly below.

Targeted Chromosomal Cleavage and Mutagenesis of the CCR5 Gene in a Human Cell Line

A 32-nucleotide deletion (Δ32) within the β-chemokine receptor 5 (CCR5) gene has been identified in individuals who remain uninfected despite extensive exposure to HIV-1. CCR5 is the major co-receptor on $CD4^+$ cells through which HIV gains entry into the cells. The virus is unable to infect $CD4^+$ lymphocytes and macrophages that are homozygous for the A32 mutant CCR5 allele. The goal here is to induce targeted mutagenesis of the CCR5 gene in primate or human cell lines by using engineered ZFN. We have identified two zinc finger target sites near the A32 locus of the CCR5 gene. Wr have engineered ZFN to target and cleave one of these sites.

In vitro studies indicate that the engineered ZFN bind and cleave the target site encoded in a plasmid as expected. Experiments are underway to test this con-struct in cell culture studies. Our long-term goal is to induce targeted A32 deletion at the chromosomal locus encoding the CCR5 gene in hematopoietic stem cells (CD34+ cells) of individuals who are at high risk for HIV infection. The HIV-1 resistant autologous cells will then be amplified and expanded in cell culture. These cells will then be used to reperfuse the bone marrow of these individuals, thereby making their CD4+ lymphocytes and macrophages resistant to HIV-1 infection.

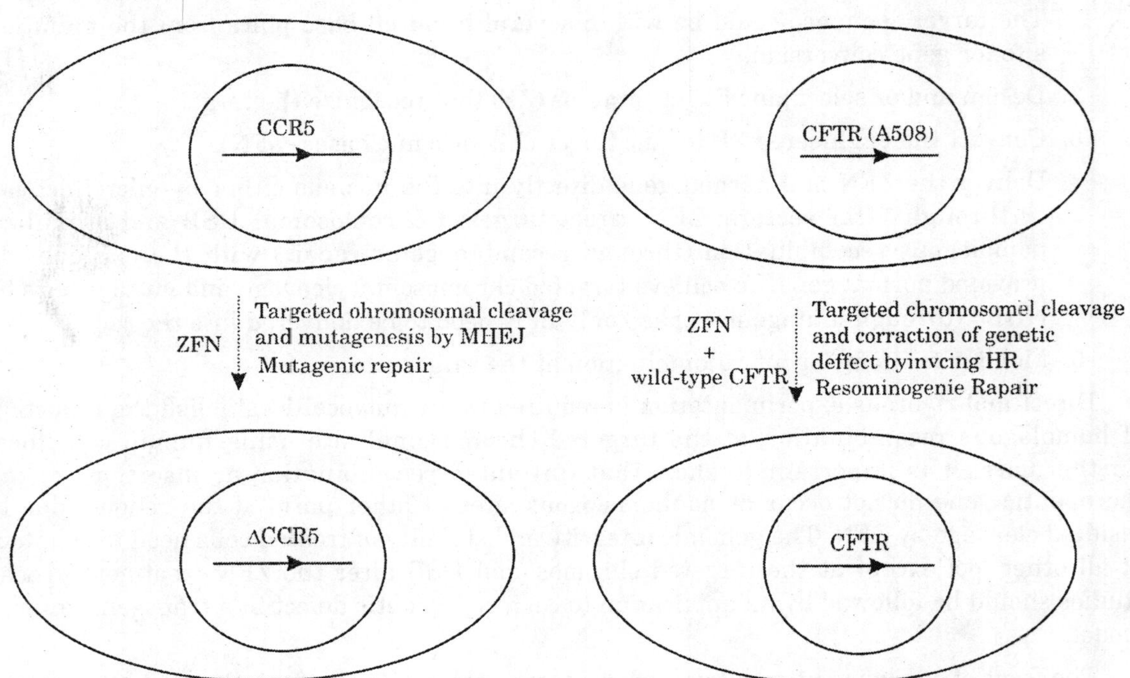

Fig. 8.4. Gene target ing in a human cell line using ZFN. A Targeted chromosomal cleavage and mutagenesis by nonhomologous end joining (NHEJ) through mutagenic repair using engineered chimeric nucleases. In this experiment, only ZFN are introduced into cells. B Targeted chromosomal cleavage and correction of a genetic defect by inducing homologous recombination *(HR)* through recombinogenic repair. In this experiment, both ZFN and the normal gene are introduced into cells.

Targeted Correction of a CFTR Genetic Defect in a Human Cell Line Using ZFN

Cystic fibrosis (CF) is a lethal autosomal recessive hereditary disorder characterized by chronic obstruction and infection of the airway, pancreatic insufficiency, and abnormally high electrolyte concentration in sweat. Although CF is a disease that affects multiple organs, respiratory disease is the major cause of morbidity and mortality: 95% of CF patients die of respiratory failure. Normal efflux of chloride ions occurs across respiratory epithelial cell membranes in response to elevated cAMP (cyclic adenosine monophosphate). This is lacking in cells derived from CF individuals, which in turn limits water secretion from epithelial cells.

As a result, dehydration and excessively thick airway mucus ultimately results in mucous obstruction, infection and inflammation, followed by damage to the airway epithelium. CF is caused by mutations of the cystic fibrosis transmembrane conductance regulator gene (CFTR gene). The most common mutation is a 3-bp deletion leading to the loss of phenylalanine at position 508 in the CFTR protein, known as AF508 mutation. Our goal is to attempt correction of AF508 mutation using ZFN in a human cell line. We have identified target sites within the CFTR gene near the AF508 mutation. We have designed ZFN to bind and cleave these targets. These are being tested in vitro for DNA binding and sequence-specific cleavage of the target.

Targeted correction of AF508 mutation involves somatic cells unlike the CCR5 experiment. The latter is to be done using hematopoietic stem cells that can be expanded in culture and reperfused into patients. The delivery of the CFTR gene to the target site and correction of the gene defect needs to occur in a sufficient number of cells in order to be helpful to the patients. One might have to use lentiviral vectors for efficient delivery of the therapeutic gene.

Potential Limitations of the Chimeric Nuclease Technology

Several things need to be considered when applying this technology in gene therapy.

These include the following : (1) ZF modules that recognize specifically each and ever) one of all possible triplets have not yet been identified. Only ZF modules for 5′-GNN-3′ and 5′-ANN-3′ triplets are known. In principle, this is likely to limit the number of human genes that could be targeted by ZFN approach. However, because the target sequence could be within several hundred bp from the mutation site for gene conversion, this may not pose a problem. (2) Targeting a single site within a gene of the human genome may not be very efficient. The identified target, if present in multiple copies within the gene of interest, will be an ideal locus for efficient cleavage. (3) Controlled and regulated expression of ZFN is necessary and highly preferable since continued expression of ZFN within cells is toxic and cell death by apoptosis ensues.

Ideally, one would like to express ZFN for a short while within cells, where hopefully they will find their target and induce a DSB, and then disappear via protein degradation pathways. Therefore, methods are needed for controlled and regulated expression of ZFN within mammalian cells. (4) In most cases, both alleles of autosomal recessive genes need to be targeted and mutated to observe a phenotypic effect. This is true of CCR5 gene targeting experiments as well. Heterozygous cells for CCR5A32 mutant allele confer only partial protection for about 1-2 years against HIV-1 infection before all cells eventually get infected (Liu et al. 1996).

Homozygous cells for CCR5A32 mutant allele are resistant to HIV 1 infection. While it may be feasible to target both copies of the CCR5 gene within cells using ZFN, it may not be a very robust process and may prove to be laborious and cumbersome because multiple rounds of selection may be needed. As an alternate route, one could envision targeted insertion of an inhibitor sequence like for a small interfering RNA (siRNA) within one allele of CCR5, and thereby, control the gene expression from the other homolog.

Incidentally, one can use a similar strategy to simultaneously control HIV-1 infection by both M-tropic and T-tropic viruses, which use CCR5 and CXCR4 (α-chemokine receptor 4) co-receptors, respectively. Furthermore, it appears that both copies of CCR5 gene can be mutated without any adverse side effects to individuals. The simplest explanation for this is that there are alternate mechanisms available within cells to compensate for the loss of CCR5 function. This makes CCR5 an ideal locus within the human genome for targeted insertion of other genes for controlled and regulated expression within cells.

Future Outlook

Gene targeting using chimeric nucleases is an emerging new technology. Recent advances now make it possible to design and/or select zinc finger proteins capable of recognizing virtually any 18-bp target sequence. This is long enough to specify a unique address within mammalian genomes. In principle, chimeric nucleases (ZFN) that combine the nonspecific cleavage domain of FokI endonuclease with zinc finger proteins (ZF) offer a general way to deliver site-specific DSb to the genome, and stimulate homologous recombination at that site. In the use of chimeric nucleases, binding of two ZFN each recognizing a 9-bp inverted site is necessary, since dimerization of the FokI cleavage domain is requited to produce a DSB. Therefore, ZFN effectively have an 18-bp recognition Site. Recent experiments from several labs indicate that ZFN find and cleave their chromosomal target; and as expected, they induce local homologous recombination at the site of cleavage.

These studies provide a glimpse of potential future therapeutic applications of ZFN in modifying and rewiring the human genome itself. This is a great time to be a biologist. There is an excited anticipation that this novel emerging technology will make targeted correction of a genetic defect feasible, especially in treating monogenic diseases. Several areas of reseaiv a appear to be converging and coalescing to make gene therapy a reality. These include the availability of the first draft sequence of the human genome. The *complete nudeotide sequence of the genome* is soon to follow this technical feat. Also, great strides and progress are being made in *stem cell* research. The first applications of chimeric nuclease technology using ZFN to treat human diseases are likely to occur in *ex-vivo gene therapy using stem celh*. Here, the desired cells can be identified through selection, expanded in cuiiurc and replenished into patients. Thus, the availability of chimeric nucleases, a new type of engineered molecular scissors that target a unique site within the human genome, will contribute and greatly aid the feasibility of genome engineering that is targeted rewiring of the genome.

Ethical issues aside, it is not unreasonable to expect that in a decade or two, all the technical problems associated with gene delivery will be overcome; and that gene therapy will be routinely used in a clinical setting, finally signifying a paradigm shift in the treatment of human diseases. Only then, the full impact of the Human Genome Project will be felt, its full potential and goals realized, and its promises fulfilled.

Index